De Gruyter Expositions in Mathematics 50

Yevhen G. Zelenyuk

Ultrafilters and Topologies on Groups

De Gruyter

Mathematical Subject Classification 2010: 22A05, 22A15, 54D80, 54G05.

ISBN 978-3-11-020422-3
e-ISBN 978-3-11-021322-5
ISSN 0938-6572

Library of Congress Cataloging-in-Publication Data

Zelenyuk, Yevhen G.
 Ultrafilters and topologies on groups / by Yevhen G. Zelenyuk.
   p. cm. − (De Gruyter expositions in mathematics ; 50)
 Includes bibliographical references and index.
 ISBN 978-3-11-020422-3 (alk. paper)
   1. Topological group theory.   2. Ultrafilters (Mathematics)
 I. Title.
   QA166.195.Z45   2011
   512'.55−dc22

                                              2010050782

Bibliographic information published by the Deutsche Nationalbibliothek

The Deutsche Nationalbibliothek lists this publication in the Deutsche Nationalbibliografie;
detailed bibliographic data are available in the Internet at http://dnb.d-nb.de.

© 2011 Walter de Gruyter GmbH & Co. KG, Berlin/New York

Typesetting: Da-TeX Gerd Blumenstein, Leipzig, www.da-tex.de
Printing and binding: Hubert & Co. GmbH & Co. KG, Göttingen
∞ Printed on acid-free paper

Printed in Germany

www.degruyter.com

Preface

This book presents the relationship between ultrafilters and topologies on groups. It shows how ultrafilters are used in constructing topologies on groups with extremal properties and how topologies on groups serve in deriving algebraic results about ultrafilters.

The contents of the book fall naturally into three parts. The first, comprising Chapters 1 through 5, introduces to topological groups and ultrafilters insofar as the semigroup operation on ultrafilters is not required. Constructions of some important topological groups are given. In particular, that of an extremally disconnected topological group based on a Ramsey ultrafilter. Also one shows that every infinite group admits a nondiscrete zero-dimensional topology in which all translations and the inversion are continuous.

In the second part, Chapters 6 through 9, the Stone–Čech compactification βG of a discrete group G is studied. For this, a special technique based on the concepts of a local left group and a local homomorphism is developed. One proves that if G is a countable torsion free group, then βG contains no nontrivial finite groups. Also the ideal structure of βG is investigated. In particular, one shows that for every infinite Abelian group G, βG contains $2^{2^{|G|}}$ minimal right ideals.

In the third part, using the semigroup βG, almost maximal topological and left topological groups are constructed and their ultrafilter semigroups are examined. Projectives in the category of finite semigroups are characterized. Also one shows that every infinite Abelian group with finitely many elements of order 2 is absolutely ω-resolvable, and consequently, can be partitioned into ω subsets such that every coset modulo infinite subgroup meets each subset of the partition.

The book concludes with a list of open problems in the field.

Some familiarity with set theory, algebra and topology is presupposed. But in general, the book is almost self-contained. It is aimed at graduate students and researchers working in topological algebra and adjacent areas.

Johannesburg, November 2010   Yevhen Zelenyuk

Contents

Chapter 1
Topological Groups

In this chapter some basic concepts and results about topological groups are presented. The largest group topology in which a given filter converges to the identity is described. As an application Markov's Criterion of topologizability of a countable group is derived. Another application is computing the minimum character of a nondiscrete group topology on a countable group which cannot be refined to a nondiscrete metrizable group topology. We conclude by proving Arnautov's Theorem on topologizability of a countably infinite ring.

1.1 The Notion of a Topological Group

Definition 1.1. A group G endowed with a topology is a *topological group* if the multiplication

$$\mu : G \times G \ni (x, y) \mapsto xy \in G$$

and the inversion

$$\iota : G \ni x \mapsto x^{-1} \in G$$

are continuous mappings. A topology which makes a group into a topological group is called a *group topology*.

The continuity of the multiplication and the inversion is equivalent to the continuity of the function

$$\mu' : G \times G \ni (x, y) \mapsto xy^{-1} \in G.$$

Indeed, $\mu'(x, y) = \mu(x, \iota(y))$, $\iota(x) = \mu'(1, x)$ and $\mu(x, y) = \mu'(x, \iota(y))$.

The continuity of μ' means that whenever $a, b \in G$ and U is a neighborhood of ab, there are neighborhoods V and W of a and b, respectively, such that

$$VW^{-1} \subseteq U.$$

It follows that whenever $a_1, \ldots, a_n \in G$, $k_1, \ldots, k_n \in \mathbb{Z}$ and U is a neighborhood of $a_1^{k_1} \ldots a_n^{k_n} \in G$, there are neighborhoods $V_1, \ldots, V_n$ of $a_1, \ldots, a_n$, respectively, such that

$$V_1^{k_1} \cdots V_n^{k_n} \subseteq U.$$

Another immediate property of a topological group G is that the translations and the inversion of G are homeomorphisms. Indeed, for each $a \in G$, the *left translation*

$$\lambda_a : G \ni x \mapsto ax \in G$$

and the *right translation*

$$\rho_a : G \ni x \mapsto xa \in G$$

are continuous mappings, being restrictions of the multiplication. The inversion ι is continuous by the definition. Since we have also that $(\lambda_a)^{-1} = \lambda_{a^{-1}}$, $(\rho_a)^{-1} = \rho_{a^{-1}}$ and $\iota^{-1} = \iota$, all of them are homeomorphisms.

A topological space X is called *homogeneous* if for every $a, b \in X$, there is a homeomorphism $f : X \to X$ such that $f(a) = b$. If G is a topological group and $a, b \in G$, then $\lambda_{ba^{-1}} : G \to G$ is a homeomorphism and $\lambda_{ba^{-1}}(a) = ba^{-1}a = b$. Thus, we have that

Lemma 1.2. *The space of a topological group is homogeneous.*

Now we establish some separation properties of topological groups.

Lemma 1.3. *Every topological group satisfying the T_0 separation axiom is regular and hence Hausdorff.*

In this book, by a *regular* space one means a T_3-space.

Proof. Let G be a T_0 topological group. We first show that G is a T_1-space. Since G is homogeneous, it suffices to show that for every $x \in G \setminus \{1\}$, there is a neighborhood U of 1 not containing x. By T_0, there is an open set U containing exactly one of two points $1, x$. If $1 \in U$, we are done. Otherwise xU^{-1} is a neighborhood of 1 not containing x.

Now we show that for every neighborhood U of 1, there is a closed neighborhood of 1 contained in U. Choose a neighborhood V of 1 such that $VV^{-1} \subseteq U$. Then for every $x \in G \setminus U$, one has $xV \cap V = \emptyset$. Indeed, otherwise $xa = b$ for some $a, b \in V$, which gives us that $x = ba^{-1} \in VV^{-1} \subseteq U$, a contradiction. Hence cl $V \subseteq U$. $\square$

In fact, the following stronger statement holds.

Theorem 1.4. *Every Hausdorff topological group is completely regular.*

Proof. See [55, Theorem 10]. $\square$

Theorem 1.4 is the best possible general separation result. However, for countable topological groups, it can be improved.

A space is *zero-dimensional* if it has a base of clopen (= both closed and open) sets. Note that if a T_0-space is zero-dimensional, then it is completely regular.

Proposition 1.5. *Every countable regular space is normal and zero-dimensional.*

Proof. Let X be a countable regular space and let A and B be disjoint closed subsets of X. Enumerate A and B as

$$A = \{a_n : n < \omega\} \quad \text{and} \quad B = \{b_n : n < \omega\}.$$

Inductively, for each $n < \omega$, choose neighborhoods U_n and V_n of a_n and b_n respectively such that

(a) cl $U_n \cap B = \emptyset$ and $A \cap$ cl $V_n = \emptyset$,

(b) $U_n \cap (\bigcup_{i<n} V_i) = \emptyset$ and $(\bigcup_{i<n} U_i) \cap V_n = \emptyset$, and

(c) $U_n \cap V_n = \emptyset$.

(a) is needed to satisfy (b). Conjunction of (b) and (c) is equivalent to

$$\left(\bigcup_{i \leq n} U_i\right) \cap \left(\bigcup_{i \leq n} V_i\right) = \emptyset.$$

It follows that

$$U = \bigcup_{n < \omega} U_n \quad \text{and} \quad V = \bigcup_{n < \omega} V_n$$

are disjoint neighborhoods of A and B, respectively.

Now, having established that X is normal, let U be an open neighborhood of a point $x \in X$. Without loss of generality one may suppose that $U \neq X$. Then by Urysohn's Lemma, there is a continuous function $f : X \to [0, 1]$ such that $f(x) = 0$ and $f(X \setminus U) = \{1\}$. Since X is countable, there is $r \in [0, 1] \setminus f(X)$. Then $f^{-1}([0, r)) = f^{-1}([0, r])$ is a clopen neighborhood of x contained in U. $\square$

It follows from Lemma 1.3 and Proposition 1.5 that

Corollary 1.6. *Every countable Hausdorff topological group is normal and zero-dimensional.*

Note that every first countable Hausdorff topological group is also normal. (A space is *first countable* if every point has a countable neighborhood base.) This is immediate from the fact that every metric space is normal and the following result.

Theorem 1.7. *A Hausdorff topological group is metrizable if and only if it is first countable. In this case, the metric can be taken to be left invariant.*

Proof. See [34, Theorem 8.3]. $\square$

Starting from Chapter 5, all topological groups are assumed to be Hausdorff.

1.2   The Neighborhood Filter of the Identity

For every set X,

$$\mathcal{P}(X) = \{Y : Y \subseteq X\}.$$

Definition 1.8. Let X be a nonempty set. A *filter* on X is a family $\mathcal{F} \subseteq \mathcal{P}(X)$ with the following properties:

(1) $X \in \mathcal{F}$ and $\emptyset \notin \mathcal{F}$,

(2) if $A, B \in \mathcal{F}$, then $A \cap B \in \mathcal{F}$, and

(3) if $A \in \mathcal{F}$ and $A \subseteq B \subseteq X$, then $B \in \mathcal{F}$.

In other words, a filter on X is a nonempty family of nonempty subsets of X closed under finite intersections and supersets. A classic example of a filter is the set $\mathcal{N}_x$ of all neighborhoods of a point x in a topological space X called the *neighborhood filter* of x. By a neighborhood of x one means any set whose interior contains x. The system $\{\mathcal{N}_x : x \subset X\}$ of all neighborhood filters on X is called the *neighborhood system* of X.

Theorem 1.9. *Let X be a space and let $\{\mathcal{N}_x : x \in X\}$ be the neighborhood system of X. Then*

(i) *for every $x \in X$ and $U \in \mathcal{N}_x$, $x \in U$, and*

(ii) *for every $x \in X$ and $U \in \mathcal{N}_x$, $\{y \in X : U \in \mathcal{N}_y\} \in \mathcal{N}_x$.*

Conversely, given a set X and a system $\{\mathcal{N}_x : x \in X\}$ of filters on X satisfying conditions (i)–(ii), there is a unique topology $\mathcal{T}$ on X for which $\{\mathcal{N}_x : x \in X\}$ is the neighborhood system.

Proof. That the neighborhood system $\{\mathcal{N}_x : x \in X\}$ of a space X satisfies (i)–(ii) is obvious. We need to prove the converse.

Define the operator int on the subsets of X by putting for every $A \subseteq X$

$$\text{int } A = \{x \in X : A \in \mathcal{N}_x\}.$$

We claim that it satisfies the following conditions:

(a) int $X = X$,

(b) int $A \subseteq A$,

(c) int (int A) = int A, and

(d) int $(A \cap B)$ = (int A) $\cap$ (int B).

Indeed, for every $x \in X$, $X \in \mathcal{N}_x$, consequently $x \in \operatorname{int} X$, and so (a) is satisfied.
For (b), if $x \in \operatorname{int} A$, then $A \in \mathcal{N}_x$, and so by (i), $x \in A$.

To check (c), let $x \in \operatorname{int} A$. Then $A \in \mathcal{N}_x$. Applying (ii) we obtain that $\operatorname{int} A \in \mathcal{N}_x$.
It follows that $x \in \operatorname{int}(\operatorname{int} A)$. Hence $\operatorname{int} A \subseteq \operatorname{int}(\operatorname{int} A)$. The converse inclusion
follows from (b).

To check (d), let $x \in (\operatorname{int} A) \cap (\operatorname{int} B)$. Then $A \in \mathcal{N}_x$ and $B \in \mathcal{N}_x$, so $A \cap B \in \mathcal{N}_x$.
It follows that $x \in \operatorname{int}(A \cap B)$. Hence $(\operatorname{int} A) \cap (\operatorname{int} B) \subseteq \operatorname{int}(A \cap B)$. Conversely,
let $x \in \operatorname{int}(A \cap B)$. Then $A \cap B \in \mathcal{N}_x$, consequently $A \in \mathcal{N}_x$ and $B \in \mathcal{N}_x$. It follows
that $x \in (\operatorname{int} A) \cap (\operatorname{int} B)$. Hence $\operatorname{int}(A \cap B) \subseteq (\operatorname{int} A) \cap (\operatorname{int} B)$.

It follows from (a)–(d) that there is a unique topology $\mathcal{T}$ on X such that int is the
interior operator for $(X, \mathcal{T})$. We have that a subset $U \subseteq X$ is a neighborhood of a
point $x \in X$ in $\mathcal{T}$ if and only if $x \in \operatorname{int} U$, and so if and only if $U \in \mathcal{N}_x$. Hence,
$\{\mathcal{N}_x : x \in X\}$ is the neighborhood system for $(X, \mathcal{T})$. $\square$

In a topological group, the neighborhood system is completely determined by the
neighborhood filter of the identity.

Lemma 1.10. *Let G be a topological group and let $\mathcal{N}$ be the neighborhood filter of 1.
Then for every $a \in G$, $a\mathcal{N} = \mathcal{N}a$ is the neighborhood filter of a.*

Here,

$$a\mathcal{N} = \{aB : B \in \mathcal{N}\} \quad \text{and} \quad \mathcal{N}a = \{Ba : B \in \mathcal{N}\}.$$

Proof. Since both λ_a and ρ_a are homeomorphisms and $\lambda_a(1) = \rho_a(1) = a$,

$$a\mathcal{N} = \lambda_a(\mathcal{N}) = \rho_a(\mathcal{N}) = \mathcal{N}a$$

is the neighborhood filter of a. $\square$

The next theorem characterizes the neighborhood filter of the identity of a topological group.

Theorem 1.11. *Let $(G, \mathcal{T})$ be a topological group and let $\mathcal{N}$ be the neighborhood
filter of 1. Then*

(1) *for every $U \in \mathcal{N}$, there is $V \in \mathcal{N}$ such that $VV \subseteq U$,*

(2) *for every $U \in \mathcal{N}$, $U^{-1} \in \mathcal{N}$, and*

(3) *for every $U \in \mathcal{N}$ and $x \in G$, $xUx^{-1} \in \mathcal{N}$.*

*Conversely, given a group G and a filter $\mathcal{N}$ on G satisfying conditions (1)–(3), there
is a unique group topology $\mathcal{T}$ on G for which $\mathcal{N}$ is the neighborhood filter of 1. The
topology $\mathcal{T}$ is Hausdorff if and only if*

(4) $\bigcap \mathcal{N} = \{1\}$.

Note that conditions (2) and (3) in Theorem 1.11 are equivalent, respectively, to

(2′) $\mathcal{N}^{-1} = \mathcal{N}$, and

(3′) for every $x \in G$, $x\mathcal{N}x^{-1} = \mathcal{N}$,

where $\mathcal{N}^{-1} = \{A^{-1} : A \in \mathcal{N}\}$ and $x\mathcal{N}x^{-1} = \{xAx^{-1} : A \in \mathcal{N}\}$.

Proof. That the neighborhood filter of 1 satisfies (1)–(3) follows from the continuity of the multiplication $\mu(x, y)$ at $(1, 1)$ and the mappings $\iota(x)$ and $\lambda_x(\rho_{x^{-1}}(y))$ at 1. To prove the converse, consider the system $\{x\mathcal{N} : x \in G\}$. We claim that it satisfies the conditions of Theorem 1.9.

To check (i), let $x \in G$ and $U \in \mathcal{N}$. It follows from (1)–(2) that there is $V \in \mathcal{N}$ such that $VV^{-1} \subseteq U$. Then $x \in xVV^{-1} \subseteq xU$.

To check (ii), let $x \in G$ and $U \in \mathcal{N}$. It follows from (1) that there is $V \in \mathcal{N}$ such that $VV \subseteq U$. For every $y \in xV$, $yV \subseteq xVV \subseteq xU$, consequently

$$xV \subseteq \{y \in G : xU \in y\mathcal{N}\},$$

and so

$$\{y \in G : xU \in y\mathcal{N}\} \in x\mathcal{N}.$$

Now by Theorem 1.9, there is a unique topology $\mathcal{T}$ on G such that for each $x \in G$, $x\mathcal{N}$ is the neighborhood filter of x, that is, the neighborhoods of x are of the form xU, where U is a neighborhood of 1. To see that $\mathcal{T}$ is a group topology, let $a, b \in G$ be given and let U be a neighborhood of 1. Using conditions (1)–(3) choose a neighborhood V of 1 such that $bVV^{-1}b^{-1} \subseteq U$. Then

$$aV(bV)^{-1} = aVV^{-1}b^{-1} = ab^{-1}bVV^{-1}b^{-1} \subseteq ab^{-1}U.$$

Since $\mathcal{T}$ is a group topology, it is Hausdorff if and only if it is a T_1-topology, and so if and only if $\bigcap \mathcal{N} = \{1\}$.   □

The notion of a filter is closely related to that of a filter base.

Definition 1.12. Let X be a nonempty set. A *filter base* on X is a nonempty family $\mathcal{B} \subseteq \mathcal{P}(X)$ with the following properties:

(1) $\emptyset \notin \mathcal{B}$, and

(2) for every $A, B \in \mathcal{B}$ there is $C \in \mathcal{B}$ such that $C \subseteq A \cap B$.

Equivalently, $\mathcal{B} \subseteq \mathcal{P}(X)$ is a filter base if

$$\mathcal{F} = \{A \subseteq X : A \supseteq B \text{ for some } B \in \mathcal{B}\}$$

is a filter, and in this case we say that $\mathcal{B}$ is a *base* for $\mathcal{F}$. Note that if $\mathcal{F}$ is a filter, then $\mathcal{B} \subseteq \mathcal{F}$ is a base for $\mathcal{F}$ if and only if for every $A \in \mathcal{F}$ there is $B \in \mathcal{B}$ such that $B \subseteq A$.

If X is a topological space and $x \in X$, then a base for the neighborhood filter of x is called a *neighborhood base* at x.

As a consequence we obtain from Theorem 1.11 the following.

Corollary 1.13. *Let $\mathcal{B}$ be a filter base on G satisfying the following conditions:*

(1) *for every $U \in \mathcal{B}$, there is $V \in \mathcal{B}$ such that $VV \subseteq U$,*

(2) *for every $U \in \mathcal{B}$, $U^{-1} \in \mathcal{B}$, and*

(3) *for every $U \in \mathcal{B}$ and $x \in G$, $xUx^{-1} \in \mathcal{B}$.*

Then there is a unique group topology $\mathcal{T}$ on G for which $\mathcal{B}$ is a neighborhood base at 1. The topology $\mathcal{T}$ is Hausdorff if and only if

(4) $\bigcap \mathcal{B} = \{1\}$.

1.3 The Topology $\mathcal{T}(\mathcal{F})$

Definition 1.14. For every filter $\mathcal{F}$ on a group G, let $\mathcal{T}(\mathcal{F})$ denote the largest group topology on G in which $\mathcal{F}$ converges to 1.

Definition 1.14 is justified by the fact that for every family $\{\mathcal{T}_i : i \in I\}$ of group topologies on G, the least upper bound $\bigvee_{i \in I} \mathcal{T}_i$ taken in the lattice of all topologies on G is a group topology.

Definition 1.15. For every filter $\mathcal{F}$ on a group G, let $\tilde{\mathcal{F}}$ denote the filter with a base consisting of subsets of the form

$$\bigcup_{x \in G} x(A_x \cup A_x^{-1} \cup \{1\})x^{-1},$$

where for each $x \in G$, $A_x \in \mathcal{F}$.

Lemma 1.16. *For every filter $\mathcal{F}$ on a group G, $\tilde{\mathcal{F}}$ is the largest filter contained in $\mathcal{F}$ such that*

(i) $1 \in \bigcap \tilde{\mathcal{F}}$,

(ii) $\tilde{\mathcal{F}}^{-1} = \tilde{\mathcal{F}}$*, and*

(iii) *for every $x \in G$, $x\tilde{\mathcal{F}}x^{-1} = \tilde{\mathcal{F}}$.*

Proof. That $\tilde{\mathcal{F}}$ satisfies (i) is obvious. To check (ii) and (iii), let $A_x \in \mathcal{F}$ for each $x \in G$. Then

$$\left(\bigcup_{x \in G} x(A_x \cup A_x^{-1} \cup \{1\})x^{-1} \right)^{-1} = \bigcup_{x \in G} x(A_x \cup A_x^{-1} \cup \{1\})x^{-1}.$$

Consequently, $\tilde{\mathscr{F}}^{-1} = \tilde{\mathscr{F}}$. Next, for every $y \in G$,

$$y\left(\bigcup_{x \in G} x(A_x \cup A_x^{-1} \cup \{1\})x^{-1}\right)y^{-1} = \bigcup_{x \in G} yx(A_x \cup A_x^{-1} \cup \{1\})(yx)^{-1}$$

$$= \bigcup_{x \in G} x(A_{y^{-1}x} \cup A_{y^{-1}x}^{-1} \cup \{1\})x^{-1}$$

$$= \bigcup_{x \in G} x(B_x \cup B_x^{-1} \cup \{1\})x^{-1},$$

where $B_x = A_{y^{-1}x}$ for each $x \in G$. It follows that $y\tilde{\mathscr{F}}y^{-1} = \tilde{\mathscr{F}}$.

To see that $\tilde{\mathscr{F}}$ is the largest filter on G contained in $\mathscr{F}$ and satisfying (i)–(iii), let $\mathscr{G}$ be any such filter and let $A \in \mathscr{G}$. Then $1 \in A$ and for each $x \in G$, there is $A_x \in \mathscr{G}$ such that $x(A_x \cup A_x^{-1})x^{-1} \subseteq A$. Since $\mathscr{G} \subseteq \mathscr{F}$, $A_x \in \mathscr{F}$ for each $x \in G$. Define $B \in \tilde{\mathscr{F}}$ by

$$B = \bigcup_{x \in G} x(A_x \cup A_x^{-1} \cup \{1\})x^{-1}.$$

Then $B \subseteq A$, and so $A \in \tilde{\mathscr{F}}$.   □

For every $n \in \mathbb{N}$, let S_n denote the group of all permutations on $\{1, \ldots, n\}$. The next theorem describes the topology $\mathscr{T}(\mathscr{F})$.

Theorem 1.17. *For every filter $\mathscr{F}$ on a group G, the neighborhood filter of 1 in $\mathscr{T}(\mathscr{F})$ has a base consisting of subsets of the form*

$$\bigcup_{n=1}^{\infty} \bigcup_{\pi \in S_n} \prod_{i=1}^{n} B_{\pi(i)},$$

where $(B_n)_{n=1}^{\infty}$ is a sequence of members of $\tilde{\mathscr{F}}$.

Proof. It is clear that these subsets form a filter base on G. In order to show that this is the neighborhood filter of 1 in a group topology, it suffices to check conditions (1)–(3) of Corollary 1.13. Let $(B_n)_{n=1}^{\infty}$ be any sequence of members of $\tilde{\mathscr{F}}$.

To check (1), define the sequence $(C_n)_{n=1}^{\infty}$ in $\tilde{\mathscr{F}}$ by $C_n = B_{2n} \cap B_{2n-1}$. Then for for every $n \in \mathbb{N}$ and $\pi, \rho \in S_n$,

$$\prod_{i=1}^{n} C_{\pi(i)} \prod_{i=1}^{n} C_{\rho(i)} \subseteq \prod_{i=1}^{n} B_{2\pi(i)-1} \prod_{i=1}^{n} B_{2\rho(i)} = \prod_{j=1}^{2n} B_{\sigma(j)}$$

where $\sigma \in S_{2n}$ is defined by

$$\sigma(j) = \begin{cases} 2\pi(j) - 1 & \text{if } j \leq n \\ 2\rho(j - n) & \text{if } j > n. \end{cases}$$

It follows that

$$\left(\bigcup_{n=1}^{\infty}\bigcup_{\pi\in S_n}\prod_{i=1}^{n}C_{\pi(i)}\right)\left(\bigcup_{n=1}^{\infty}\bigcup_{\pi\in S_n}\prod_{i=1}^{n}C_{\pi(i)}\right)\subseteq\bigcup_{n=1}^{\infty}\bigcup_{\pi\in S_n}\prod_{i=1}^{n}B_{\pi(i)}.$$

To check (2), define the sequence $(C_n)_{n=1}^{\infty}$ in $\tilde{\mathcal{F}}$ by $C_n = B_n^{-1}$ (Lemma 1.16). Then for every $n \in \mathbb{N}$ and $\pi \in S_n$,

$$\left(\prod_{i=1}^{n}B_{\pi(i)}\right)^{-1}=\prod_{i=1}^{n}B_{\rho(i)}^{-1}=\prod_{i=1}^{n}C_{\rho(i)}$$

where $\rho \in S_n$ is defined by $\rho(i) = \pi(n+1-i)$. Consequently,

$$\left(\bigcup_{n=1}^{\infty}\bigcup_{\pi\in S_n}\prod_{i=1}^{n}B_{\pi(i)}\right)^{-1}=\bigcup_{n=1}^{\infty}\bigcup_{\pi\in S_n}\prod_{i=1}^{n}C_{\pi(i)}.$$

To check (3), let $x \in G$. Define the sequence $(C_n)_{n=1}^{\infty}$ in $\tilde{\mathcal{F}}$ by $C_n = xB_nx^{-1}$ (Lemma 1.16). Then for every $n \in \mathbb{N}$ and $\pi \in S_n$,

$$x\left(\prod_{i=1}^{n}B_{\pi(i)}\right)x^{-1}=\prod_{i=1}^{n}xB_{\pi(i)}x^{-1}=\prod_{i=1}^{n}C_{\pi(i)}.$$

Consequently,

$$x\left(\bigcup_{n=1}^{\infty}\bigcup_{\pi\in S_n}\prod_{i=1}^{n}B_{\pi(i)}\right)x^{-1}=\bigcup_{n=1}^{\infty}\bigcup_{\pi\in S_n}\prod_{i=1}^{n}C_{\pi(i)}.$$

Now let G be endowed with any group topology in which $\mathcal{F}$ converges to 1 and let U be a neighborhood of 1. Note that every neighborhood of 1 is a member of $\tilde{\mathcal{F}}$ (Lemma 1.16). Choose inductively a sequence $(V_n)_{n=0}^{\infty}$ of neighborhoods of 1 such that $V_0 = U$ and for every n,

$$V_{n+1}V_{n+1}V_{n+1}\subseteq V_n.$$

Then whenever $n_1,\ldots,n_k$ are distinct numbers in $\mathbb{N}$, one has

$$V_{n_1}\cdots V_{n_k}\subseteq V_n,$$

where $n = \min\{n_1,\ldots,n_k\} - 1$. (To see this, pick $i \in \{1,\ldots,k\}$ such that $n_i = \min\{n_1,\ldots,n_k\}$ and write $V_{n_1}\cdots V_{n_k}$ as $(V_{n_1}\cdots V_{n_{i-1}})V_{n_i}(V_{n_{i+1}}\cdots V_{n_k})$.) It follows that

$$\bigcup_{n=1}^{\infty}\bigcup_{\pi\in S_n}\prod_{i=1}^{n}V_{\pi(i)}\subseteq U,$$

and so U is a neighborhood of 1 in $\mathcal{T}(\mathcal{F})$. $\qquad\square$

1.4   Topologizing a Group

Definition 1.18. Let G be a countably infinite group. Enumerate G as $\{g_n : n < \omega\}$ without repetitions and with $g_0 = 1$.

(i) For every infinite sequence $(a_n)_{n=1}^{\infty}$ in G, define $U((a_n)_{n=1}^{\infty}) \subseteq G$ by

$$U((a_n)_{n=1}^{\infty}) = \bigcup_{n=1}^{\infty} \bigcup_{\pi \in S_n} \prod_{i=1}^{n} B_{\pi(i)},$$

where $B_i = \bigcup_{j=0}^{\infty} g_j \{1, a_{i+j}^{\pm 1}, a_{i+j+1}^{\pm 1}, \ldots\} g_j^{-1}$.

(ii) For every finite sequence $a_1, \ldots, a_n$ in G, define $U(a_1, \ldots, a_n) \subseteq G$ by

$$U(a_1, \ldots, a_n) = \bigcup_{\pi \in S_n} \prod_{i=1}^{n} B_{\pi(i)}^n,$$

where $B_i^n = \bigcup_{j=0}^{n-i} g_j \{1, a_{i+j}^{\pm 1}, a_{i+j+1}^{\pm 1}, \ldots, a_n^{\pm 1}\} g_j^{-1}$. That is, $U(a_1, \ldots, a_n)$ consists of all elements of the form

$$g_{j_1} c_1 g_{j_1}^{-1} \cdots g_{j_n} c_n g_{j_n}^{-1},$$

where $j_i \in \{0, \ldots, n - \pi(i)\}$ and $c_i \in \{1, a_{\pi(i)+j_i}^{\pm 1}, \ldots, a_n^{\pm 1}\}$ for each $i = 1, \ldots, n$, and $\pi \in S_n$. In particular, $U(a_1) = \{1, a_1^{\pm 1}\}$. Also put $U(\emptyset) = \{1\}$.

(iii) For every finite sequence $a_1, \ldots, a_{n-1}$ in G, let $T(a_1, \ldots, a_{n-1}, x)$ denote the set of group words $f(x)$ in the alphabet $G \cup \{x\}$ in which variable x occurs and which have the form

$$f(x) = g_{j_1} c_1 g_{j_1}^{-1} \cdots g_{j_n} c_n g_{j_n}^{-1},$$

where $j_i \in \{0, \ldots, n - \pi(i)\}$ and $c_i \in \{1, a_{\pi(i)+j_i}^{\pm 1}, \ldots, a_{n-1}^{\pm 1}, x^{\pm 1}\}$ for each $i = 1, \ldots, n$, and $\pi \in S_n$. In particular, $T(x)$ consists of two group words x and x^{-1}.

Of course, in the case where G is Abelian, all these definitions look simpler. In particular,

$$B_i^n = \{0, \pm a_i, \ldots, \pm a_n\} \quad \text{and} \quad U(a_1, \ldots, a_n) = \sum_{i=1}^{n} B_i^n.$$

Theorem 1.19. *For every sequence $(a_n)_{n=1}^{\infty}$ in G, the following statements hold:*

(1) $U((a_n)_{n=1}^{\infty})$ *is a neighborhood of* 1 *in* $\mathcal{T}((a_n)_{n=1}^{\infty})$,

(2) $U((a_n)_{n=1}^{\infty}) = \bigcup_{n=1}^{\infty} U(a_1, \ldots, a_n)$,

(3) $U(a_1, \ldots, a_n) = U(a_1, \ldots, a_{n-1}) \cup \{f(a_n) : f(x) \in T(a_1, \ldots, a_{n-1}, x)\}$ *for every* $n \in \mathbb{N}$, *and*

(4) *for every* $n \in \mathbb{N}$ *and* $f(x) \in T(a_1, \ldots, a_{n-1}, x)$, $f(1) \in U(a_1, \ldots, a_{n-1})$.

Proof. (1) follows from Theorem 1.17, and (2)–(4) from Definition 1.18. □

Theorem 1.19 gives us a method of topologizing a countable group. We illustrate it by proving Markov's Criterion.

We say that a group is *topologizable* if it admits a nondiscrete Hausdorff group topology.

An *inequality* over a group G is any expression of the form $f(x) \neq b$, where $f(x)$ is a group word in the alphabet $G \cup \{x\}$ and $b \in G$.

Theorem 1.20 (Markov's Criterion). *A countable group G is topologizable if and only if every finite system of inequalities over G having a solution has also another solution.*

Proof. Necessity is obvious. Indeed, let $\mathcal{T}$ be a nondiscrete Hausdorff group topology on G. Consider any finite system of inequalities over G, say $f_i(x) \neq b_i$, where $i = 1, \dots, n$, having a solution, say $a \in G$, that is, $f_i(a) \neq b_i$ for each $i = 1, \dots, n$. Since $\mathcal{T}$ is a Hausdorff group topology, there is a neighborhood U of a in $\mathcal{T}$ such that $b_i \notin f_i(U)$ for each $i = 1, \dots, n$. Then every element of U is a solution of the system, and since $\mathcal{T}$ is nondiscrete, $U \setminus \{a\} \neq \emptyset$.

The proof of sufficiency is based on Theorem 1.19. It is enough to construct a sequence $(a_n)_{n=1}^{\infty}$ in $G \setminus \{1\}$ such that

$$g_i \notin U(a_i, a_{i+1}, \dots, a_n)$$

for each $n \in \mathbb{N}$ and $i = 1, \dots, n$. This implies that

$$g_i \notin U((a_n)_{n=i}^{\infty})$$

for each $i \in \mathbb{N}$. Then the topology $\mathcal{T}((a_n)_{n=1}^{\infty})$ would be nondiscrete and Hausdorff.

At the first step pick any $a_1 \in G \setminus \{1, g_1^{\pm 1}\}$. Then $g_1 \notin U(a_1) = \{1, a_1^{\pm 1}\}$.

Now fix $n \in \mathbb{N}$ and suppose that elements $a_1, \dots, a_n \in G$ have already been chosen so that $g_i \notin U(a_i, \dots, a_n)$ for each $i = 1, \dots, n$. We need to find $a_{n+1} \in G \setminus \{1\}$ such that $g_i \notin U(a_i, \dots, a_{n+1})$ for each $i = 1, \dots, n+1$. Since

$$U(a_i, \dots, a_{n+1}) = U(a_i, \dots, a_n) \cup \{f(a_{n+1}) : f(x) \in T(a_i, \dots, a_n, x)\},$$

it follows that $g_i \notin U(a_i, \dots, a_{n+1})$ for each $i = 1, \dots, n+1$ if and only if a_{n+1} is a solution of the system of inequalities

$$f(x) \neq g_i,$$

where $i = 1, \dots, n+1$ and $f(x) \in T(a_i, \dots, a_n, x)$. Since $f(1) \in U(a_i, \dots, a_n)$ for each $i = 1, \dots, n+1$ and $f(x) \in T(a_i, \dots, a_n, x)$, 1 is a solution of this system. Hence, there is a solution $a_{n+1} \neq 1$. □

In connection with Theorem 1.20, A. Markov asked whether every infinite group is topologizable. The next two theorems show that for infinite Abelian groups and for free groups the answer is "Yes".

Theorem 1.21. *For every Abelian group G and for every $a \in G \setminus \{0\}$, there is a homomorphism $f : G \to \mathbb{T}$ such that $f(a) \neq 1$.*

Proof. We first define a homomorphism $f_0 : \langle a \rangle \to \mathbb{T}$ such that $f_0(a) \neq 1$. If a has finite order, say n, define

$$f(ka) = e^{\frac{2\pi i k}{n}}$$

for every $k = 1, \ldots, n$. If a has infinite order, pick any $x \in \mathbb{T} \setminus \{1\}$ and put

$$f(ka) = x^k$$

for every $k \in \mathbb{Z}$.

Now consider the set of all pairs (H, g), where H is a subgroup of G containing $\langle a \rangle$ and $g : H \to \mathbb{T}$ is a homomorphism extending f_0, ordered by

$$(H_1, g_1) \leq (H_2, g_2) \quad \text{if and only if} \quad H_1 \subseteq H_2 \text{ and } g_2|_{H_1} = g_1.$$

By Zorn's Lemma, there is a maximal pair (H, f). We claim that $H = G$.

Indeed, assume on the contrary that there is $c \in G \setminus H$. To derive a contradiction, let $H' = H + \langle c \rangle$ and define a homomorphism $f' : H' \to \mathbb{T}$ extending f. If there is $n \in \mathbb{N}$ with $nc \in H$, choose the smallest such n and then $z \in \mathbb{T}$ such that $z^n = f(nc)$, and if there is no such n, choose arbitrary $z \in \mathbb{T}$. Define

$$f'(b + kc) = f(b) \cdot z^k$$

for every $b \in H$ and $k \in \mathbb{Z}$.   □

A topological group G is called *totally bounded* if it is Hausdorff and for every nonempty open $U \subseteq G$ there is a finite $F \subseteq G$ such that $FU = G$. A topological group is totally bounded if and only if it can be topologically and algebraically embedded into a compact group (see [4, Corollary 3.7.17]). By a *compact group* one means a compact Hausdorff topological group.

Corollary 1.22. *Every infinite Abelian group admits a totally bounded group topology.*

Proof. Let G be an infinite Abelian group. By Theorem 1.21, for every $a \in G \setminus \{0\}$, there is a homomorphism $f_a : G \to \mathbb{T}$ with $f_a(a) \neq 1$. Let $K = \prod_{a \in G \setminus \{0\}} \mathbb{T}_a$ where $\mathbb{T}_a = \mathbb{T}$. Define $f : G \to K$ by $(f(x))_a = f_a(x)$. Clearly f is an injective homomorphism. Being a subgroup of a compact group, $f(G)$ is totally bounded. Hence, the topology on G consisting of subsets $f^{-1}(U)$, where U ranges over open subsets of $f(G)$, is as required.   □

Theorem 1.23. *Let X be a set and let F be the free group generated by X. Then for every $w \in F \setminus \{\emptyset\}$, there exist $n \in \mathbb{N}$ and a homomorphism $f : F \to S_{n+1}$ such that $f(w) \neq \iota$, where ι is the identity permutation.*

Proof. Write w as $x_1^{\varepsilon_1} x_2^{\varepsilon_2} \dots x_n^{\varepsilon_n}$ where each $\varepsilon_i = \pm 1$ and $\varepsilon_i = \varepsilon_{i+1}$ whenever $x_i = x_{i+1}$. It suffices to define a mapping $X \ni x \mapsto \pi_x \in S_{n+1}$ such that

$$\pi_{x_1}^{\varepsilon_1} \circ \pi_{x_2}^{\varepsilon_2} \circ \cdots \circ \pi_{x_n}^{\varepsilon_n} \neq \iota.$$

Given $x \in X$ and $\varepsilon = \pm 1$, let

$$D_\varepsilon(x) = \{i \in \{1, \dots, n\} : x_i = x \text{ and } \varepsilon_i = \varepsilon\}.$$

Note that $D_{-1}(x) \cap (D_1(x) + 1) = \emptyset$ and $(D_{-1}(x) + 1) \cap D_1(x) = \emptyset$, so

$$(D_{-1}(x) \cup (D_{-1}(x) + 1)) \cap (D_1(x) \cup (D_1(x) + 1)) = \emptyset.$$

First define π_x on $D_{-1}(x) \cup (D_1(x) + 1)$ by

$$\pi_x(i) = \begin{cases} i + 1 & \text{if } i \in D_{-1}(x) \\ i - 1 & \text{if } i \in D_1(x) + 1. \end{cases}$$

Then extend it in any way to a member $\pi_x \in S_{n+1}$.

Now we claim that for each $i = 1, \dots, n$, $\pi_{x_i}^{\varepsilon_i}(i + 1) = i$.

Indeed, if $\varepsilon_i = 1$, then $\pi_{x_i}^{\varepsilon_i}(i + 1) = \pi_{x_i}(i + 1) = i$. If $\varepsilon_i = -1$, then $\pi_{x_i}(i) = i + 1$, and consequently, $\pi_{x_i}^{\varepsilon_i}(i + 1) = i$.

It follows that $\pi_{x_1}^{\varepsilon_1} \circ \pi_{x_2}^{\varepsilon_2} \circ \cdots \circ \pi_{x_n}^{\varepsilon_n}(n + 1) = 1$ and hence $\pi_{x_1}^{\varepsilon_1} \circ \pi_{x_2}^{\varepsilon_2} \circ \cdots \circ \pi_{x_n}^{\varepsilon_n} \neq \iota$. $\quad\square$

Corollary 1.24. *Every free group admits a totally bounded group topology.*

Markov's question had been open for a long time. However, eventually it was solved in the negative.

Example 1.25. Let $m, n \in \mathbb{N}$, $m \geq 2$, $n \geq 665$ and n is odd. Consider the *Adian* group $A(m, n)$. This is a torsion free m-generated group whose center is an infinite cyclic group $\langle c \rangle$ and the quotient $A(m, n)/\langle c \rangle$ is an infinite group of period n. More precisely, $A(m, n)/\langle c \rangle$ is the *Burnside* group $B(m, n)$, the largest group on m generators satisfying the identity $x^n = 1$.

Let $x \in A(m, n) \setminus \langle c \rangle$. It is clear that $x^n \in \langle c \rangle$, because $A(m, n)/\langle c \rangle$ has period n. We claim that $x^n \notin \langle c^n \rangle$.

To see this, assume the contrary. So $x^n = (c^n)^k = (c^k)^n$ for some $k \in \mathbb{Z}$. Let $z = xc^{-k}$. Then $z \notin \langle c \rangle$ and $z^n = x^n c^{-kn} = 1$, since $\langle c \rangle$ is the center. But this contradicts that $A(m, n)$ is torsion free.

Now let $G = A(m,n)/\langle c^n \rangle$ and let $D = \langle c \rangle / \langle c^n \rangle$. Then G is infinite,

$$D = \{1, d_1, \ldots, d_{n-1}\} \subset G$$

and for every $x \in G \setminus D$, one has $x^n \in \{d_1, \ldots, d_{n-1}\}$. It follows that for every T_1-topology on G in which 1 is not an isolated point, the mapping

$$G \ni x \mapsto x^n \in G$$

is discontinuous at 1. Hence G is nontopologizable.

1.5 Metrizable Refinements

The *character* of a space is the minimum cardinal κ such that every point has a neighborhood base of cardinality $\leq \kappa$.

Definition 1.26. For every countable topologizable group G, let $\mathfrak{p}_G$ denote the minimum character of a nondiscrete Hausdorff group topology on G which cannot be refined to a nondiscrete metrizable group topology.

Equivalently, $\mathfrak{p}_G$ is the supremum of all cardinals κ such that every nondiscrete Hausdorff group topology on G of character $< \kappa$ has a nondiscrete metrizable refinement.

In this section we show that the cardinal $\mathfrak{p}_G$ is equal to a well-known cardinal invariant of the continuum.

We say that a family $\mathcal{F} \subseteq \mathcal{P}(X)$ has the *strong finite intersection property* if for every finite $\mathcal{H} \subseteq \mathcal{F}$, $\bigcap \mathcal{H}$ is infinite. A subset $A \subseteq X$ is a *pseudo-intersection* of a family $\mathcal{F} \subseteq \mathcal{P}(X)$ if $A \setminus B$ is finite for all $B \in \mathcal{F}$.

Definition 1.27. The *pseudo-intersection* number $\mathfrak{p}$ is the minimum cardinality of a family $\mathcal{F} \subseteq \mathcal{P}(\omega)$ having the strong finite intersection property but no infinite pseudo-intersection.

We show that

Theorem 1.28. *For every countable topologizable group G, $\mathfrak{p}_G = \mathfrak{p}$.*

Before proving Theorem 1.28 we establish several auxiliary statements.

Lemma 1.29. *Let $\mathcal{T}$ be a nondiscrete group topology on G, let $(V_n)_{n=1}^{\infty}$ be a sequence of neighborhoods of 1, and let $A \subseteq G$ be such that $1 \in \mathrm{cl}_{\mathcal{T}} A \setminus A$. Then there is a sequence $(a_n)_{n=1}^{\infty}$ in A such that*

$$U((a_n)_{n=i}^{\infty}) \subseteq V_i$$

for every $i \in \mathbb{N}$.

Proof. Construct inductively a sequence $(a_n)_{n=1}^{\infty}$ in A such that

$$U(a_i, \ldots, a_n) \subseteq V_i$$

for every $n \in \mathbb{N}$ and $i = 1, \ldots, n$.

Without loss of generality one may assume that all V_n are open. Pick $a_1 \in A \cap V_1 \cap V_1^{-1}$. Then $U(a_1) \subseteq V_1$.

Now fix $n > 1$ and suppose that elements $a_1, \ldots, a_{n-1} \in A$ have already been chosen so that

$$U(a_i, \ldots, a_{n-1}) \subseteq V_i$$

for every $i = 1, \ldots, n - 1$. Choose a neighborhood W_n of 1 such that for every $i = 1, \ldots, n - 1$ and $f(x) \in T(a_i, \ldots, a_{n-1}, x)$, one has $f(W_n) \subseteq V_i$. This can be done because $f(1) \in U(a_i, \ldots, a_{n-1})$. Pick $a_n \in A \cap V_n \cap V_n^{-1} \cap W_n$. Then $U(a_n) \subseteq V_n$ and for every $i = 1, \ldots, n - 1$,

$$U(a_i, \ldots, a_n) = U(a_i, \ldots, a_{n-1}) \cup \{f(a_n) : f(x) \in T(a_i, \ldots, a_{n-1}, x)\} \subseteq V_i.$$

$\square$

Lemma 1.30. *Let $\mathcal{T}$ be a nondiscrete Hausdorff group topology on G and let $A \subseteq G$ be such that $1 \in \mathrm{cl}_{\mathcal{T}} A \setminus A$. Then there is a sequence $(a_n)_{n=1}^{\infty}$ in A such that*

(i) *$a_n \notin U(a_1, \ldots, a_{n-1})$ for every $n \in \mathbb{N}$, and*

(ii) *whenever $k \in \mathbb{N}$ and $1 \le n_1 < \cdots < n_k < \omega$,*

$$U(a_{n_1}, \ldots, a_{n_k}) \setminus U(a_{n_1}, \ldots, a_{n_{k-1}})$$
$$\subseteq U(a_1, a_2, \ldots, a_{n_k}) \setminus U(a_1, a_2, \ldots, a_{n_k - 1}).$$

Proof. Construct inductively a sequence $(a_n)_{n=1}^{\infty}$ in A such that for every $n \in \mathbb{N}$ and $f(x) \in T(a_1, \ldots, a_{n-1}, x)$, either $f(a_n) = f(1)$ or $f(a_n) \notin U(a_1, \ldots, a_{n-1})$. Then (i) is satisfied because $x \in T(a_1, \ldots, a_{n-1}, x)$ and $1 \notin A$. To check (ii), let

$$g \in U(a_{n_1}, \ldots, a_{n_k}) \setminus U(a_{n_1}, \ldots, a_{n_{k-1}}).$$

Then $g = f(a_{n_k})$ for some $f(x) \in T(a_{n_1}, \ldots, a_{n_{k-1}}, x)$. Since

$$T(a_{n_1}, \ldots, a_{n_{k-1}}, x) \subseteq T(a_1, \ldots, a_{n_k - 1}, x),$$

$f(x) \in T(a_1, \ldots, a_{n_k - 1}, x)$, and since $g \notin U(a_{n_1}, \ldots, a_{n_{k-1}})$, $f(a_{n_k}) \ne f(1)$. Hence by the construction, $g = f(a_{n_k}) \notin U(a_1, \ldots, a_{n_k - 1})$.

$\square$

Lemma 1.31. *Let $\mathcal{T}$ be a nondiscrete Hausdorff group topology on G and let $\mathcal{U} \subseteq \mathcal{T}$. Then there is a Hausdorff group topology $\mathcal{T}'$ on G such that $\mathcal{U} \subseteq \mathcal{T}' \subseteq \mathcal{T}$ and the character of $\mathcal{T}'$ does not exceed $\max\{\omega, |\mathcal{U}|\}$.*

Proof. It suffices to show that for every $U \in \mathcal{U}$, there is a Hausdorff metrizable group topology $\mathcal{T}_U$ on G such that $U \in \mathcal{T}_U \subseteq \mathcal{T}$. Then the topology $\mathcal{T}' = \bigvee_{U \in \mathcal{U}} \mathcal{T}_U$ would be as required.

Let $\mathcal{V} = \{Ux^{-1} : x \in U\}$. Enumerate $\mathcal{V}$ and $G \setminus \{1\}$ as $\{V_n : n \in \mathbb{N}\}$ and $\{x_n : n \in \mathbb{N}\}$, respectively. Construct inductively a sequence $(W_n)_{n=0}^{\infty}$ of neighborhoods of 1 in $\mathcal{T}$ such that $W_0 = G$ and for every $n \in \mathbb{N}$ the following conditions are satisfied:

(a) $W_n \subseteq V_n$,

(b) $x_n \notin W_n$,

(c) $W_n W_n^{-1} \subseteq W_{n-1}$, and

(d) $x_k W_n x_k^{-1} \subseteq W_{n-1}$ for all $k = 1, \ldots, n$.

It follows from (b)–(d) that there is a group topology $\mathcal{T}_U$ on G for which $\{W_n : n \in \mathbb{N}\}$ is a neighborhood base at 1. Then $\mathcal{T}_U$ is metrizable, $\mathcal{T}_U \subseteq \mathcal{T}$ and by (a), $U \in \mathcal{T}_U$. □

For every group topology $\mathcal{T}_0$ on G, let $\mathcal{T}_0^{\triangle}$ denote the lattice of all group topologies $\mathcal{T}$ such that that $\mathcal{T}_0 \subseteq \mathcal{T}$. Let $\mathcal{F}_0$ denote the Fréchet filter on $\mathbb{N}$ and $\mathcal{F}_0^{\triangle}$ the lattice of all filters $\mathcal{F}$ such that $\mathcal{F}_0 \subseteq \mathcal{F}$.

Theorem 1.32. *Let $\mathcal{T}'$ be a nondiscrete metrizable group topology on G. Then there is a metrizable group topology $\mathcal{T}_0 \supseteq \mathcal{T}'$ and a mapping $h : G \to \mathbb{N}$ such that*

$$\mathcal{T}_0^{\triangle} \ni \mathcal{T} \mapsto h(\mathcal{N}(\mathcal{T})) \in \mathcal{F}_0^{\triangle}$$

is a surjective order preserving mapping.

Here, $\mathcal{N}(\mathcal{T})$ denotes the neighborhood filter of 1 in $\mathcal{T}$, and $h(\mathcal{N}(\mathcal{T}))$ the filter on $\mathbb{N}$ with a base consisting of subsets $h(U)$ where $U \in \mathcal{N}(\mathcal{T})$.

Proof. Pick any sequence $(b_n)_{n=1}^{\infty}$ in $G \setminus \{1\}$ converging to 1 in $\mathcal{T}'$ and let $A = \{b_n : n \in \mathbb{N}\}$. By Lemma 1.30, there is a sequence $(a_n)_{n=1}^{\infty}$ in A such that

(i) $a_n \notin U(a_1, \ldots, a_{n-1})$ for every $n \in \mathbb{N}$, and

(ii) whenever $k \in \mathbb{N}$ and $1 \leq n_1 < \cdots < n_k < \omega$,

$$U(a_{n_1}, \ldots, a_{n_k}) \setminus U(a_{n_1}, \ldots, a_{n_{k-1}})$$
$$\subseteq U(a_1, a_2, \ldots, a_{n_k}) \setminus U(a_1, a_2, \ldots, a_{n_k-1}).$$

Clearly, $(a_n)_{n=1}^{\infty}$ converges to 1 in $\mathcal{T}'$. By Lemma 1.31, there is a metrizable group topology $\mathcal{T}_0$ on G such that

$$\mathcal{T}' \subseteq \mathcal{T}_0 \subseteq \mathcal{T}((a_n)_{n=1}^{\infty}).$$

Define a function $h : G \to \mathbb{N}$ by the condition that for every $n \in \mathbb{N}$ and $x \in U(a_1, \ldots, a_n) \setminus U(a_1, \ldots, a_{n-1})$, one has $h(x) = n$.

It is clear that the mapping

$$\mathcal{T}_0^{\Delta} \ni \mathcal{T} \mapsto h(\mathcal{N}(\mathcal{T})) \in \mathcal{F}_0^{\Delta}$$

is order preserving. We need to show that it is surjective.

Let I be an infinite subset of $\mathbb{N}$. Write I as $\{n_k : k \in \mathbb{N}\}$, where $(n_k)_{k=1}^{\infty}$ is an increasing sequence in $\mathbb{N}$. Define the group topology $\mathcal{T}_I$ on G by

$$\mathcal{T}_I = \mathcal{T}((a_{n_k})_{k=1}^{\infty}).$$

Then $h(\mathcal{N}(\mathcal{T}_I))$ is the filter on $\mathbb{N}$ consisting of all subsets $J \subseteq \mathbb{N}$ with finite $I \setminus J$. Indeed, for every $x \in U(a_{n_1}, \ldots, a_{n_k}) \setminus U(a_{n_1}, \ldots, a_{n_{k-1}})$, one has $h(x) = n_k$, and $a_{n_k} \in U(a_{n_1}, \ldots, a_{n_k}) \setminus U(a_{n_1}, \ldots, a_{n_{k-1}})$.

Now let $\mathcal{F}$ be any filter on $\mathbb{N}$ containing $\mathcal{F}_0$. Define the group topology $\mathcal{T}$ on G by

$$\mathcal{T} = \bigvee_{I \in \mathcal{F}} \mathcal{T}_I.$$

Then $h(\mathcal{N}(\mathcal{T})) = \mathcal{F}$.   □

Now we are in a position to prove Theorem 1.28.

Proof of Theorem 1.28. To prove that $\mathfrak{p}_G \geq \mathfrak{p}$, let $\mathcal{T}'$ be a nondiscrete Hausdorff group topology on G of character $\kappa < \mathfrak{p}$. We need to find a metrizable group topology $\mathcal{T}_0 \supseteq \mathcal{T}'$. Let $\{V_{\alpha} : \alpha < \kappa\}$ be a neighborhood base at 1 in $\mathcal{T}'$. Since $\kappa < \mathfrak{p}$, there is an infinite pseudo-intersection of the neighborhood base, and consequently, a one-to-one sequence $(a_n)_{n=1}^{\infty}$ in G converging to 1. Using Lemma 1.29 and that $\kappa < \mathfrak{p}$, inductively for each $\alpha \leq \kappa$, construct a subsequence $(a_n^{\alpha})_{n=1}^{\infty}$ of $(a_n)_{n=1}^{\infty}$ such that

(i) for each $\alpha < \kappa$, $U((a_n^{\alpha})_{n=1}^{\infty}) \subseteq V_{\alpha}$, and

(ii) for each $\alpha \leq \kappa$ and $\gamma < \alpha$, $\{a_n^{\alpha} : n \in \mathbb{N}\} \setminus \{a_n^{\gamma} : n \in \mathbb{N}\}$ is finite.

Then $(a_n^{\kappa})_{n=1}^{\infty}$ is a subsequence of $(a_n)_{n=1}^{\infty}$ with the following property: for each $\alpha < \kappa$, there is $i(\alpha) \in \mathbb{N}$ such that $U((a_n^{\kappa})_{n=i(\alpha)}^{\infty}) \subseteq V_{\alpha}$. By Lemma 1.31, the topology $\mathcal{T}((a_n^{\kappa})_{n=1}^{\infty})$ can be weakened to a metrizable group topology $\mathcal{T}_0$ in which for each $i \in \mathbb{N}$, $U((a_n^{\kappa})_{n=i}^{\infty})$ remains a neighborhood of 1. It then follows that $\mathcal{T}' \subseteq \mathcal{T}_0$.

To prove that $\mathfrak{p} \geq \mathfrak{p}_G$, let $\mathcal{A}$ be a family of subsets of $\mathbb{N}$ having the strong finite intersection property and with $|\mathcal{A}| = \kappa < \mathfrak{p}_G$. We need to find an infinite pseudo-intersection of $\mathcal{A}$. Define the filter $\mathcal{F}$ on $\mathbb{N}$ by

$$\mathcal{F} = \left\{A \subseteq \mathbb{N} : A \supseteq \bigcap \mathcal{B} \text{ for some finite } \mathcal{B} \subseteq \mathcal{A}\right\}.$$

Without loss of generality one may suppose that $\mathcal{F}$ contains the Fréchet filter $\mathcal{F}_0$. Since G is topologizable, there is a nondiscrete Hausdorff group topology $\mathcal{T}'$ on G, and by Lemma 1.31, $\mathcal{T}'$ can be chosen to be metrizable. By Theorem 1.28, there is a metrizable group topology $\mathcal{T}_0 \supseteq \mathcal{T}'$ and a mapping $h : G \to \mathbb{N}$ such that

$$\mathcal{T}_0^{\triangle} \ni \mathcal{T} \mapsto h(\mathcal{N}(\mathcal{T})) \in \mathcal{F}_0^{\triangle}$$

is a surjective order preserving mapping. Pick $\mathcal{T} \in \mathcal{T}_0^{\triangle}$ such that $h(\mathcal{N}(\mathcal{T})) = \mathcal{F}$. Choosing a family $\mathcal{U}$ of open neighborhoods of 1 in $\mathcal{T}$ such that $|\mathcal{U}| = \kappa$ and $\{h(U) : U \in \mathcal{U}\}$ is a base for $\mathcal{F}$ and applying Lemma 1.31, one may suppose that $\mathcal{T}$ has character κ. Since $\kappa < \mathfrak{p}_G$, there is a nondiscrete metrizable group topology $\mathcal{T}_1 \supseteq \mathcal{T}$. Then the filter $\mathcal{F}_1 = h(\mathcal{N}(\mathcal{T}_1))$ has a countable base, and consequently, an infinite pseudo-intersection $A \subseteq \mathbb{N}$. Since $\mathcal{F}_1 \supseteq \mathcal{F} \supseteq \mathcal{A}$, A is also a pseudo-intersection of $\mathcal{A}$.   □

1.6 Topologizability of a Countably Infinite Ring

A ring is *topologizable* if it admits a nondiscrete Hausdorff *ring topology*, that is, a topology in which the addition, the additive inversion, and the multiplication are continuous.

In this section we show that

Theorem 1.33 (Arnautov's Theorem). *Every countably infinite ring is topologizable.*

Note that the ring in Theorem 1.33 is not assumed to be associative.

The proof of Theorem 1.33 is based on two auxiliary results. First is the ring version of Theorem 1.20.

Proposition 1.34. *A countable ring R is topologizable if and only if every finite system of inequalities over R having a solution has also another solution.*

By an *inequality* over R one means any expression of the form $f(x) \neq b$, where $f(x)$ is a ring word in the alphabet $R \cup \{x\}$ and $b \in R$.

The proof of Proposition 1.34 is similar to that of Theorem 1.20. First of all we need the following lemma.

Lemma 1.35. *A filter $\mathcal{N}$ on a ring R is the neighborhood filter of 0 in a ring topology if and only if the following conditions are satisfied:*

(1) *for every $U \in \mathcal{N}$, there is $V \in \mathcal{N}$ such that $V + V \subseteq U$,*

(2) *for every $U \in \mathcal{N}$, $-U \in \mathcal{N}$,*

(3) *for every $U \in \mathcal{N}$, there is $V \in \mathcal{N}$ such that $VV \subseteq U$,*

(4) *for every $U \in \mathcal{N}$ and $x \in R$, there is $V \in \mathcal{N}$ such that $xV \subseteq U$ and $Vx \subseteq U$.*

Proof. Necessity is obvious. We need to check sufficiency. It follows from (1), (2) and Theorem 1.11 that there is a topology $\mathcal{T}$ on R in which, for each $x \in R$, $x + \mathcal{N}$ is the neighborhood filter of x, and the addition and the additive inversion are continuous mappings.

To see that the multiplication is continuous, let $x, y \in R$ and let U_{xy} be a neighborhood of $xy \in R$. Put $U = -xy + U_{xy}$. Then U is a neighborhood of 0 and $U_{xy} = xy + U$. Choose a neighborhood V of 0 such that $V + V + V \subseteq U$. Applying (3) and (4) we obtain that there is a neighborhood W of 0 such that $WW \subseteq V$, $xW \subseteq V$ and $Wy \subseteq V$. Then $W_x = x + W$ is a neighborhood of x, $W_y = y + W$ is a neighborhood of y and

$$W_x W_y = (x + W)(y + W)$$
$$= xy + xW + Wy + WW \subseteq xy + V + V + V \subseteq xy + U = U_{xy}. \quad \square$$

Proof of Proposition 1.34. Necessity is obvious. We have to prove sufficiency. Assume that every finite system of inequalities over R having a solution has also another solution. Enumerate R without repetitions as $\{b_n : n < \omega\}$ with $b_0 = 0$. Define inductively a set P of ring words in the alphabet $\{x_1, x_2, \ldots\}$ and to each $f = f(x_1, \ldots, x_k) \in P$ assign a vector $\vec{r}_f = (r_1, \ldots, r_k) \in \mathbb{N}^k$ as follows:

(i) $x_1 \in P$ and $\vec{r}_{x_1} = (1)$,

(ii) if $f \in P$ and $\vec{r}_f = (r_1, \ldots, r_k)$, then

$$g = g(x_1, \ldots, x_k) = -f(x_1, \ldots, x_k) \in P$$

and $\vec{r}_g = (r_1, \ldots, r_k)$,

(iii) if $f, g \in P$, $\vec{r}_f = (r_1, \ldots, r_k)$ and $\vec{r}_g = (s_1, \ldots, s_l)$, then

$$h = h(x_1, \ldots, x_{k+l}) = f(x_1, \ldots, x_k) + g(x_{k+1}, \ldots, x_{k+l}) \in P,$$
$$t = t(x_1, \ldots, x_{k+l}) = f(x_1, \ldots, x_k) \cdot g(x_{k+1}, \ldots, x_{k+l}) \in P,$$

and $\vec{r}_h = \vec{r}_t = (2r_1, \ldots, 2r_k, 2s_1, \ldots, 2s_l)$,

(iv) if $f \in P$ and $\vec{r}_f = (r_1, \ldots, r_k)$, then for each $m \in \mathbb{N}$,

$$h = h(x_1, \ldots, x_k) = b_m \cdot f(x_1, \ldots, x_k) \in P,$$
$$t = t(x_1, \ldots, x_k) = f(x_1, \ldots, x_k) \cdot b_m \in P,$$

and $\vec{r}_h = \vec{r}_t = (r_1 + m, \ldots, r_k + m)$.

Lemma 1.36. *For every filter $\mathcal{F}$ on R, the subsets*

$$\bigcup_{f \in P} f(A_{r_1}, \ldots, A_{r_k}),$$

where $(r_1, \ldots, r_k) = \vec{r}_f$ and $(A_n)_{n=1}^{\infty}$ is a sequence of members of $\mathcal{F}$, form a neighborhood base at 0 in the largest ring topology on R in which $\mathcal{F}$ converges to 0.

Proof. Let $(A_n)_{n=1}^\infty$ be any sequence of members of $\mathscr{F}$ and let

$$U = \bigcup_{f \in P} f(A_{r_1}, \dots, A_{r_k}).$$

By (ii), $U = -U$. Define the sequence $(B_n)_{n=1}^\infty$ of members of $\mathscr{F}$ by $B_n = A_{2n}$ and let

$$V = \bigcup_{f \in P} f(B_{r_1}, \dots, B_{r_k}).$$

Then by (iii), $V + V \subseteq U$. Indeed,

$$f(B_{r_1}, \dots, B_{r_k}) + g(B_{s_1}, \dots, B_{s_l}) = f(A_{2r_1}, \dots, A_{2r_k}) + g(A_{2s_1}, \dots, A_{2s_l})$$
$$= h(A_{2r_1}, \dots, A_{2r_k}, A_{2s_1}, \dots, A_{2s_l}).$$

Similarly, $VV \subseteq U$. Next, for each $m \in \mathbb{N}$, define the sequence $(C_n)_{n=1}^\infty$ of members of $\mathscr{F}$ by $C_n = A_{n+m}$ and let

$$W = \bigcup_{f \in P} f(C_{r_1}, \dots, C_{r_k}).$$

Then by (iv), $b_m W \subseteq U$ and $W b_m \subseteq U$.

Applying Lemma 1.35, we obtain that there is a ring topology $\mathscr{T}$ on R in which the above subsets form a neighborhood base at 0, and so $\mathscr{F}$ converges to 0. It remains to check that $\mathscr{T}$ is the largest such topology.

Let $\mathscr{T}'$ be any ring topology on R in which $\mathscr{F}$ converges to 0 and let U be any neighborhood of 0 in $\mathscr{T}'$. We need to show that U is a neighborhood of 0 in $\mathscr{T}$. Choose a sequence $(U_n)_{n=0}^\infty$ of neighborhoods of 0 in $\mathscr{T}'$ with $U_0 = -U_0 \subseteq U$ such that for each $n \in \mathbb{N}$ and $m = 1, \dots, n$,

$$U_n = -U_n, \ U_n + U_n \subseteq U_{n-1}, \ U_n U_n \subseteq U_{n-1}, \ b_m U_n \subseteq U_{n-1} \text{ and } U_n b_m \subseteq U_{n-1}.$$

We show that for every $f \in P$ with $\vec{r}(f) = (r_1, \dots, r_k)$ and for every $j = 0, 1, \dots,$ one has

$$f(U_{r_1+j}, \dots, U_{r_k+j}) \subseteq U_j,$$

in particular, $f(U_{r_1}, \dots, U_{r_k}) \subseteq U_0$.

Suppose that the statement holds for some $f, g \in P$ with $\vec{r}_f = (r_1, \dots, r_k)$ and $\vec{r}_g = (s_1, \dots, s_l)$. Consider the words

$$h = h(x_1, \dots, x_{k+l}) = f(x_1, \dots, x_k) + g(x_{k+1}, \dots, x_{k+l}) \in P,$$
$$t = t(x_1, \dots, x_k) = b_m f(x_1, \dots, x_k) \in P$$

with $\vec{r}_h = (2r_1, \ldots, 2r_k, 2s_1, \ldots, 2s_l)$ and $\vec{r}_t = (r_1 + m, \ldots, r_k + m)$. We have that

$$h(U_{2r_1+j}, \ldots, U_{2s_l+j}) = f(U_{2r_1+j}, \ldots, U_{2r_k+j}) + g(U_{2s_1+j}, \ldots, U_{2s_l+j})$$
$$\subseteq f(U_{r_1+j+1}, \ldots, U_{r_k+j+1}) + g(U_{s_1+j+1}, \ldots, U_{s_l+j+1})$$
$$\subseteq U_{j+1} + U_{j+1} \subseteq U_j$$

and

$$t(U_{r_1+m+j}, \ldots, U_{r_k+m+j}) = b_m f(U_{r_1+m+j}, \ldots, U_{r_k+m+j})$$
$$\subseteq b_m U_{m+j} \subseteq U_{m+j-1} \subseteq U_j.$$

Considering the words $f(x_1, \ldots, x_k) g(x_{k+1}, \ldots, x_{k+l})$ and $f(x_1, \ldots, x_k) b_m$ is similar, and the case of the word $-f(x_1, \ldots, x_k)$ is trivial. $\square$

Now for every $n \in \mathbb{N}$, denote by P_n the subset of P consisting of all $f \in P$ such that $\max\{r_1, \ldots, r_k\} \leq n$, where $(r_1, \ldots, r_k) = \vec{r}_f$, in particular, $P_1 = \{x_1, -x_1\}$. Note that P_n is finite. For every finite sequence $a_1, \ldots, a_n$ in R, define the subset $U(a_1, \ldots, a_n) \subseteq R$ by

$$U(a_1, \ldots, a_n) = \{0\} \cup \bigcup_{f \in P_n} f(\{a_{r_1}, \ldots, a_n\}, \ldots, \{a_{r_k}, \ldots, a_n\}),$$

where $(r_1, \ldots, r_k) = \vec{r}_f$. We then obtain that for every infinite sequence $(a_n)_{n=1}^{\infty}$ in R,

$$\bigcup_{n=1}^{\infty} U(a_1, \ldots, a_n) = \bigcup_{f \in P} f(A_{r_1}, \ldots, A_{r_k}),$$

where $A_i = \{a_i, a_{i+1}, \ldots\}$ for all $i \in \mathbb{N}$. Consequently by Lemma 1.36,

$$\bigcup_{n=1}^{\infty} U(a_1, \ldots, a_n)$$

is a neighborhood of 0 in the largest ring topology on R in which $(a_n)_{n=1}^{\infty}$ converges to 0. Hence, in order to prove that R is topologizable, it suffices to construct a sequence $(a_n)_{n=1}^{\infty}$ in R such that $a_n \neq 0$ and $b_i \notin U(a_i, a_{i+1}, \ldots, a_n)$ for all $n \in \mathbb{N}$ and $i = 1, \ldots, n$.

To this end, for every finite sequence $a_1, \ldots, a_{n-1}$ in R, denote by $T(a_1, \ldots, a_{n-1}, x)$ the set of ring words in the alphabet $\{a_1, \ldots, a_{n-1}, x\}$ in which variable x occurs and which are obtained from words $f \in P_n$, say $f = f(x_1, \ldots, x_k)$ with $\vec{r}_f = (r_1, \ldots, r_k)$, by substituting $x_i \in \{a_{r_i}, \ldots, a_{n-1}, x\}$ into $f(x_1, \ldots, x_k)$ for each $i = 1, \ldots, k$, in particular, $T(x) = \{x, -x\}$. Then for every finite sequence $a_1, \ldots, a_n$ in R,

(a) $U(a_1, \ldots, a_n) = U(a_1, \ldots, a_{n-1}) \cup \{ f(a_n) : f(x) \in T(a_1, \ldots, a_{n-1}, x) \}$ and $U(a_1) = \{0\} \cup \{ f(a_1) : f(x) \in T(x) \}$,

(b) for every $f(x) \in T(a_1, \ldots, a_{n-1}, x)$, $f(0) \in U(a_1, \ldots, a_{n-1})$, and for $f(x) \in T(x)$, $f(0) = 0$.

Statement (a) is obvious and (b) follows from the next lemma.

Lemma 1.37. *Let $f \in P$ and let $\vec{r}_f = (r_1, \ldots, r_k)$. Then*

(1) $f(0, \ldots, 0) = 0$, *and*

(2) *whenever $\emptyset \neq \{i_1, \ldots, i_l\} \subset \{1, \ldots, k\}$ and $g(x_{i_1}, \ldots, x_{i_l})$ is the word obtained from $f(x_1, \ldots, x_k)$ by substituting $x_i = 0$ for each $i \in \{1, \ldots, k\} \setminus \{i_1, \ldots, i_k\}$, there exists $h \in P$ with $\vec{r}_h = (s_1, \ldots, s_l)$ such that $s_j \leq r_{i_j}$ for all $j = 1, \ldots, l$ and $g(c_1, \ldots, c_l) = h(c_1, \ldots, c_l)$ for all $c_1, \ldots, c_l \in R$.*

Proof. Suppose that the lemma holds for some $f_1, f_2 \in P$ with $\vec{r}_{f_1} = (r'_1, \ldots, r'_{k_1})$ and $\vec{r}_{f_2} = (r''_1, \ldots, r''_{k_2})$. Let $k = k_1 + k_2$ and consider the word

$$f = f(x_1, \ldots, x_k) = f_1(x_1, \ldots, x_{k_1}) + f_2(x_{k_1+1}, \ldots, x_k) \in P$$

with $\vec{r}_f = (2r'_1, \ldots, 2r'_{k_1}, 2r''_1, \ldots, 2r''_{k_2})$. Clearly

$$f(0, \ldots, 0) = f_1(0, \ldots, 0) + f_2(0, \ldots, 0) = 0 + 0 = 0.$$

Let $\emptyset \neq \{i_1, \ldots, i_l\} \subset \{1, \ldots, k\}$ and let $g(x_{i_1}, \ldots, x_{i_l})$ be the word obtained from $f(x_1, \ldots, x_k)$ by substituting $x_i = 0$ for each $i \in \{1, \ldots, k\} \setminus \{i_1, \ldots, i_k\}$. Without loss of generality one may assume that $\{i_1, \ldots, i_l\} \cap \{1, \ldots, k_1\} \neq \emptyset$. Let

$$\{i_1, \ldots, i_l\} \cap \{1, \ldots, k_1\} = \{i_1, \ldots, i_{l_1}\}$$

and let $g_1(x_{i_1}, \ldots, x_{i_{l_1}})$ be the word obtained from $f_1(x_1, \ldots, x_{k_1})$ by substituting $x_i = 0$ for each $i \in \{1, \ldots, k_1\} \setminus \{i_1, \ldots, i_{l_1}\}$. By the hypothesis, there exists $h_1 \in P$ with $\vec{r}_{h_1} = (s'_1, \ldots, s'_{l_1})$ such that $s'_j \leq r'_{i_j}$ for all $j = 1, \ldots, l_1$ and $g_1(c_1, \ldots, c_{l_1}) = h_1(c_1, \ldots, c_{l_1})$ for all $c_1, \ldots, c_{l_1} \in R$.

Suppose first that $l_1 < l$. Let

$$\{i_1, \ldots, i_l\} \cap \{k_1+1, \ldots, k\} = \{k_1+q_1, \ldots, k_1+q_{l_2}\}$$

and let $g_2(x_{q_1}, \ldots, x_{q_{l_2}})$ be the term obtained from $f_2(x_1, \ldots, x_{k_2})$ by substituting $x_q = 0$ for each $q \in \{1, \ldots, k_2\} \setminus \{q_1, \ldots, q_{l_2}\}$. By the hypothesis, there exists $h_2 \in P$ with $\vec{r}_{h_2} = (s''_1, \ldots, s''_{l_2})$ such that $s''_j \leq r''_{q_j}$ for all $j = 1, \ldots, l_2$ and $g_2(c_1, \ldots, c_{l_2}) = h_2(c_1, \ldots, c_{l_2})$ for all $c_1, \ldots, c_{l_2} \in R$. Then the word

$$h = h(x_1, \ldots, x_l) = h_1(x_1, \ldots, x_{l_1}) + h_2(x_{l_1+1}, \ldots, x_l) \in P$$

with $\vec{r}_h = (2s'_1, \ldots, 2s'_{l_1}, 2s''_1, \ldots, 2s''_{l_2})$ is as required. Indeed, whenever $c_1, \ldots, c_l \in R$, we have that

$$
\begin{aligned}
g(c_1, \ldots, c_l) &= g_1(c_1, \ldots, c_{l_1}) + g_2(c_{l_1+1}, \ldots, c_l) \\
&= h_1(c_1, \ldots, c_{l_1}) + h_2(c_{l_1+1}, \ldots, c_l) \\
&= h(c_1, \ldots, c_l)
\end{aligned}
$$

and $2s'_j \leq 2r'_{i_j}$ for all $j = 1, \ldots, l_1$ and $2s''_j \leq 2r''_{q_j}$ for all $j = 1, \ldots, l_2$.

Suppose now that $l_1 = l$. Then the word

$$
h = h(x_1, \ldots, x_l) = h_1(x_1, \ldots, x_{l_1}) \in P
$$

with $\vec{r}_h = (s'_1, \ldots, 2s'_{l_1})$ is as required. Indeed, whenever $c_1, \ldots, c_l \in R$, we have that

$$
\begin{aligned}
g(c_1, \ldots, c_l) &= g_1(c_1, \ldots, c_{l_1}) + f_2(0, \ldots, 0) \\
&= h_1(c_1, \ldots, c_{l_1}) + 0 \\
&= h(c_1, \ldots, c_l)
\end{aligned}
$$

and $s'_j \leq r'_{i_j}$ for all $j = 1, \ldots, l_1$.

Considering the words $f_1(x_1, \ldots, x_{k_1}) f_2(x_{k_1+1}, \ldots, x_k)$, $b_m f_1(x_1, \ldots, x_{k_1})$ and $f_1(x_1, \ldots, x_{k_1}) b_m$ is similar, and the case of the word $-f(x_1, \ldots, x_k)$ is trivial.   □

Using (a), (b) and our assumption, the required sequence $(a_n)_{n=1}^{\infty}$ can be constructed in the same way as in the proof of Theorem 1.20.   □

Another result that we need is Hindman's Theorem.

Definition 1.38. Given a set X, $\mathcal{P}_f(X)$ is the set of finite nonempty subsets of X.

Definition 1.39. Let S be a semigroup. Given an infinite sequence $(x_n)_{n=1}^{\infty}$ in S, the set of *finite products* of the sequence is defined by

$$
\mathrm{FP}((x_n)_{n=1}^{\infty}) = \left\{ \prod_{n \in F} x_n : F \in \mathcal{P}_f(\mathbb{N}) \right\}.
$$

Given a finite sequence $(x_n)_{n=1}^{m}$ in S,

$$
\mathrm{FP}((x_n)_{n=1}^{m}) = \left\{ \prod_{n \in F} x_n : F \in \mathcal{P}_f(\{1, \ldots, m\}) \right\}.
$$

If S is an additive semigroup, we write FS instead of FP and say *finite sums* instead of finite products.

We state Hindman's Theorem in the form involving product subsystem.

Definition 1.40. Let S be a semigroup and let $(x_n)_{n=1}^{\infty}$ be a sequence in S. A sequence $(y_n)_{n=1}^{\infty}$ is a *product subsystem* of $(x_n)_{n=1}^{\infty}$ if there is a sequence $(H_n)_{n=1}^{\infty}$ in $\mathcal{P}_f(\mathbb{N})$ such that $\max H_n < \min H_{n+1}$ and $y_n = \prod_{i \in H_n} x_i$ for each $n \in \mathbb{N}$. If S is an additive semigroup, we say *sum subsystem* instead of product subsystem.

Theorem 1.41. *Let S be a semigroup and let $(x_n)_{n=1}^{\infty}$ be a sequence in S. Whenever $\mathrm{FP}((x_n)_{n=1}^{\infty})$ is partitioned into finitely many subsets, there exists a product subsystem $(y_n)_{n=1}^{\infty}$ of $(x_n)_{n=1}^{\infty}$ such that $\mathrm{FP}((y_n)_{n=1}^{\infty})$ is contained in one subset of the partition.*

We shall prove Theorem 1.41 in Chapter 6. Now we need the following.

Corollary 1.42. *Let S be an infinite cancellative semigroup and let $A \subset S$ with $|A| < |S|$. Whenever $S \setminus A$ is partitioned into finitely many subsets, there exists a sequence $(y_n)_{n=1}^{\infty}$ such that $\mathrm{FP}((y_n)_{n=1}^{\infty})$ is contained in one subset of the partition.*

Proof. Construct inductively a sequence $(x_n)_{n=1}^{\infty}$ in S such that $\mathrm{FP}((x_n)_{n=1}^{\infty}) \cap A = \emptyset$. Then apply Theorem 1.41.   □

Given a ring R, let $R[x]$ denote the set of all ring words in the alphabet $R \cup \{x\}$. Using ring identities every $f(x) \in R[x]$ can be rewritten as a noncommutative polynomial. We denote by $\deg f(x)$ the degree of that polynomial.

Lemma 1.43. *For every $f(x) \in R[x]$ with $\deg f(x) > 0$, there is $g(x) \in R[x]$ such that $\deg g(x) < \deg f(x)$ and*

$$g(a) = f(a + b) - f(a) - f(b)$$

for all $a, b \in R$.

Proof. Rewriting $f(x + b)$ as a polynomial of x we obtain that

$$f(x + b) = f(x) + f(b) + g(x),$$

where $\deg g(x) < \deg f(x)$.   □

Now we are in a position to prove Theorem 1.33.

Proof of Theorem 1.33. Let R be a countably infinite ring and assume on the contrary that R is not topologizable. By Proposition 1.34, there is a finite sequence

$$f_1(x), \ldots, f_m(x)$$

in $R[x]$ with the following properties:

(i) $f_i(0) \neq 0$ for each $i = 1, \ldots, m$, and

(ii) for every $a \in R \setminus \{0\}$, there is $i = 1, \ldots, m$ such that $f_i(a) = 0$.

Let $k = \max\{\deg f_i(x) : i = 1, \ldots, m\}$. By Theorem 1.41, applied to the additive group of R, there exist a sequence $(a_n)_{n=1}^{k+1}$ in $R \setminus \{0\}$ and $i = 1, \ldots, m$ such that

$$f_i(a) = 0 \quad \text{for every } a \in FS((a_n)_{n=1}^{k+1}).$$

Let $f(x) = f_i(x)$. Then by Lemma 1.43, there is $g(x) \in R[x]$ with $\deg g(x) < \deg f(x)$ such that
$$g(a) = f(a + b) - f(a) - f(b)$$

for all $a, b \in R$. It follows that

$$g(0) = f(0 + 0) - f(0) - f(0) = -f(0) \neq 0$$

and for every $a \in FS((a_n)_{n=1}^{k})$,

$$g(a) = f(a + a_{k+1}) - f(a) - f(a_{k+1}) = 0.$$

After at most k such reductions we obtain a ring word $h(x)$ with $\deg h(x) = 0$ such that $h(0) = \pm f(0) \neq 0$ and $h(a_1) = 0$, which is a contradiction. $\qquad\square$

References

The standard references for topological groups are [55] and [34]. A great deal of information about topological groups can be found also in [7], [11], and [4]. Theorem 1.17 is from [62], a result of collaboration with I. Protasov. Its Abelian case was proved in [83]. Theorem 1.20 is due to A. Markov [48]. The first example of a nontopologizable group was produced by S. Shelah [68]. It was an uncountable group and its construction used the Continuum Hypothesis CH. G. Hesse [32] showed that CH can be dropped in Shelah's construction. Example 1.25 is due to A. Ol'šanskiĭ [53]. For the Adian group see [1]. That the Burnside group $B(m, n)$ is infinite for $m \geq 2$ and for n sufficiently large and odd was proved by P. Novikov and S. Adian [51]. Theorem 1.28 is from [110]. For more information about p and other cardinal invariants of the continuum see [79]. Theorem 1.33 is due to V. Arnautov [5]. Theorem 1.41 is due to N. Hindman [35]. Our proof of Theorem 1.33, as well as Theorem 1.20, is based on the treatment in [62].

Chapter 2
Ultrafilters

This chapter contains some basic facts about ultrafilters and the Stone–Čech compactification of a discrete space. We also discuss Ramsey ultrafilters, P-points, and countably complete ultrafilters.

2.1 The Notion of an Ultrafilter

Let D be a nonempty set. Recall that a *filter* on D is a family $\mathcal{F} \subseteq \mathcal{P}(D)$ with the following properties:

(1) $D \in \mathcal{F}$ and $\emptyset \notin \mathcal{F}$,

(2) if $A, B \in \mathcal{F}$, then $A \cap B \in \mathcal{F}$, and

(3) if $A \in \mathcal{F}$ and $A \subseteq B \subseteq D$, then $B \in \mathcal{F}$.

Definition 2.1. An *ultrafilter* on D is a filter on D which is not properly contained in any other filter on D.

In other words, an ultrafilter is a maximal filter.

Definition 2.2. A family $\mathcal{A} \subseteq \mathcal{P}(D)$ has the *finite intersection property* if $\bigcap \mathcal{B} \neq \emptyset$ for every finite $\mathcal{B} \subseteq \mathcal{A}$.

It is clear that every filter has the finite intersection property. Conversely, every family $\mathcal{A} \subseteq \mathcal{P}(D)$ with the finite intersection property generates a filter $\mathrm{flt}(\mathcal{A})$ on D by

$$\mathrm{flt}(\mathcal{A}) = \{A \subseteq D : A \supseteq \bigcap \mathcal{B} \text{ for some finite } \mathcal{B} \subseteq \mathcal{A}\}.$$

Proposition 2.3. *Let* $\mathcal{A} \subseteq \mathcal{P}(D)$. *Then the following statements are equivalent:*

(1) $\mathcal{A}$ *is an ultrafilter,*

(2) $\mathcal{A}$ *is a maximal family with the finite intersection property,*

(3) $\mathcal{A}$ *is a filter and for every* $A \subseteq D$, *either* $A \in \mathcal{A}$ *or* $D \setminus A \in \mathcal{A}$.

Proof. $(1) \Rightarrow (3)$ Consider two cases.

Case 1: there is $B \in \mathcal{A}$ such that $B \cap A = \emptyset$. Then $B \subseteq D \setminus A$ and, since $\mathcal{A}$ is closed under supersets, $D \setminus A \in \mathcal{A}$.

Case 2: for every $B \in \mathcal{A}$, $B \cap A \neq \emptyset$. Then

$$\{B \cap A : B \in \mathcal{A}\}$$

is a family of nonempty sets closed under finite intersections, because $\mathcal{A}$ is closed under finite intersections. Consequently,

$$\mathcal{F} = \{C \subseteq D : C \supseteq B \cap A \text{ for some } B \in \mathcal{A}\}$$

is a filter. Clearly $\mathcal{A} \subseteq \mathcal{F}$ and $A \in \mathcal{F}$. Since $\mathcal{A}$ is an ultrafilter, $\mathcal{A} = \mathcal{F}$. Hence $A \in \mathcal{A}$.

(3) $\Rightarrow$ (2) Let $A \subseteq D$ and $A \notin \mathcal{A}$. Then $D \setminus A \in \mathcal{A}$. Since $A \cap (D \setminus A) = \emptyset$, $\mathcal{A} \cup \{A\}$ has no finite intersection property.

(2) $\Rightarrow$ (1) Since $\mathcal{A}$ has the finite intersection property, so does flt($\mathcal{A}$), and since $\mathcal{A}$ is maximal, flt($\mathcal{A}$) $= \mathcal{A}$. It follows that $\mathcal{A}$ is a filter, and consequently, an ultrafilter.

□

It is obvious that for every $a \in D$, $\{A \subseteq D : a \in A\}$ is an ultrafilter. Such ultrafilters are called *principal*. Ultrafilters which are not principal are called *nonprincipal*.

Lemma 2.4. *Let $\mathcal{U}$ be an ultrafilter on D. Then the following statements are equivalent:*

(1) *$\mathcal{U}$ is a nonprincipal ultrafilter,*

(2) *$\bigcap \mathcal{U} = \emptyset$,*

(3) *for every $A \in \mathcal{U}$, $|A| \geq \omega$.*

Proof. (1) $\Rightarrow$ (2) If $a \in \bigcap \mathcal{U}$, then $\mathcal{U} \subseteq \{A \subseteq D : a \in A\}$, and since $\mathcal{U}$ is a maximal filter, $\mathcal{U} = \{A \subseteq D : a \in A\}$.

(2) $\Rightarrow$ (3) Suppose that some $A \in \mathcal{U}$ is finite. Then, applying Proposition 2.3, we obtain that there is $a \in A$ such that $\{a\} \in \mathcal{U}$. It follows that $\mathcal{U} = \{A \subseteq D : a \in A\}$.

(3) $\Rightarrow$ (1) is obvious.

□

The existence of nonprincipal ultrafilters involves the Axiom of Choice.

Proposition 2.5 (Ultrafilter Theorem). *Every filter on D can be extended to an ultrafilter on D.*

Proof. Let $\mathcal{F}$ be a filter on D and let

$$P = \{\mathcal{G} \subseteq \mathcal{P}(D) : \mathcal{F} \subseteq \mathcal{G} \text{ and } \mathcal{G} \text{ is a filter on } D\}.$$

Given any chain C in P, $\bigcup C$ is a filter, and so $\bigcup C$ is an upper bound of C. Hence, by Zorn's Lemma, P has a maximal element $\mathcal{U}$. Clearly, $\mathcal{U}$ is a maximal filter. □

An ultrafilter $\mathcal{U}$ on D is *uniform* if for every $A \in \mathcal{U}$, $|A| = |D|$.

Corollary 2.6. *There are uniform ultrafilters on any infinite set.*

Proof. Let D be any infinite set and let

$$\mathcal{F} = \{A \subseteq D : |D \setminus A| < |D|\}.$$

By Proposition 2.3, there is an ultrafilter $\mathcal{U}$ on D containing $\mathcal{F}$. If $A \subseteq D$ and $|A| < |D|$, then $D \setminus A \in \mathcal{F} \subseteq \mathcal{U}$, so $A \notin \mathcal{U}$. Hence $\mathcal{U}$ is uniform. $\square$

Definition 2.7. Let $\mathcal{F}$ be a filter on D. If $C \subseteq D$ and $C \cap A \neq \emptyset$ for every $A \in \mathcal{F}$, then

$$\mathcal{F}|_C = \{C \cap A : A \in \mathcal{F}\}$$

is a filter on C called the *trace* of $\mathcal{F}$ on C. If $f : D \to E$, then

$$f(\mathcal{F}) = \{f(A) : A \in \mathcal{F}\}$$

is a filter base on E called the *image* of $\mathcal{F}$ with respect to f.

Note that if f is surjective, then $f(\mathcal{F})$ is a filter.

It is clear that if $\mathcal{F}$ is an ultrafilter, so is $\mathcal{F}|_C$.

Lemma 2.8. *If $\mathcal{F}$ is an ultrafilter, $f(\mathcal{F})$ is an ultrafilter base.*

Proof. Let $B \subseteq E$ and let $A = f^{-1}(B)$. Since $\mathcal{F}$ is an ultrafilter, either $A \in \mathcal{F}$ or $D \setminus A \in \mathcal{F}$. Then either $B \supseteq f(A) \in f(\mathcal{F})$ or $E \setminus B \supseteq f(D \setminus A) \in f(\mathcal{F})$. $\square$

Recall that a space X is called *compact* if every open cover of X has a finite subcover. Equivalently, X is compact if every family of closed subsets of X with the finite intersection property has a nonempty intersection. A filter base $\mathcal{B}$ on a space X *converges* to a point $x \in X$ if for every neighborhood U of $x \in X$, there is $A \in \mathcal{B}$ such that $A \subseteq U$. Note that in the case where $\mathcal{B}$ is a filter, $\mathcal{B}$ converges to x if and only if $\mathcal{B}$ contains the neighborhood filter of x.

Proposition 2.9. *A space X is compact if and only if every ultrafilter on X is convergent.*

Proof. Let X be a compact space and let $\mathcal{U}$ be an ultrafilter on X. Assume on the contrary that for every point $x \in X$, there is a neighborhood U_x of x such that $U_x \notin \mathcal{U}$. Clearly one can choose U_x to be open. Then, since $\mathcal{U}$ is an ultrafilter, $X \setminus U_x \in \mathcal{U}$. The open sets U_x, where $x \in X$, cover X. Since X is compact, there is a finite subcover $\{U_{x_i} : i < n\}$ of $\{U_x : x \in X\}$. But then $\emptyset = \bigcap_{i<n}(X \setminus U_{x_i}) \in \mathcal{U}$, a contradiction.

Conversely, suppose that every ultrafilter on X is convergent and let $\mathcal{A}$ be a family of closed subsets of X with the finite intersection property. Assume on the contrary that $\bigcap \mathcal{A} = \emptyset$. By Proposition 2.5, there is an ultrafilter $\mathcal{U}$ on X such that $\mathcal{A} \subseteq \mathcal{U}$. We claim that $\mathcal{U}$ is not convergent. Indeed, let $x \in X$. Since $\bigcap \mathcal{A} = \emptyset$, there is $F_x \in \mathcal{A}$ such that $x \notin F_x$. Then $U_x = X \setminus F_x$ is a neighborhood of x and $U_x \notin \mathcal{U}$. Hence $\mathcal{U}$ is not convergent, a contradiction. $\square$

2.2 The Space βD

Definition 2.10. Let D be a nonempty set and let βD denote the set of all ultrafilters on D. For every $A \subseteq D$, define $\overline{A} \subseteq \beta D$ by

$$\overline{A} = \{p \in \beta D : A \in p\}.$$

Thus, for every $A \subseteq D$ and $p \in \beta D$, $p \in \overline{A}$ if and only if $A \in p$. We have that

$$\overline{A} \cap \overline{B} = \overline{A \cap B}$$

for all $A, B \subseteq D$, so the family $\{\overline{A} : A \in \mathcal{P}(D)\}$ is closed under finite intersections. Consequently, this family is a base for a topology on βD. For every pair of different $p, q \in \beta D$, there are $A \in p$ and $B \in q$ such that $A \cap B = \emptyset$, and so $\overline{A} \cap \overline{B} = \emptyset$. Hence, this topology is Hausdorff. Note also that the set of principal ultrafilters on D is a discrete open dense subset of βD with respect to this topology.

Definition 2.11. Given a discrete space D, βD is the space of ultrafilters on D with the topology generated by taking as a base the subsets $\overline{A}$, where $A \in \mathcal{P}(D)$. The principal ultrafilters are being identified with the points of D. For every $A \subseteq D$, $A^* = \overline{A} \setminus A$.

We have that

$$\overline{D \setminus A} = \beta D \setminus \overline{A}$$

for every $A \subseteq D$. Consequently, the subset $\overline{A} \subseteq \beta D$ is closed as well as open. Since also A is dense in $\overline{A}$, it follows that

$$\operatorname{cl}_{\beta D} A = \overline{A}.$$

Definition 2.12. The *Stone–Čech compactification* of a discrete space D is a compact Hausdorff space Y containing D as a dense subspace such that every mapping $f : D \to Z$ from D into any compact Hausdorff space Z can be extended to a continuous mapping $\overline{f} : Y \to Z$.

It is easy to see that the Stone–Čech compactification of D is unique up to a homeomorphism leaving D pointwise fixed.

Theorem 2.13. *βD is the Stone–Čech compactification of D.*

Before proving Theorem 2.13, let us consider the following general situation.

Suppose that X is a dense subset of a space Y and $f : X \to Z$ is a continuous mapping from X into a Hausdorff space Z. For every $p \in Y$, denote by $\mathcal{F}_p$ the trace of the neighborhood filter of $p \in Y$ on X and let $\lim_{X \ni x \to p} f(x)$ denote the limit of the filter base $f(\mathcal{F}_p)$ in Z, if exists. It is clear that if f has a continuous extension $\overline{f} : Y \to Z$, then it is unique and

$$\overline{f}(p) = \lim_{X \ni x \to p} f(x)$$

for every $p \in Y$.

Lemma 2.14. *Suppose that Z is regular and for every $p \in Y$, there is $\lim_{X \ni x \to p} f(x)$. Define $\overline{f} : Y \to Z$ by*

$$\overline{f}(p) = \lim_{X \ni x \to p} f(x).$$

Then $\overline{f}$ is the continuous extension of f.

Proof. It follows from the definition of $\overline{f}$ that for every $A \subseteq X$,

$$\overline{f}(\mathrm{cl}_Y A) \subseteq \mathrm{cl}_Z f(A).$$

To see this, let $q \in \mathrm{cl}_Y A$ and let W be a neighborhood of $\overline{f}(q) \in Z$. Since $\overline{f}(q) = \lim_{x \to q} f(x)$, there is $B \in \mathcal{F}_q$ such that $f(B) \subseteq W$. We have that $B \cap A \neq \emptyset$ and $f(B \cap A) \subseteq W$. Hence, $\overline{f}(q) \in \mathrm{cl}_Z f(A)$.

Now to see that $\overline{f}$ is continuous, let $p \in Y$ and let U be a neighborhood of $\overline{f}(p) \in Z$. Since Z is regular, one may suppose that U is closed. Choose $A \in \mathcal{F}_p$ such that $f(A) \subseteq U$. Put $V = \mathrm{cl}_Y A$. Then V is a neighborhood of $p \in Y$ and $\overline{f}(V) \subseteq U$. □

Now we are ready to prove Theorem 2.13.

Proof of Theorem 2.13. Since $\overline{D \setminus A} = \beta D \setminus \overline{A}$, the sets $\overline{A}$ form also a base for the closed sets (that is, every closed set in βD is the intersection of some sets $\overline{A}$). Consequently, in order to show that βD is compact, it suffices to show that every family $\mathcal{A}$ of sets of the form $\overline{A}$ with the finite intersection property has a nonempty intersection. Let

$$\mathcal{B} = \{A \subseteq D : \overline{A} \in \mathcal{A}\}$$

Clearly $\mathcal{B}$ has the finite intersection property. Then by Proposition 2.5, there is $p \in \beta D$ such that $\mathcal{B} \subseteq p$. But then $p \in \bigcap \overline{A}$.

Now let $f : D \to Z$ be any mapping from D into any compact Hausdorff space Z and let $p \in \beta D$. The trace of the neighborhood filter of $p \in \beta D$ on D is the ultrafilter

p. Since Z is compact, the ultrafilter base $f(p)$ converges, so $\lim_{D \ni x \to p} f(x)$ exists. Hence by Lemma 2.14,

$$\overline{f} : \beta D \ni p \mapsto \lim_{D \ni x \to p} f(x) \in Z$$

is the continuous extension of f.   □

Definition 2.15. For every mapping $f : D \to Z$ of a discrete space D into a compact Hausdorff space Z, let $\overline{f} : \beta D \to Z$ denote the continuous extension of f.

Note that if $f : D \to E \subset \beta E$ and $p \in \beta D$, then $\overline{f}(p)$ is the ultrafilter on E with a base $f(p)$.

Theorem 2.16. *Let $f : D \to D$ be a mapping with no fixed points. Then there is a 3-partition $\{A_i : i < 3\}$ of D such that $A_i \cap f(A_i) = \emptyset$ for each $i < 3$.*

Proof. Consider the set $\mathcal{G}$ of all functions g such that

(i) $\operatorname{dom}(g) \subseteq D$ and $\operatorname{ran}(g) \subseteq \{0, 1, 2\}$,

(ii) $f(\operatorname{dom}(g)) \subseteq \operatorname{dom}(g)$, and

(iii) $g(a) \neq g(f(a))$ for each $a \in \operatorname{dom}(g)$,

ordered by $\subseteq$. Note that $\mathcal{G}$ is nonempty, and if $\mathcal{C}$ is a chain in $\mathcal{G}$, then $\bigcup \mathcal{C} \in \mathcal{G}$. Consequently by Zorn's Lemma, there is a maximal element $g \in \mathcal{G}$. We claim that $\operatorname{dom}(g) = D$.

Let $D_g = \operatorname{dom}(g)$ and assume on the contrary that there is $b \in D \setminus D_g$. To obtain a contradiction, it suffices to define an extension h of g such that

(a) $\operatorname{dom}(h) = D_g \cup \{f^n(b) : n < \omega\}$, where $f^0(b) = b$ and $f^{n+1}(b) = f(f^n(b))$, and

(b) $g(f^n(b)) \neq g(f^{n+1}(b))$ for each $n < \omega$.

We distinguish between three cases.

Case 1: there is $n < \omega$ such that $f^k(b) \notin D_g$ for all $k \leq n$ and $f^{n+1}(b) \in D_g$. Let $i = g(f^{n+1}(b))$ and pick $j < 2$ different from i. For each $k \leq n$, put

$$h(f^{n-k}(b)) = \begin{cases} j & \text{if } k \text{ is even} \\ i & \text{otherwise.} \end{cases}$$

Case 2: all elements $f^n(b)$, where $n < \omega$, are different and do not belong to D_g. For each $n < \omega$, put

$$h(f^n(b)) = \begin{cases} 0 & \text{if } n \text{ is even} \\ 1 & \text{otherwise.} \end{cases}$$

Case 3: there are $m < n < \omega$ such that all elements $f^k(b)$, where $k \le n$, are different, do not belong to D_g, and $f^{n+1}(b) = f^m(b)$. If $n - m$ is odd, for each $k \le n$, put

$$h(f^k(b)) = \begin{cases} 0 & \text{if } k \text{ is even} \\ 1 & \text{otherwise.} \end{cases}$$

If $n - m$ is even, we correct the latter definition at $f^n(b)$ by putting $h(f^n(b)) = 2$.

Having established that $\mathrm{dom}(g) = D$, let $A_i = g^{-1}(i)$ for each $i < 3$.   □

Corollary 2.17. *If* $f : D \to D$ *has no fixed points, neither does* $\overline{f} : \beta D \to \beta D$.

Proof. By Theorem 2.16, there is a partition $\{A_i : i < 3\}$ of D such that $A_i \cap f(A_i) = \emptyset$ for each $i < 3$. Let $p \in \beta D$. Then $A_j \in p$ for some $j < 3$. Consequently $f(A_j) \in \overline{f}(p)$. We have that $p \in \overline{A_j}$, $\overline{f}(p) \in \overline{f(A_j)}$ and $\overline{A_j} \cap \overline{f(A_j)} = \emptyset$. Hence $\overline{f}(p) \ne p$.   □

Corollary 2.18. *Let* $f : D \to D$ *and let* $p \in \beta D$. *Then* $\overline{f}(p) = p$ *if and only if*

$$\{a \in D : f(a) = a\} \in p.$$

Proof. Denote $F = \{a \in D : f(a) = a\}$ and $A = D \setminus F$.

Let $\overline{f}(p) = p$ and assume on the contrary that $F \notin p$. Then $A \in p$. Pick any $b \in A$ and define $g : D \to D$ by

$$g(x) = \begin{cases} f(x) & \text{if } x \in A \\ b & \text{if } x \in F. \end{cases}$$

Then g has no fixed points and $g|_A = f|_A$, so $\overline{g}|_{\overline{A}} = \overline{f}|_{\overline{A}}$. Since $p \in \overline{A}$, we obtain that $\overline{g}(p) = \overline{f}(p) = p$. But by Corollary 2.17, $\overline{g}(p) \ne p$, a contradiction.

Conversely, let $F \in p$. Then $p \in \overline{F}$ and $\overline{f}(x) = x$ for all $x \in \overline{F}$. Hence $\overline{f}(p) = p$.   □

Definition 2.19. A space is *extremally disconnected* if the closure of an open set is open.

Equivalently, a space is extremally disconnected if the closures of disjoint open sets are disjoint.

Lemma 2.20. βD *is extremally disconnected.*

Proof. Let U be an open subset of βD. Put $A = U \cap D$. Then $U \subseteq \mathrm{cl}_{\beta D} A$. Hence $\mathrm{cl}_{\beta D} U = \overline{A}$.   □

Remark 2.21. Although βD is extremally disconnected, D^* is not (see [22, Example 6.2.31]).

A space is called σ-*compact* if it can be represented as a countable union of compact subsets.

Theorem 2.22. *Let X be a regular extremally disconnected space and let A and B be σ-compact subsets of X. If $(\mathrm{cl}\ A) \cap B = A \cap (\mathrm{cl}\ B) = \emptyset$, then $(\mathrm{cl}\ A) \cap (\mathrm{cl}\ B) = \emptyset$.*

Proof. Write $A = \bigcup_{n<\omega} A_n$ and $B = \bigcup_{n<\omega} B_n$ where A_n and B_n are compact for each $n < \omega$. Inductively, for each $n < \omega$, choose open neighborhoods U_n and V_n of A_n and B_n respectively such that

(a) $(\mathrm{cl}\ U_n) \cap B = \emptyset$ and $A \cap (\mathrm{cl}\ V_n) = \emptyset$,

(b) $U_n \cap \left(\bigcup_{i<n} V_i\right) = \emptyset$ and $\left(\bigcup_{i<n} U_i\right) \cap V_n = \emptyset$, and

(c) $U_n \cap V_n = \emptyset$.

(a) is needed to satisfy (b). Conjunction of (b) and (c) is equivalent to

$$\left(\bigcup_{i\leq n} U_i\right) \cap \left(\bigcup_{i\leq n} V_i\right) = \emptyset.$$

It follows that

$$U = \bigcup_{n<\omega} U_n \quad \text{and} \quad V = \bigcup_{n<\omega} V_n$$

are disjoint open neighborhoods of A and B respectively. Since X is extremally disconnected, $(\mathrm{cl}\ U) \cap (\mathrm{cl}\ V) = \emptyset$ and so $(\mathrm{cl}\ A) \cap (\mathrm{cl}\ B) = \emptyset$. $\qquad\square$

Corollary 2.23. *Let A and B be countable subsets of βD. If $(\mathrm{cl}\ A) \cap B = A \cap (\mathrm{cl}\ B) = \emptyset$, then $(\mathrm{cl}\ A) \cap (\mathrm{cl}\ B) = \emptyset$.*

Corollary 2.24. *Let $X = \bigcup_{n<\omega} X_n$ and $Y = \bigcup_{n<\omega} Y_n$ be subsets of βD. If $(\mathrm{cl}\ X) \cap (\mathrm{cl}\ Y_n) = (\mathrm{cl}\ X_n) \cap (\mathrm{cl}\ Y) = \emptyset$ for every $n < \omega$, then $(\mathrm{cl}\ X) \cap (\mathrm{cl}\ Y) = \emptyset$.*

Proof. Let $A = \bigcup_{n<\omega} \mathrm{cl}\ X_n$ and $B = \bigcup_{n<\omega} \mathrm{cl}\ Y_n$. We have that $X \subseteq A$, $Y \subseteq B$, $\mathrm{cl}\ X = \mathrm{cl}\ A$, and $\mathrm{cl}\ Y = \mathrm{cl}\ B$. It then follows that $(\mathrm{cl}\ A) \cap B = A \cap (\mathrm{cl}\ B) = \emptyset$. Since also both A and B are σ-compact, we can apply Theorem 2.22. Hence $(\mathrm{cl}\ A) \cap (\mathrm{cl}\ B) = \emptyset$, and so $(\mathrm{cl}\ X) \cap (\mathrm{cl}\ Y) = \emptyset$. $\qquad\square$

Definition 2.25. For every infinite set D, let $U(D)$ denote the subset of βD consisting of uniform ultrafilters.

Theorem 2.26. *Let D be an infinite set of cardinality κ. Then $|U(D)| = |\beta D| = 2^{2^\kappa}$.*

Proof. Since every ultrafilter is a member of $\mathcal{P}(\mathcal{P}(D))$, $|\beta D| \leq 2^{2^\kappa}$. Let $U = U(D)$. In order to show that $|U| \geq 2^{2^\kappa}$, it suffices to construct a mapping of U onto a set of cardinality 2^{2^κ}.

Let Z be the product of 2^κ copies of the discrete space $\{0, 1\}$. Then $|Z| = 2^{2^\kappa}$. By the Hewitt–Marczewski–Pondiczery Theorem (see [22, Theorem 2.3.15]), Z has a dense subset E of cardinality κ. Enumerate E as $\{q_\alpha : \alpha < \kappa\}$. Now let $\{A_\alpha : \alpha < \kappa\}$ be a partition of D into subsets of cardinality κ. For each $\alpha < \kappa$, there is $p_\alpha \in U$ with $A_\alpha \in p$. Define $f : D \to E$ by $f(A_\alpha) = \{q_\alpha\}$. Then $\overline{f}(p_\alpha) = q_\alpha$. Since $U \subseteq \beta D$ is closed, it follows that $\overline{f}(U)$ is a compact subset of Z containing E. Hence $\overline{f}(U) = Z$. $\square$

We conclude this section by establishing a one-to-one correspondence between nonempty closed subsets of βD and filters on D.

Definition 2.27. Given a family $\mathcal{A} \subseteq \mathcal{P}(D)$, define $\overline{\mathcal{A}} \subseteq \beta D$ by

$$\overline{\mathcal{A}} = \bigcap_{A \in \mathcal{A}} \overline{A}.$$

Lemma 2.28. *For every filter $\mathcal{F}$ on D, $\overline{\mathcal{F}}$ is a nonempty closed subset of βD consisting of all $p \in \beta D$ such that $\mathcal{F} \subseteq p$. Conversely, for every nonempty closed subset $X \subseteq \beta D$, the intersection of all ultrafilters from X is a filter $\mathcal{F}$ on D such that $\overline{\mathcal{F}} = X$.*

Proof. The first part of the lemma is obvious, so it suffices to prove the second. It is clear that $\mathcal{F}$ is a filter and $X \subseteq \overline{\mathcal{F}}$. To see the reverse inclusion, let $p \in \overline{\mathcal{F}}$. Assume on the contrary that $p \notin X$. Then there is $A \in p$ such that $\overline{A} \cap X = \emptyset$. It follows that for each $q \in X$, $D \setminus A \in q$. Hence $D \setminus A \in \mathcal{F} \subseteq p$, a contradiction. $\square$

2.3   Martin's Axiom

Let $P = (P, \leq)$ be a partially ordered set. A *filter* in P is a nonempty subset $G \subseteq P$ such that

   (i) for every $a, b \in G$, there is $c \in G$ such that $c \leq a$ and $c \leq b$, and

   (ii) for every $a \in G$ and $b \in P$, $a \leq b$ implies $b \in G$.

Elements $a, b \in P$ are *incompatible* if there is no $c \in P$ such that $c \leq a$ and $c \leq b$. An *antichain* in P is a subset of pairwise incompatible elements. P has the *countable chain condition* if every antichain in P is countable. A subset $D \subseteq P$ is *dense* if for every $a \in P$ there is $b \in D$ such that $b \leq a$.

Definition 2.29. MA is the following assertion: whenever $(P, \leq)$ is a partially ordered set with the countable chain condition and $\mathcal{D}$ is a family of $< 2^\omega$ dense subsets of P, there is a filter G in P such that $G \cap D \neq \emptyset$ for all $D \in \mathcal{D}$.

We first show that MA follows from the Continuum Hypothesis CH.

Lemma 2.30. *CH implies MA.*

Proof. Let $(P, \leq)$ be a partially ordered set with the countable chain condition and let $\mathcal{D}$ be a family of $< 2^\omega$ dense subsets of P. Then by CH, $\mathcal{D}$ is countable. Enumerate $\mathcal{D}$ as $\{D_n : n < \omega\}$. Construct inductively a sequence $(a_n)_{n<\omega}$ in P such that $a_n \in D_n$ and $a_{n+1} \leq a_n$. Then the filter $G = \{b \in P : b \geq a_n \text{ for some } n < \omega\}$ generated by $(a_n)_{n<\omega}$ is as required.   □

Now we derive an important combinatorial principle from MA.

Recall that the *pseudo-intersection number* $\mathfrak{p}$ is the minimum cardinality of a family $\mathcal{F} \subseteq \mathcal{P}(\omega)$ having the strong finite intersection property but no infinite pseudo-intersection. It is easy to see that $\omega_1 \leq \mathfrak{p} \leq \mathfrak{c}$.

Theorem 2.31. *MA implies $\mathfrak{p} = \mathfrak{c}$.*

Proof. Let $\mathcal{F} \subseteq \mathcal{P}(\omega)$ be a family having the strong finite intersection property with $|\mathcal{F}| < \mathfrak{c}$. We need to find an infinite $A \subseteq \omega$ such that $A \setminus B$ is finite for all $B \in \mathcal{F}$. Let P be the set of all pairs $(F, \mathcal{H})$ where F is a finite subset of ω and $\mathcal{H}$ is a finite subfamily of $\mathcal{F}$. Define the relation $\leq$ on P by

$$(F', \mathcal{H}') \leq (F, \mathcal{H}) \quad \text{if and only if} \quad F \subseteq F', \mathcal{H} \subseteq \mathcal{H}' \text{ and } F' \setminus F \subseteq \bigcap \mathcal{H}.$$

It is routine to verify that $\leq$ is a partial order. Next notice that if $(F, \mathcal{H})$ and $(F', \mathcal{H}')$ are incompatible elements of P, then $F \neq F'$. Indeed, otherwise $(F, \mathcal{H} \cup \mathcal{H}') \leq (F, \mathcal{H})$ and $(F, \mathcal{H} \cup \mathcal{H}') \leq (F, \mathcal{H}')$. Since the set of finite subsets of ω is countable, it follows that P has the countable chain condition.

For each $B \in \mathcal{F}$, let

$$D_B = \{(F, \mathcal{H}) \in P : B \in \mathcal{H}\}.$$

Given any $(F, \mathcal{H}) \in P$, one has $(F, \mathcal{H} \cup \{B\}) \in D_B$ and $(F, \mathcal{H} \cup \{B\}) \leq (F, \mathcal{H})$. So D_B is dense.

Also for each $n < \omega$, let

$$D_n = \{(F, \mathcal{H}) \in P : \max F \geq n\}.$$

Given any $(F, \mathcal{H}) \in P$, there is $m \in \bigcap \mathcal{H}$ such that $m \geq n$. Then $(F \cup \{m\}, \mathcal{H}) \in D_n$ and $(F \cup \{m\}, \mathcal{H}) \leq (F, \mathcal{H})$. So D_n is dense.

Now by MA, there is a filter G in P such that $G \cap D_B \neq \emptyset$ for each $B \in \mathcal{F}$ and $G \cap D_n \neq \emptyset$ for each $n < \omega$. Let

$$A = \bigcup_{(F,\mathcal{H}) \in G} F.$$

Since $G \cap D_n \neq \emptyset$ for each $n < \omega$, it follows that A is infinite. To show that $A \setminus B$ is finite for each $B \in \mathcal{F}$, pick $(F, \mathcal{H}) \in G \cap D_B$. We claim that $A \setminus B \subseteq F$. Indeed, let $x \in A \setminus B$. Then there is $(F', \mathcal{H}') \in G$ such that $x \in F'$. Since G is a filter, there is $(F'', \mathcal{H}'') \in G$ such that $(F'', \mathcal{H}'') \leq (F, \mathcal{H})$ and $(F'', \mathcal{H}'') \leq (F', \mathcal{H}')$. We have that $x \in F' \subseteq F''$, $F'' \setminus F \subseteq \bigcap \mathcal{H}$ and $B \in \mathcal{H}$. Hence $x \in F$.   □

2.4   Ramsey Ultrafilters and P-points

Given a set X and a cardinal k,

$$[X]^k = \{Y \subseteq X : |Y| = k\}.$$

Ramsey's Theorem says that whenever $k, r \in \mathbb{N}$ and $[\omega]^k$ is r-colored, there exists an infinite $A \subseteq \omega$ such that $[A]^k$ is monochrome (see [30]).

Definition 2.32. An ultrafilter p on ω is *Ramsey* if whenever $k, r \in \mathbb{N}$ and $[\omega]^k$ is r-colored, there exists $A \in p$ such that $[A]^k$ is monochrome.

Theorem 2.33. *Let p be an ultrafilter on ω. Then the following statements are equivalent:*

(1) *p is Ramsey,*

(2) *for every 2-coloring of $[\omega]^2$, there exists $A \in p$ such that $[A]^2$ is monochrome, and*

(3) *for every partition $\{A_n : n < \omega\}$ of ω, either $A_n \in p$ for some $n < \omega$ or there exists $A \in p$ such that $|A \cap A_n| \leq 1$ for all $n < \omega$.*

Proof. $(1) \Rightarrow (2)$ is obvious.

$(2) \Rightarrow (3)$ Define $\chi : [\omega]^2 \to \{0, 1\}$ by

$$\chi(i, j) = \begin{cases} 0 & \text{if } \{i, j\} \subseteq A_n \text{ for some } n < \omega \\ 1 & \text{otherwise.} \end{cases}$$

By (2), there is $A \in p$ be such that $[A]^2$ is monochrome. Suppose that $A_n \notin p$ for all $n < \omega$. It then follows that $\chi([A]^2) = \{1\}$, and consequently, $|A \cap A_n| \leq 1$ for all $n < \omega$.

$(3) \Rightarrow (1)$ We prove (1) for fixed r by induction on k. For $k = 1$, it is obvious.

Now assume (1) holds for some k and let $\chi : [\omega]^{k+1} \to \{0, 1, \ldots, r - 1\}$ be given. For each $i < \omega$, define $\chi_i : [\omega]^k \to \{0, 1, \ldots, r - 1\}$ by

$$\chi_i(x) = \begin{cases} \chi(\{i\} \cup x) & \text{if } \min x > i \\ 0 & \text{otherwise.} \end{cases}$$

By the inductive assumption, for each $i < \omega$, there are $B_i \in p$ and $l_i < r$ such that $\chi_i([B_i]^k) = \{l_i\}$. Clearly one may suppose that $\min B_i > i$ and the sequence $(B_i)_{i<\omega}$ is strictly decreasing, so $\{B_i \setminus B_{i+1} : i < \omega\}$ is a partition of B_0 into nonempty subsets. Applying (3), we obtain that there is a sequence $(b_i)_{i<\omega}$ with $b_i \in B_i \setminus B_{i+1}$ such that $B = \{b_i : i < \omega\} \in p$. We then have that whenever $j_0 < j_1 < \ldots < j_k < \omega$ and $j_1 \geq b_{j_0}$,

$$\chi(\{b_{j_0}, b_{j_1}, \ldots, b_{j_k}\}) = \chi_{b_{j_0}}(\{b_{j_1}, \ldots, b_{j_k}\}) = l_{b_{j_0}}$$

Define recursively an increasing sequence $(i_n)_{n<\omega}$ in ω by $i_0 = 0$ and

$$i_{n+1} = \max\{i_n + 1, \max\{b_i : i < i_n\}\}.$$

Applying once again (3), we obtain that there is $I \subseteq \omega$ such that $|I \cap [i_n, i_{n+1})| \leq 1$ for each $n < \omega$ and $\{b_i : i \in I\} \in p$. Let

$$I_0 = I \cap \bigcup_{n<\omega} [i_{2n}, i_{2n+1}) \quad \text{and} \quad I_1 = I \cap \bigcup_{n<\omega} [i_{2n+1}, i_{2n+2}).$$

There is $s < 2$ such that $\{b_i : i \in I_s\} \in p$. Then whenever $i, j \in I_s$ and $i < j$, there is $n < \omega$ such that $i < i_n < i_{n+1} \leq j$, and so $j \geq b_i$. Finally, there is $l < r$, such that

$$\{b_i : i \in I_s \text{ and } l_{b_i} = l\} \in p.$$

Denote this member of p by A. Then $\chi([A]^{k+1}) = \{l\}$. $\square$

Theorem 2.34. *Assume $\mathfrak{p} = \mathfrak{c}$. Then there exists a nonprincipal Ramsey ultrafilter on ω.*

Proof. Let $\{X_\alpha : \alpha < 2^\omega\}$ be an enumeration of $\mathcal{P}(\omega)$ and let $\{\chi_\alpha : \alpha < 2^\omega\}$ be an enumeration of 2-colorings $[\omega]^2 \to \{0, 1\}$. We shall construct inductively a 2^ω-sequence $(A_\alpha)_{\alpha<2^\omega}$ of infinite subsets of ω such that for every $\alpha < 2^\omega$, the following conditions are satisfied:

(i) for every $\beta < \alpha$, $|A_\alpha \setminus A_\beta| < \omega$,

(ii) $[A_\alpha]^2$ is monochrome with respect to χ_α, and

(iii) either $A_\alpha \subseteq X_\alpha$ or $A_\alpha \subseteq \omega \setminus X_\alpha$.

By Ramsey's Theorem, there is an infinite $C_0 \subseteq \omega$ such that $[C_0]^2$ is monochrome with respect to χ_0. If $|C_0 \cap X_0| = \omega$, put $A_0 = C_0 \cap X_0$. Otherwise put $A_0 = C_0 \cap (\omega \setminus X_0)$. Now fix $\gamma < 2^\omega$ and suppose that we have already constructed a sequence $(A_\alpha)_{\alpha<\gamma}$ satisfying conditions (i)–(iii) for all $\alpha < \gamma$. Since $\mathfrak{p} = \mathfrak{c}$, there is $B_\gamma \subseteq \omega$ such that $|B_\gamma \setminus A_\alpha| < \omega$ for all $\alpha < \gamma$. Choose an infinite $C_\gamma \subseteq B_\gamma$ such that $[C_\gamma]^2$ is monochrome with respect to χ_γ and put

$$A_\gamma = \begin{cases} C_\gamma \cap X_\gamma & \text{if } |C_\gamma \cap X_\gamma| = \omega \\ C_\gamma \cap (\omega \setminus X_\gamma) & \text{otherwise.} \end{cases}$$

Having constructed $(A_\alpha)_{\alpha < 2^\omega}$, let

$$p = \{A \subseteq \omega : |A_\alpha \setminus A| < \omega \text{ for some } \alpha < 2^\omega\}.$$

Then p is a nonprincipal Ramsey ultrafilter.   □

A point x in a topological space X is called a *P-point* if the intersection of countably many neighborhoods of x is again a neighborhood of x.

Definition 2.35. We say that a nonprincipal ultrafilter p on ω is a *P-point* if p is a P-point in ω^*.

Lemma 2.36. *Let p be a nonprincipal ultrafilter on ω. Then p is a P-point if and only if whenever $\{A_n : n < \omega\}$ is a partition of ω and $A_n \notin p$ for all $n < \omega$, there exists $A \in p$ such that $A \cap A_n$ is finite for all $n < \omega$.*

Proof. Necessity. Let $\{A_n : n < \omega\}$ be a partition of ω and $A_n \notin p$ for all $n < \omega$. Define the sequence $\{B_n : n < \omega\}$ of members of p by $B_n = \bigcup_{n < i < \omega} A_i$. Since p is a P-point, $\bigcap_{n < \omega} B_n^*$ is a neighborhood of $p \in \omega^*$. Consequently, there is $A \in p$ such that $A^* \subseteq \bigcap_{n < \omega} B_n^*$. It follows that for every $n < \omega$, $A \setminus B_n$ is finite, and so $A \cap A_n$ is finite.

Sufficiency. Let $\{U_n : n < \omega\}$ be a sequence of neighborhoods of $p \in \omega^*$. For every $n < \omega$, pick $B_n \in p$ such that $B_n^* \subseteq U_n$. Without loss of generality one may suppose that the sequence $\{B_n : n < \omega\}$ is decreasing and $\bigcap_{n < \omega} B_n = \emptyset$. Define the partition $\{A_n : n < \omega\}$ of ω by putting $A_0 = \omega \setminus B_0$ and $A_{n+1} = B_n \setminus B_{n+1}$. Clearly $A_n \notin p$ for all $n < \omega$. Consequently by the assumption, there is $A \in p$ such that $A \cap A_n$ is finite for all $n < \omega$. Then for every $n < \omega$, $A \setminus B_n$ is finite, and so $A^* \subseteq B_n^*$. We have that $p \in A^* \subseteq \bigcap_{n < \omega} B_n^* \subseteq \bigcap_{n < \omega} U_n$. Hence $\bigcap_{n < \omega} U_n$ is a neighborhood of $p \in \omega^*$.   □

Corollary 2.37. *Every nonprincipal Ramsey ultrafilter on ω is a P-point.*

By Theorem 2.34, $\mathfrak{p} = \mathfrak{c}$ implies the existence of a Ramsey ultrafilter, and consequently, a P-point. Assuming $\mathfrak{p} = \mathfrak{c}$, one can construct also a P-point not being a Ramsey ultrafilter. However, neither of such ultrafilters can be constructed in ZFC, the system of usual axioms of set theory.

Theorem 2.38. *It is consistent with ZFC that there is no P-point in ω^*.*

Proof. See [69, VI, §4].   □

2.5   Measurable Cardinals

Definition 2.39. Let κ and λ be infinite cardinals. A filter $\mathcal{F}$ on κ is λ-*complete* if $\bigcap \mathcal{G} \in \mathcal{F}$ whenever $\mathcal{G} \subseteq \mathcal{F}$ and $|\mathcal{G}| < \lambda$.

Every filter is ω-complete. An ω^+-complete filter is called *countably complete*.

Lemma 2.40. *An ultrafilter p on κ is λ-complete if and only if whenever $\mu < \lambda$ and $\{A_\alpha : \alpha < \mu\}$ is a partition of κ, there is $\alpha < \mu$ such that $A_\alpha \in p$.*

Proof. Necessity. Assume on the contrary that for every $\alpha < \mu$, one has $A_\alpha \notin p$, so $\kappa \setminus A_\alpha \in p$. Then by λ-completeness of p, $\bigcap_{\alpha < \mu} \kappa \setminus A_\alpha \in p$. But $\bigcap_{\alpha < \mu} \kappa \setminus A_\alpha = \emptyset$, a contradiction.

Sufficiency. Assume on the contrary that p is not λ-complete. Choose $\mu < \lambda$ and $\{C_\alpha : \alpha < \mu\} \subseteq p$ such that $\bigcap_{\alpha < \mu} C_\alpha \notin p$ and μ is as small as possible. Taking $C_\alpha \setminus \bigcap_{\alpha < \mu} C_\alpha$ instead of C_α, one may suppose that $\bigcap_{\alpha < \mu} C_\alpha = \emptyset$. Using minimality of μ, define $\{B_\alpha : \alpha < \mu\} \subseteq p$ by $B_0 = \kappa$ and $B_\alpha = \bigcap_{\beta < \alpha} C_\alpha$ if $\alpha > 0$. Then the sets $A_\alpha = B_\alpha \setminus B_{\alpha+1}, \alpha < \mu$, form a partition of κ and $A_\alpha \notin p$ for every $\alpha < \mu$, a contradiction. $\qquad\square$

Corollary 2.41. *An ultrafilter p on κ is countably complete if and only if whenever $\{A_n : \alpha < \omega\}$ is a partition of κ, there is $n < \omega$ such that $A_n \in p$.*

Definition 2.42. A cardinal κ is *measurable* (*Ulam-measurable*) if there is a κ-complete (countably complete) nonprincipal ultrafilter on κ.

Note that ω is a measurable cardinal, but not Ulam-measurable.

Theorem 2.43. *A cardinal is Ulam-measurable if and only if it is greater than or equal to the first uncountable measurable cardinal.*

Proof. It is clear that every uncountable measurable cardinal is Ulam-measurable, and if a cardinal is Ulam-measurable, then any greater cardinal is also Ulam-measurable. Therefore, it suffices to prove that the first Ulam-measurable cardinal is measurable.

Let κ be the first Ulam-measurable cardinal and let p be a nonprincipal countably complete ultrafilter on κ. We show that p is κ-complete. Assume the contrary. Then by Lemma 2.40, there are $\lambda < \kappa$ and a partition $\{A_\xi : \xi < \lambda\}$ of κ such that $A_\xi \notin p$ for every $\xi < \lambda$. Define $f : \kappa \to \lambda$ by $f(x) = \xi$ if $x \in A_\xi$. Now let $q = f(p)$. It is easy to see that q is a nonprincipal countably complete ultrafilter on λ, contradicting the choice of κ. $\qquad\square$

A cardinal κ is called *strongly inaccessible* if κ is regular and $2^\lambda < \kappa$ whenever $\lambda < \kappa$. Note that ω is a strongly inaccessible cardinal.

Theorem 2.44. *A measurable cardinal is strongly inaccessible.*

Proof. Let κ be a measurable cardinal and let p be a nonprincipal κ-complete ultra-filter on κ.

We first show that κ is regular. Let $\lambda = \mathrm{cf}(\kappa)$ and assume on the contrary that $\lambda < \kappa$. Pick an increasing cofinal λ-sequence $(\kappa_\xi)_{\xi<\lambda}$ of cardinals in κ. If $\kappa \setminus \kappa_\xi \in p$ for every $\xi < \lambda$, then $\emptyset = \bigcap_{\xi<\lambda} \kappa \setminus \kappa_\xi \in p$, since p is κ-complete. Consequently, $\kappa_\xi \in p$ for some $\xi < \lambda$. But then, since p is nonprincipal and κ-complete, we obtain that $\emptyset = \bigcap_{\alpha<\kappa_\xi} \kappa_\xi \setminus \{\alpha\} \in p$, a contradiction.

Now let $\lambda < \kappa$ and assume on the contrary that $\kappa \leq 2^\lambda$. Pick an injective function $F : \kappa \to {}^\lambda 2$. For every $\xi < \lambda$ and $i < 2$, let

$$X_\xi^i = \{\alpha < \kappa : F(\alpha)(\xi) = i\}.$$

Note that $\{X_\xi^i : i < 2\}$ is a partition of κ. Define a function $g : \lambda \to 2$ by $X_\xi^{g(\xi)} \in p$, and let $X = \bigcap_{\xi<\lambda} X_\xi^{g(\xi)}$. Then $X \in p$, by κ-completeness of p, and

$$
\begin{aligned}
X &= \{\alpha < \kappa : F(\alpha)(\xi) = g(\xi) \text{ for all } \xi < \lambda\} \\
&= \{\alpha < \kappa : F(\alpha) = g\} \\
&= F^{-1}(g).
\end{aligned}
$$

Since F is injective, it follows that X is a singleton. But this contradicts the assumption that p is nonprincipal.   □

Theorem 2.45. *It is consistent with ZFC that there is no uncountable strongly inaccessible cardinal.*

Proof. See [43, Corollary IV 6.9].   □

References

Ultrafilters were introduced by F. Riesz [64] and S. Ulam [76]. The Ultrafilter Theorem is equivalent to the Prime Ideal Theorem, a well-known weaker form of the Axiom of Choice AC (see [41, Section 2.3]). The existence of a nonprincipal ultrafilter on ω cannot be established in ZF [41, Problem 5.24]. The Stone–Čech compactification was produced independently by M. Stone [72] and E. Čech [82]. A great deal of information about ultrafilters and the Stone–Čech compactification can be found in [12]. Concerning $\beta\omega$ see also [81]. Theorem 2.16 goes back to N. de Bruijn and P. Erdős [16]. Theorem 2.22 is apparently due to Z. Frolík [26]. Theorem 2.31 is a result of D. Booth [6]. For more information about MA see [43] and [25]. Our proof of Theorem 2.33 is based on the treatment in [10]. P-points were invented by W. Rudin [66] who used them to show that under CH, ω^* is not homogeneous. Theorem 2.38 is due to S. Shelah [69]. Measurable cardinals were introduced by S. Ulam [77].

Chapter 3
Topological Spaces with Extremal Properties

In this chapter we discuss extremally disconnected, irresolvable, maximal and other spaces with extremal properties.

3.1 Filters and Ultrafilters on Topological Spaces

Definition 3.1. Let $\mathcal{F}$ be a filter on a topological space.

  (a) $\mathcal{F}$ is *open* (*closed*) if it has a base of open (closed) sets.

  (b) $\mathcal{F}$ is *nowhere dense* if there is $A \in \mathcal{F}$ such that int cl $A = \emptyset$.

  (a) $\mathcal{F}$ is *dense* if for every $A \in \mathcal{F}$, int cl $A \in \mathcal{F}$.

Lemma 3.2. *Every ultrafilter on a space is either dense or nowhere dense.*

Proof. Let $\mathcal{U}$ be an ultrafilter on a space X. Suppose that $\mathcal{U}$ is not dense. Then there is a closed $A \in \mathcal{F}$ such that int $A \notin \mathcal{U}$. Let $B = A \setminus \text{int } A$. It follows that $B \in \mathcal{U}$ and int cl $B = \emptyset$. Hence, $\mathcal{U}$ is nowhere dense. □

Corollary 3.3. *An ultrafilter $\mathcal{U}$ on a space is dense if and only if for every $A \in \mathcal{U}$, int cl $A \neq \emptyset$.*

Note that a filter $\mathcal{F}$ on a space is nowhere dense if and only if every ultrafilter $\mathcal{U} \supseteq \mathcal{F}$ is nowhere dense.

Lemma 3.4. *Let $\mathcal{F}$ be a filter on a space X and suppose that $\mathcal{F}$ is not nowhere dense. Define the filter $\mathcal{G}$ on X by taking as a base the subsets of the form $U \setminus Y$ where $U \in \mathcal{F}$ and Y ranges over nowhere dense subsets of X. Then for every ultrafilter $\mathcal{U} \supseteq \mathcal{F}$, $\mathcal{U}$ is dense if and only if $\mathcal{U} \supseteq \mathcal{G}$. If $\mathcal{F}$ is open, $\mathcal{G}$ is open as well.*

Proof. It follows from the definition of $\mathcal{G}$ that the ultrafilters containing $\mathcal{G}$ are precisely those of ultrafilters containing $\mathcal{F}$ which are not nowhere dense, and so by Lemma 3.2, these are all dense ultrafilters containing $\mathcal{F}$. If $\mathcal{F}$ is open, then $\mathcal{G}$ has a base consisting of subsets of the form $U \setminus Y$ where $U \in \mathcal{F}$ is open and $Y \subset X$ is closed nowhere dense, so $\mathcal{G}$ is open as well. □

We say that filters $\mathcal{F}$ and $\mathcal{G}$ are *incompatible* if $A \cap B = \emptyset$ for some $A \in \mathcal{F}$ and $B \in \mathcal{G}$.

Lemma 3.5. *Every open filter is contained in a maximal open filter. Different maximal open filters are pairwise incompatible.*

Proof. The first statement is immediate from Zorn's Lemma. To see the second, let $\mathcal{F}$ and $\mathcal{G}$ be maximal open filters on a space X. Suppose that $U \cap V \neq \emptyset$ for all $U \in \mathcal{F}$ and $V \in \mathcal{G}$. Then the sets $U \cap V \neq \emptyset$, where $U \in \mathcal{F}$ and $V \in \mathcal{G}$, form a base for a filter $\mathcal{H}$. Since $\mathcal{H}$ is open and contains $\mathcal{F}$ and $\mathcal{G}$, it follows that $\mathcal{F} = \mathcal{G}$. $\square$

Lemma 3.6. *Let $\mathcal{F}$ be a maximal open filter on a space X and let $A \subseteq X$ be such that $A \cap U \neq \emptyset$ for all $U \in \mathcal{F}$. Then int cl $A \in \mathcal{F}$.*

Proof. Since $\mathcal{F}$ is open, it suffices to show that cl $A \in \mathcal{F}$. Assume the contrary. Then $U \setminus \text{cl } A \neq \emptyset$ for all $U \in \mathcal{F}$. Consequently, the sets $U \setminus \text{cl } A$, where $U \in \mathcal{F}$, form a base for an open filter $\mathcal{G} \supseteq \mathcal{F}$. Since $\mathcal{F}$ is maximal open, $\mathcal{F} = \mathcal{G}$. It follows that $(\text{cl } A) \cap U = \emptyset$ for some $U \in \mathcal{F}$, and so $A \cap U = \emptyset$, a contradiction. $\square$

Corollary 3.7. *Every ultrafilter containing a maximal open filter is dense.*

Lemma 3.8. *Let $\mathcal{F}$ be an open filter on a space and let $\mathcal{U}$ be a dense ultrafilter containing $\mathcal{F}$. Then there is a maximal open filter $\mathcal{G}$ such that $\mathcal{U} \supseteq \mathcal{G} \supseteq \mathcal{F}$.*

Proof. For every $A \in \mathcal{U}$, int cl $A \in \mathcal{U}$. Consequently, there is a maximal open filter $\mathcal{G}$ such that int cl $A \in \mathcal{G}$ for all $A \in \mathcal{U}$. Let $V \in \mathcal{G}$ be open and let $A \in \mathcal{U}$. Then $V \cap (\text{int cl } A) \neq \emptyset$, and so $V \cap A \neq \emptyset$. It follows that $\mathcal{G} \subseteq \mathcal{U}$. $\square$

Summarizing, we obtain the following.

Proposition 3.9. *Let $\mathcal{F}$ be an open filter on a space X, let $\mathcal{G}$ be the filter with a base consisting of subsets $U \setminus Y$ where $U \in \mathcal{F}$ and Y ranges over nowhere dense subsets of X, and let $\{\mathcal{G}_i : i \in I\}$ be the family of all maximal open filters containing $\mathcal{F}$. Then*

(1) *for every ultrafilter $\mathcal{U} \supseteq \mathcal{F}$, $\mathcal{U}$ is dense if and only if $\mathcal{U} \supseteq \mathcal{G}$,*

(2) *the filters $\mathcal{G}_i$, where $i \in I$, are pairwise incompatible and $\bigcap_{i \in I} \mathcal{G}_i = \mathcal{G}$, and*

(3) *for every $i \in I$, ultrafilter $\mathcal{U} \supseteq \mathcal{G}_i$ and $A \in \mathcal{U}$, one has int cl $A \in \mathcal{G}_i$.*

3.2   Spaces with Extremal Properties

Definition 3.10. A topological space X is *nodec* if every nowhere dense subset of X is closed.

Equivalently, a space is nodec if every nowhere dense subset is discrete.

Proposition 3.11. *A T_1-space X is nodec if and only if every nonprincipal converging ultrafilter on X is dense.*

Proof. A T_1-space X contains no nonclosed nowhere dense set if and only if there is no nonprincipal converging nowhere dense ultrafilter on X, consequently by Lemma 3.2, if and only if every nonprincipal converging ultrafilter on X is dense.   □

Recall that a space is *extremally disconnected* if the closures of disjoint open sets are disjoint.

Proposition 3.12. *A space X is extremally disconnected if and only if for every $x \in X$, there is exactly one maximal open filter on X converging to x.*

Proof. Suppose that X is not extremally disconnected. Then there are disjoint open $U, V \subset X$ and $x \in X$ such that $x \in (\mathrm{cl}\, U) \cap (\mathrm{cl}\, V)$. Let $\mathscr{F}$ be the neighborhood filter of x and let

$$\mathscr{F}_U = \mathrm{flt}(\mathscr{F} \cup \{U\}) \quad \text{and} \quad \mathscr{F}_V = \mathrm{flt}(\mathscr{F} \cup \{V\}).$$

Then $\mathscr{F}_U$ and $\mathscr{F}_V$ are incompatible open filters on X converging to x. Hence, maximal open filters extending $\mathscr{F}_U$ and $\mathscr{F}_V$ (Lemma 3.5) are different.

Conversely, suppose that there are at least two different maximal open filters on X, say $\mathscr{G}$ and $\mathscr{H}$, converging to some $x \in X$. Since they are incompatible (Lemma 3.5), there are disjoint open $U \in \mathscr{G}$ and $V \in \mathscr{H}$. Clearly $x \in (\mathrm{cl}\, U) \cap (\mathrm{cl}\, V)$. Hence, X is not extremally disconnected.   □

An important property of extremally disconnected spaces is contained in the following theorem.

Theorem 3.13. *Let X be an extremally disconnected Hausdorff space and let $f : X \to X$ be a homeomorphism. Then the set $M = \{x \in X : h(x) = x\}$ of all fixed points of f is clopen.*

Proof. By Zorn's Lemma, there is a maximal open set $U \subset X$ such that $f(U) \cap U = \emptyset$. Since X is extremally disconnected, U is clopen. Let

$$F = U \cup f(U) \cup f^{-1}(U).$$

Then F is clopen and $M \cap F = \emptyset$. We claim that $M = X \setminus F$.

Indeed, assume on the contrary that $f(x) \neq x$ for some $x \in X \setminus F$. Choose an open neighborhood V of x such that $V \cap F = \emptyset$ and $f(V) \cap V = \emptyset$. It follows that $f(V) \cap U = \emptyset$, since $V \cap f^{-1}(U) = \emptyset$. But then $f(U \cup V) \cap (U \cup V) = \emptyset$, which contradicts maximality of U.   □

Definition 3.14. A space X is *strongly extremally disconnected* if for every open nonclosed $U \subset X$, there exists $x \in X \setminus U$ such that $U \cup \{x\}$ is open.

Proposition 3.15. *A space* X *is strongly extremally disconnected if and only if it is extremally disconnected and every nonempty nowhere dense subset of* X *has an isolated point.*

Proof. Necessity. To see that X is extremally disconnected, let U be an open subset of X and let $V = \text{int cl } U$. We claim that $V = \text{cl } U$. Indeed, otherwise V is nonclosed and so there is $x \in \text{cl } V \setminus V = \text{cl } U \setminus V$ such that $V \cup \{x\}$ is open, which contradicts $V = \text{int cl } U$.

Now let Y be a nonempty nowhere dense subset of X. One may suppose that Y is closed. Then $U = X \setminus Y$ is open and nonclosed. Consequently, there is $x \in Y$ such that $U \cup \{x\}$ is open. It follows that x is an isolated point of Y.

Sufficiency. Let U be an open nonclosed subset of X. Then $Y = \text{cl } U \setminus U$ is a nonempty nowhere dense subset. Consequently, there is an isolated point $x \in Y$. Also we have that cl U is open. It follows that $U \cup \{x\}$ is open. $\qquad\square$

Note that to say that every nonempty nowhere dense subset of a space has an isolated point is the same as saying that every nonempty perfect subset has a nonempty interior. A subset of a space is called *perfect* if it is closed and dense in itself.

Corollary 3.16. *A dense in itself space* X *is strongly extremally disconnected if and only if every perfect subset of* X *is open.*

Proof. Necessity. Let F be a perfect subset of X and let $U = \text{int } F$. We have to show that $U = F$. Assume on the contrary that $F \setminus U \neq \emptyset$. Then U is nonclosed. Indeed, otherwise $F \setminus U$ is perfect, so by Proposition 3.15, int $(F \setminus U) \neq \emptyset$, which contradicts $U = \text{int } F$. Since U is nonclosed, there is $x \in F \setminus U$ such that $U \cup \{x\}$ is open, a contradiction with $U = \text{int } F$.

Sufficiency. By Proposition 3.15, we have to check only that X is extremally disconnected. Let U be a nonempty open subset of X. Since X is dense in itself, so is U. Consequently, cl U is perfect, and so it is open. $\qquad\square$

Definition 3.17. A space is *resolvable* or *irresolvable* depending on whether or not it can be partitioned into two dense subsets. A space X is *open-hereditarily irresolvable* if every nonempty open subset of X is irresolvable.

Proposition 3.18. *A space* X *is open-hereditarily irresolvable if and only if every (converging) maximal open filter on* X *is an ultrafilter.*

Proof. Let $\mathcal{F}$ be a maximal open filter on X and suppose that $\mathcal{F}$ is not an ultrafilter. Then there are two different ultrafilters $\mathcal{U}$ and $\mathcal{V}$ on X containing $\mathcal{F}$. Pick pairwise disjoint $A \in \mathcal{U}$ and $B \in \mathcal{V}$. By Corollary 3.7, both $\mathcal{U}$ and $\mathcal{V}$ are dense, so int cl $A \in \mathcal{F}$ and int cl $B \in \mathcal{F}$. Define an open $U \in \mathcal{F}$ by

$$U = (\text{int cl } A) \cap (\text{int cl } B).$$

Then both $A \cap U$ and $B \cap U$ are dense in U, so U is resolvable. Hence, X is not open-hereditarily irresolvable.

Conversely, suppose that every converging maximal open filter on X is an ultrafilter. Let U be any nonempty open subset of X and let $\{A_0, A_1\}$ be a partition of U. Pick any converging maximal open filter $\mathcal{F}$ on X containing U. By the assumption $\mathcal{F}$ is an ultrafilter, so $A_i \in \mathcal{F}$ for some $i < 2$, say $A_0 \in \mathcal{F}$. Since $\mathcal{F}$ is open, int $A_0 \neq \emptyset$. Consequently, A_1 is not dense in U. Hence, X is not open-hereditarily irresolvable. $\qquad\square$

Definition 3.19. A space is *submaximal* if every dense subset is open.

Proposition 3.20. *A space is submaximal if and only if it is open-hereditarily irresolvable and nodec.*

Proof. Suppose that X is submaximal.

To see that X is open-hereditarily irresolvable, let U be a nonempty open subset of X and let A be a dense subset of U. Then $(X \setminus U) \cup A$ is dense in X, and consequently, open. It follows that $U \setminus A$ is is not dense in U. Hence, U is irresolvable.

To see that X is nodec, let Y be a nowhere dense subset of X. Then $X \setminus Y$ is dense, and consequently, open. So Y is closed.

Now suppose that X is open-hereditarily irresolvable and nodec. To show that X is submaximal, let A be a dense subset of X. Since X is open-hereditarily irresolvable, int A is dense in X. Consequently, $X \setminus$ int A is nowhere dense, and so discrete, since X is nodec. It follows from this that int $A = A$. $\qquad\square$

Definition 3.21. A space $(X, \mathcal{T})$ is *maximal* if $\mathcal{T}$ is maximal among all dense in itself topologies on X.

Equivalently, a space is maximal if it has no isolated point but it does have an isolated point in any stronger topology.

Theorem 3.22. *For a dense in itself Hausdorff space X, the following statements are equivalent:*

(1) *X is maximal,*

(2) *X is extremally disconnected, open-hereditarily irresolvable and nodec,*

(3) *for each $x \in X$, there is exactly one nonprincipal ultrafilter on X converging to x.*

Proof. Let $X = (X, \mathcal{T})$.

(1) $\Rightarrow$ (2) By Proposition 3.20, it suffices to show that that X is extremally disconnected and submaximal.

Assume that X is not extremally disconnected. Then there is an open subset $V \subset X$ and $x \in \operatorname{cl} V \setminus V$ such that $V \cup \{x\}$ is not open. Define the topology on X by taking as a base

$$\mathcal{T} \cup \{U \cap (V \cup \{x\}) : U \in \mathcal{T}\}.$$

This topology is dense in itself and stronger than $\mathcal{T}$. Hence, $\mathcal{T}$ is not maximal, a contradiction.

Now assume that X is not submaximal. So there is a dense nonopen subset $A \subset X$. Define the topology on X by taking as a base

$$\mathcal{T} \cup \{U \cap A : U \in \mathcal{T}\}.$$

Clearly, this topology is dense in itself and stronger than $\mathcal{T}$, a contradiction.

$(2) \Rightarrow (3)$ It follows from extremal disconnectedness of X that there is exactly one maximal open filter $\mathcal{F}$ on X converging to x (Proposition 3.12). Since X is nodec, every nonprincipal ultrafilter on X converging to x is dense (Proposition 3.11) and so contains $\mathcal{F}$ (Proposition 3.9). Finally, open-hereditary irresolvability of X implies that $\mathcal{F}$ is an ultrafilter (Proposition 3.18).

$(3) \Rightarrow (1)$ Let $\mathcal{T}_1$ be any topology on X stronger than $\mathcal{T}$. Then there is $x \in X$ and a nonprincipal ultrafilter $\mathcal{U}$ on X converging to x in $\mathcal{T}$ but not converging to x in $\mathcal{T}_1$. It follows that x is an isolated point in $\mathcal{T}_1$. Hence, $\mathcal{T}$ is maximal. $\square$

The equivalence $(1) \Leftrightarrow (3)$ in Theorem 3.22 justifies the following definition.

Definition 3.23. We say that a space X is *almost maximal* if it is dense in itself and for every $x \in X$ there are only finitely many ultrafilters on X converging to x.

3.3   Irresolvability

The notion of irresolvability (Definition 3.17) naturally generalizes as follows.

Definition 3.24. Given a cardinal $\kappa \geq 2$, a space is *κ-resolvable* or *κ-irresolvable* depending on whether or not it can be partitioned into κ-many dense subsets. A space X is *hereditarily κ-irresolvable (open-hereditarily κ-irresolvable)* if every nonempty subset of X (every nonempty open subset of X) is κ-irresolvable.

The next lemma contains the main simple properties of κ-resolvability.

Lemma 3.25. *For every $\kappa \geq 2$, the following statements hold:*

(1) *an open subset of a κ-resolvable space is κ-resolvable,*

(2) *the closure of a κ-resolvable subset of a space is κ-resolvable, and*

(3) *the union of a family of κ-resolvable subsets of a space is κ-resolvable.*

Proof. (1) and (2) are obvious, so it suffices to show (3). Let $\mathcal{R}$ be a family of κ-resolvable subsets of a space X and let $Y = \bigcup \mathcal{R}$. We have to show that Y is κ-resolvable.

Consider the set P of all κ-sequences $\mathcal{A} = \{A_\alpha : \alpha < \kappa\}$ consisting of pairwise disjoint nonempty subsets $A_\alpha \subseteq Y$ dense in $\bigcup \mathcal{A}$. Define the order on P by

$$\mathcal{A} \leq \mathcal{B} \quad \text{if and only if} \quad A_\alpha \subseteq B_\alpha \text{ for all } \alpha < \kappa.$$

Every chain $(\mathcal{A}_i)_{i \in I}$ in P, where $\mathcal{A}_i = \{A_\alpha^i : \alpha < \kappa\}$, has an upper bound

$$\left\{ \bigcup_{i \in I} A_\alpha^i : \alpha < \kappa \right\}.$$

Consequently by Zorn's Lemma, there is a maximal element $\mathcal{A} = \{A_\alpha : \alpha < \kappa\}$ in P. Let $Z = \bigcup_{\alpha < \kappa} A_\alpha$. Obviously, Z is κ-resolvable. It is clear also that Z is closed. Indeed, otherwise the element $\mathcal{B} = \{B_\alpha : \alpha < \kappa\}$ in P defined by

$$B_\alpha = \begin{cases} A_0 \cup \operatorname{cl} Z \setminus Z & \text{if } \alpha = 0 \\ A_\alpha & \text{otherwise} \end{cases}$$

is greater than $\mathcal{A}$. We claim that $Z = Y$. To see this, assume the contrary. Then there is $R \in \mathcal{R}$ such that $R \setminus Z \neq \emptyset$. Since Z is closed, $R \setminus Z \in \mathcal{R}$ as well. It follows that there is $\mathcal{C} = \{C_\alpha : \alpha < \kappa\}$ in P such that $\bigcup \mathcal{C} \cap Z = \emptyset$. Then the element $\mathcal{D} = \{D_\alpha : \alpha < \kappa\}$ in P defined by

$$D_\alpha = A_\alpha \cup C_\alpha$$

is greater than $\mathcal{A}$, a contradiction. $\qquad\square$

Lemma 3.25 leads to the following construction.

Lemma 3.26. *Given a space X and $\kappa \geq 2$, let $R_\kappa(X)$ denote the union of all κ-resolvable subsets of X. Then*

(i) *$R_\kappa(X)$ is the largest κ-resolvable subset of X,*

(ii) *$R_\kappa(X)$ is closed,*

(iii) *X is κ-resolvable if and only if $R_\kappa(X) = X$, and*

(iv) *if X is κ-irresolvable, then $X \setminus R_\kappa(X)$ is hereditarily κ-irresolvable.*

Proof. It is immediate from Lemma 3.25. $\qquad\square$

Corollary 3.27. *A homogeneous space is κ-irresolvable if and only if it is hereditarily κ-irresolvable.*

Proof. Let X be a homogeneous κ-irresolvable space. Assume on the contrary that X is not hereditarily κ-irresolvable. Then $R_\kappa(X)$ is a proper subset of X. Pick $x \in R_\kappa(X)$ and $y \in X \setminus R_\kappa(X)$. Let $f : X \to X$ be a homeomorphism with $f(x) = y$. Then $f(R_\kappa(X))$ is a κ-resolvable subset of X and $f(R_\kappa(X)) \setminus R_\kappa(X) \neq \emptyset$, which is a contradiction. □

Of special interest are finitely and ω-irresolvable spaces.

Recall that for every filter $\mathcal{F}$ on a set X, $\overline{\mathcal{F}}$ denotes the set of all ultrafilters on X containing $\mathcal{F}$.

Proposition 3.28. *A space X is open-hereditarily n-irresolvable if and only if for every (converging) maximal open filter $\mathcal{F}$ on X, $|\overline{\mathcal{F}}| < n$.*

Proof. The same as that of Proposition 3.18. □

Lemma 3.29. *Let X be a space and suppose that there is an open filter $\mathcal{F}$ on X such that $|\overline{\mathcal{F}}| < n$ for some $n < \omega$. Then X is n-irresolvable.*

Proof. Assume on the contrary that there is a partition of X into n dense sets A_i, where $i < n$. Then there is $j < n$ such that $\bigcup_{j \neq i < n} A_i \in \mathcal{F}$. Since $\mathcal{F}$ is open, it follows that A_j is not dense, a contradiction. □

Corollary 3.30. *Every almost maximal space is ω-irresolvable.*

Theorem 3.31. *A space X is n-irresolvable if and only if there is a (converging) open filter $\mathcal{F}$ on X with $|\overline{\mathcal{F}}| < n$.*

Proof. Necessity. By Lemma 3.26, the subspace $Y = X \setminus R_n(X)$ of X is open and hereditarily n-irresolvable. Pick a converging maximal open filter $\mathcal{F}$ on X such that $Y \in \mathcal{F}$. Then by Proposition 3.28, $|\overline{\mathcal{F}}| < n$.

Sufficiency is immediate from Lemma 3.29. □

Corollary 3.32. *A space X is irresolvable if and only if there is a (converging) open ultrafilter on X.*

The next theorem tells us that in a homogeneous regular ω-irresolvable space every countable discrete subset is closed.

Theorem 3.33. *Let X be a homogeneous regular space and suppose that X contains a countable discrete nonclosed set. Then X is ω-resolvable.*

Before proving Theorem 3.33, we establish the following simple fact.

A subset D of a space X is called *strongly discrete* if for every $x \in D$, there is a neighborhood U_x of $x \in X$ such that the subsets U_x, where $x \in D$, are pairwise disjoint.

Lemma 3.34. *Every countable discrete subset of a regular space is strongly discrete.*

Proof. Let X be a regular space and let D be a countably infinite discrete subset of X. Enumerate D without repetitions as $\{x_n : n < \omega\}$. For each $n < \omega$, pick a closed neighborhood U_n of $x_n \in X$ such that $U_n \cap D = \{x_n\}$. For each $n < \omega$, define a neighborhood V_n of $x_n \in X$ by

$$V_n = U_n \setminus \bigcup_{i<n} U_i.$$

Then the subsets V_n, where $n < \omega$, are pairwise disjoint. $\square$

Proof of Theorem 3.33. Let D be a countable discrete nonclosed subset of X and let $a \in \operatorname{cl} D \setminus D$. For each $x \in X$, there is a homeomorphism $f_x : X \to X$ with $f_x(a) = x$. Construct inductively a sequence $(D_n)_{n=1}^{\infty}$ of discrete subsets of X such that

$$D_n \subseteq \operatorname{cl} D_{n+1} \setminus D_{n+1}.$$

Put $D_1 = D$. Fix $k \in \mathbb{N}$ and suppose that the sequence $(D_n)_{n=1}^{k}$ has already been constructed satisfying that condition. Denote

$$Y_{k-1} = \begin{cases} \bigcup_{n=1}^{k-1} \operatorname{cl} D_n & \text{if } k > 1 \\ \varnothing & \text{otherwise.} \end{cases}$$

It follows from the condition that $D_k \cap Y_{k-1} = \varnothing$. By Lemma 3.34, for each $x \in D_k$, there is an open neighborhood U_x of $x \in X \setminus Y_{k-1}$ such that the sets U_x, where $x \in D_k$, are pairwise disjoint. Then the set

$$D_{k+1} = \bigcup_{x \in D_k} f_x(D \cap f_x^{-1}(U_x))$$

is as required.

Now define $Z \subseteq X$ by

$$Z = \bigcup_{n=1}^{\infty} D_n.$$

We claim that Z is ω-resolvable.

Indeed, let $\{I_n : n \in \mathbb{N}\}$ be any partition of $\mathbb{N}$ into infinite sets. Define the partition $\{A_n : n \in \mathbb{N}\}$ of Z by

$$A_n = \bigcup_{i \in I_n} D_i.$$

Then each A_n is dense in Z.

Finally, since X contains a nonempty ω-resolvable subset, X itself is ω-resolvable by Corollary 3.27. $\square$

We conclude this section by showing that ω-irresolvable spaces are in fact finitely irresolvable.

Theorem 3.35. *If a space is n-resolvable for every $n < \omega$, then it is ω-resolvable.*

To prove Theorem 3.35, we need two lemmas.

Lemma 3.36. *Given a space X, let $W = W(X)$ denote the union of all open sets in X containing a dense hereditarily irresolvable subset. Then*

  (i) *W is the largest open set in X containing a dense open-hereditarily irresolvable subset, and*

  (ii) *every dense subset of $X \setminus \mathrm{cl}\, W$ is resolvable, and consequently, ω-resolvable.*

Proof. By Zorn's Lemma, there is a maximal family $\mathcal{U}$ of pairwise disjoint open sets in X containing a dense hereditarily irresolvable subset. Note that $\bigcup \mathcal{U}$ is dense in W. For each $U \in \mathcal{U}$, pick a dense hereditarily irresolvable $D_U \subseteq U$. Then $D = \bigcup_{U \in \mathcal{U}} D_U$ is a dense open-hereditarily irresolvable subset of W.

Now let A be a dense subset of $X \setminus \mathrm{cl}\, W$. Assume on the contrary that A is irresolvable. It then follows from Lemma 3.26 that there is a nonempty open set V in $X \setminus \mathrm{cl}\, W$ such that $V \cap A$ is hereditarily irresolvable. Clearly, $V \cap A$ is also dense in V. But this contradicts maximality of $\mathcal{U}$. $\square$

Lemma 3.37. *Let D be an open-hereditarily irresolvable subset of a space X and suppose that X is $(n + 1)$-resolvable for some n. Then $X \setminus D$ is dense in X and n-resolvable.*

Proof. By Lemma 3.26, it suffices to show that for every nonempty open set $U \subseteq X$, $U \setminus D$ contains a nonempty n-resolvable subset. One may suppose that $D \cap U \neq \emptyset$. Partition X into dense subsets $A_1, \ldots, A_{n+1}$. Then $D \cap U$ is partitioned into subsets $A_1 \cap D \cap U, \ldots, A_{n+1} \cap D \cap U$. Since no space is the union of finitely many nowhere dense sets, at least one of those sets, say $A_{n+1} \cap D \cap U$, is not nowhere dense in $D \cap U$. So there is an open $U_0 \subseteq U$ such that

$$\emptyset \neq D \cap U_0 \subseteq \mathrm{cl}_{D \cap U}(A_{n+1} \cap D \cap U).$$

It follows that $A_{n+1} \cap D \cap U_0$ is dense in $D \cap U_0$. Then, since $D \cap U_0$ is irresolvable, $(D \cap U_0) \setminus A_{n+1}$ is not dense in $D \cap U_0$. Consequently, there is an open $U_1 \subseteq U_0$ such that $\emptyset \neq D \cap U_1 \subseteq A_{n+1} \cap D \cap U_0$, and so $\emptyset \neq D \cap U_1 \subseteq A_{n+1} \cap U_1$. It follows that $A_1 \cap U_1, \ldots, A_n \cap U_1$ are pairwise disjoint dense subsets of $U_1 \setminus D$. $\square$

Proof of Theorem 3.35. By Lemma 3.26, it suffices to show that every nonempty open set $X_0 \subseteq X$ contains a nonempty ω-resolvable subset. Being open subset of X, X_0 is

n-resolvable for each n. Consider the subset $W(X_0) \subseteq X_0$ given by Lemma 3.36. If it is not dense, then $X_0 \setminus \mathrm{cl}\, W(X_0)$ is ω-resolvable and we are done. Otherwise pick a dense open-hereditarily irresolvable subset $D_1 \subset W(X_0)$ and put $X_1 = W(X_0) \setminus D_1$. Being an open subset of X_0, $W(X_0)$ is n-resolvable for each n, and then by Lemma 3.37, X_1 is n-resolvable for each n. If $W(X_1) \subseteq X_1$ is not dense, then $X_1 \setminus \mathrm{cl}\, W(X_1)$ is ω-resolvable. Otherwise pick a dense open-hereditarily irresolvable subset $D_2 \subset W(X_1)$ and put $X_2 = W(X_1) \setminus D_2$, and so on.

If at some m, $W(X_m) \subseteq X_m$ is not dense, then $X_m \setminus \mathrm{cl}\, W(X_m)$ is the required nonempty ω-resolvable subset of X_0. Otherwise we construct an infinite sequence $(D_m)_{m=1}^{\infty}$ of pairwise disjoint dense subsets of X_0, and then X_0 itself is ω-resolvable.

$\square$

References

The study of maximal and irresolvable spaces was initiated by E. Hewitt [33]. Theorem 3.13 is due to Z. Frolík [27]. Theorem 3.22 is from [99]. Another characterizations of maximal spaces can be found in [80]. Corollary 3.32 is a result of A. El'kin [18]. Theorem 3.35 is due to A. Illanes [40].

Chapter 4

Left Invariant Topologies and Strongly Discrete Filters

In this chapter left invariant topologies on semigroups are studied. We describe the largest left invariant topology in which a given filter converges to the identity. A special attention is paid to the case where the filter is strongly discrete, then the topology possesses certain important properties. We conclude by showing that every infinite group admits a nondiscrete zero-dimensional Hausdorff invariant topology.

4.1   Left Topological Semigroups

Definition 4.1. A semigroup (group) S endowed with a topology is a *left topological semigroup (group)* and the topology itself a *left invariant topology* if for every $a \in S$, the left translation

$$\lambda_a : S \ni x \mapsto ax \in S$$

is continuous.

A topology $\mathcal{T}$ on a semigroup S is left invariant if and only if for every $a \in S$ and $U \in \mathcal{T}, a^{-1}U \in \mathcal{T}$ where

$$a^{-1}U = \lambda_a^{-1}(U) = \{x \in S : ax \in U\}.$$

Note that in a left topological group left translations are homeomorphisms. Consequently, if S is a group, a left invariant topology on S is completely determined by the neighborhood filter $\mathcal{N}$ of 1: for every $a \in S$, the neighborhood filter of a is $a\mathcal{N} = \{aU : U \in \mathcal{N}\}$. Topologies with this property on semigroups are characterized by the following lemma.

Lemma 4.2. *Let S be a semigroup with identity, let $\mathcal{T}$ be a topology on S, and let $\mathcal{N}$ be the neighborhood filter of 1 in $\mathcal{T}$. Then the following statements are equivalent:*

(a) *for every $a \in S$, $a\mathcal{N}$ is a neighborhood base at a,*

(b) *for every $a \in S$, the left translation $\lambda_a : S \ni x \mapsto ax \in S$ is continuous and open, and*

(c) *for every $a \in S$ and $U \in \mathcal{T}$, both $a^{-1}U \in \mathcal{T}$ and $aU \in \mathcal{T}$.*

Proof. (a) $\Rightarrow$ (b) To see that λ_a is continuous, let $b \in S$ and let U be a neighborhood of $\lambda_a(b) = ab$. Pick $V \in \mathcal{N}$ such that $abV \subseteq U$. Then bV is a neighborhood of b and $\lambda_a(bV) = abV \subseteq U$.

To see that λ_a is open, let $b \in S$ and let U be a neighborhood of b. Pick $V \in \mathcal{N}$ such that $bV \subseteq U$. Then abV is a neighborhood of $\lambda_a(b)$ and $\lambda_a(U) \supseteq \lambda_a(bV) = abV$.

(b) $\Rightarrow$ (c) Since λ_a is continuous, $a^{-1}U = \lambda_a^{-1}(U) \in \mathcal{T}$, and since λ_a is open, $aU = \lambda_a(U) \in \mathcal{T}$.

(c) $\Rightarrow$ (a) If U is an open neighborhood of 1, then $a \in aU \in \mathcal{T}$, so aU is an open neighborhood of a. Conversely, let V be an open neighborhood of a and let $U = a^{-1}V$. Then U is an open neighborhood of 1 and $aU \subseteq V$. $\square$

The next theorem characterizes the neighborhood filter of the identity of a left topological semigroup.

Theorem 4.3. *Let S be a left topological semigroup with identity and let $\mathcal{N}$ be the neighborhood filter of 1. Then*

(1) *for every $U \in \mathcal{N}$, $1 \in U$, and*

(2) *for every $U \in \mathcal{N}$, $\{x \in S : x^{-1}U \in \mathcal{N}\} \in \mathcal{N}$.*

Conversely, given a semigroup S with identity and a filter $\mathcal{N}$ on S satisfying conditions (1)–(2), there is a left invariant topology on S in which for each $a \in S$, $a\mathcal{N}$ is a neighborhood base at a.

Note that condition (2) in Theorem 4.3 is equivalent to

(2$'$) *for every $U \in \mathcal{N}$, there are $V \in \mathcal{N}$ and $V \ni x \mapsto W_x \in \mathcal{N}$ such that $xW_x \subseteq U$.*

Proof. Suppose that S is a left topological semigroup with identity and let $U \in \mathcal{N}$. Clearly $1 \in U$. Putting $V = \operatorname{int} U$, we obtain that $V \in \mathcal{N}$ and for every $x \in V$, $x^{-1}U \in \mathcal{N}$.

Conversely, suppose that $\mathcal{N}$ is a filter on S satisfying conditions (1)–(2). For every $x \in S$, let $\mathcal{N}_x$ be the filter with a base $x\mathcal{N}$. We show that, whenever $x \in S$ and $U \in \mathcal{N}_x$, one has

(i) $x \in U$, and

(ii) $\{y \in S : U \in \mathcal{N}_y\} \in \mathcal{N}_x$.

(i) follows from (1) and definition of $\mathcal{N}_x$. To show (ii), let $U_0 = x^{-1}U$,

$$V_0 = \{z \in S : U_0 \in \mathcal{N}_z\},$$

and $V = xV_0$. Clearly $U_0 \in \mathcal{N}$, so by (2), $V_0 \in \mathcal{N}$, and consequently $V \in \mathcal{N}_x$. We claim that for every $y \in V$, $U \in \mathcal{N}_y$. To see this, write $y = xz$ for some $z \in V_0$. Then $x^{-1}U = U_0 \in \mathcal{N}_z = z\mathcal{N}$, and so $U \in xz\mathcal{N} = \mathcal{N}_y$.

Now by Theorem 1.9, there is a topology $\mathcal{T}$ on S for which $\{\mathcal{N}_x : x \in S\}$ is the neighborhood system. By Lemma 4.2, $\mathcal{T}$ is left invariant. $\square$

A semigroup is called *cancellative* (*left cancellative*, *right cancellative*) if all translations (left translations, right translations) are injective.

Note that whenever $f : X \rightarrow Y$ and $\mathcal{B} \subseteq \mathcal{P}(X)$, one has $f(\bigcap \mathcal{B}) \subseteq \bigcap f(\mathcal{B})$, and if f is injective, then $f(\bigcap \mathcal{B}) = \bigcap f(\mathcal{B})$.

From this and Theorem 4.3 we obtain as a consequence the following.

Corollary 4.4. *Let S be a semigroup with identity and let $\mathcal{B}$ be a filter base on S satisfying the following conditions:*

(1) *for every $U \in \mathcal{B}$, $1 \in U$, and*

(2) *for every $U \in \mathcal{B}$ and $a \in U$, there is $V \in \mathcal{B}$ such that $aV \subseteq U$.*

Then there is a left invariant topology $\mathcal{T}$ on S in which for each $a \in S$, $a\mathcal{B}$ is an open neighborhood base at a. Furthermore, if S is left cancellative, then $\mathcal{T}$ is a T_1-topology if and only if

(3) $\bigcap \mathcal{B} = \{1\}$.

Theorem 4.3 gives us also one more characterization of topologies from Lemma 4.2.

Corollary 4.5. *Let S be a semigroup with identity, let $\mathcal{T}$ be a topology on S, and let $\mathcal{N}$ be the neighborhood filter of 1 in $\mathcal{T}$. Then the following statements are equivalent:*

(a) *for every $a \in S$, $a\mathcal{N}$ is a neighborhood base at a, and*

(d) *$\mathcal{T}$ is the largest left invariant topology on S in which $\mathcal{N}$ is the neighborhood filter of 1.*

Proof. (a) $\Rightarrow$ (d) That $\mathcal{T}$ is left invariant follows from Lemma 4.2. To see that it is the largest one, let $\mathcal{T}'$ be any left invariant topology on S in which $\mathcal{N}$ is the neighborhood filter of 1. Then whenever $a \in S$ and U is a neighborhood of a in $\mathcal{T}'$, there is $V \in \mathcal{N}$ such that $aV \subseteq U$. Hence $\mathcal{T}' \subseteq \mathcal{T}$.

(d) $\Rightarrow$ (a) By Theorem 4.3, there is a left invariant topology $\mathcal{T}'$ on S in which for each $a \in S$, $a\mathcal{N}$ is a neighborhood base at a. Applying (a)$\Rightarrow$(d) to $\mathcal{T}'$, we obtain that $\mathcal{T}'$ is the largest left invariant topology on S in which $\mathcal{N}$ is the neighborhood filter of 1. Hence $\mathcal{T}' = \mathcal{T}$. $\qquad\qquad\square$

4.2   The Topology $\mathcal{T}[\mathcal{F}]$

Definition 4.6. Let S be a semigroup with identity. For every filter $\mathcal{F}$ on S, let $\mathcal{T}[\mathcal{F}]$ denote the largest left invariant topology on S in which $\mathcal{F}$ converges to 1.

Note that by Corollary 4.5, if $\mathcal{N}$ is the neighborhood filter of 1 in $\mathcal{T}[\mathcal{F}]$, then for every $a \in S$, $a\mathcal{N}$ is a neighborhood base at a, and by Lemma 4.2, the left translations in $\mathcal{T}[\mathcal{F}]$ are continuous and open.

Definition 4.7. For every mapping $M : S \to \mathcal{P}(S)$ and $a \in S$, define $[M]_a \in \mathcal{P}(S)$ by

$$[M]_a = \{x_0 x_1 \cdots x_n : n < \omega, x_0 = a \text{ and } x_{i+1} \in M(x_0 \cdots x_i) \text{ for each } i < n\}.$$

The next theorem describes the topology $\mathcal{T}[\mathcal{F}]$.

Theorem 4.8. *Let S be a semigroup with identity and let $\mathcal{F}$ be a filter on S. Then for every $a \in S$, the subsets $[M]_a$, where $M : S \to \mathcal{F}$, form an open neighborhood base at a in $\mathcal{T}[\mathcal{F}]$.*

Proof. The subsets $[M]_a$, where $a \in S$ and $M : S \to \mathcal{F}$, possess the following properties:

(i) for every $a \in S$ and $M, N : S \to \mathcal{F}$, there is $K : S \to \mathcal{F}$ such that $[K]_a \subseteq [M]_a \cap [N]_a$,

(ii) for every $a \in S$ and $M : S \to \mathcal{F}$, $a \in [M]_a$,

(iii) for every $a \in S$, $M : S \to \mathcal{F}$ and $x \in [M]_a$, one has $[M]_x \subseteq [M]_a$, and

(iv) for every $a \in S$ and $M : S \to \mathcal{F}$, there is $N : S \to \mathcal{F}$ such that $[M]_a = a[N]_1$.

For (i), define $K : S \to \mathcal{F}$ by $K(x) = M(x) \cap N(x)$. (ii) and (iii) are obvious. To see (iv), define $N : S \to \mathcal{F}$ by $N(x) = M(ax)$. Then

$$[M]_a = \{x_0 x_1 \cdots x_n : n < \omega, x_0 = a \text{ and } x_{i+1} \in M(x_0 \cdots x_i) \text{ for each } i < n\}$$
$$= a\{x_0 x_1 \cdots x_n : n < \omega, x_0 = 1 \text{ and } x_{i+1} \in M(ax_0 \cdots x_i) \text{ for each } i < n\}$$
$$= a\{x_0 x_1 \cdots x_n : n < \omega, x_0 = 1 \text{ and } x_{i+1} \in N(x_0 \cdots x_i) \text{ for each } i < n\}$$
$$= a[N]_1.$$

By (i), the subsets $[M]_1$, where $M : S \to \mathcal{F}$, form a filter base $\mathcal{B}$ on S, and by (ii), $1 \in U$ for all $U \in \mathcal{B}$. (iii) and (iv) give us that for every $U \in \mathcal{B}$ and $a \in U$, there is $V \in \mathcal{B}$ such that $aV \subseteq U$. Hence by Corollary 4.4, there is a left invariant topology $\mathcal{T}$ on S in which for every $a \in S$, $a\mathcal{B}$ is an open neighborhood base at a. (iii) and (iv) give us also that for every $a \in S$ and $M : S \to \mathcal{F}$, $[M]_a$ is an open neighborhood of a in $\mathcal{T}$. On the other hand, let U be an open neighborhood of $a \in S$ in any left invariant topology on S in which $\mathcal{F}$ converges to 1. Define $M : S \to \mathcal{F}$ by

$$M(x) = \begin{cases} x^{-1}U & \text{if } x \in U \\ S & \text{otherwise.} \end{cases}$$

Then $[M]_a \subseteq U$. It follows that the subsets $[M]_a$, where $M : S \to \mathcal{F}$, form an open neighborhood base at a in $\mathcal{T}$ and $\mathcal{T} = \mathcal{T}[\mathcal{F}]$. $\square$

Corollary 4.9. *If $A \in \mathcal{F}$ and for every $x \in A$, U_x is a neighborhood of 1 in $\mathcal{T}[\mathcal{F}]$, then $\left(\bigcup_{x \in A} xU_x\right) \cup \{1\}$ is a neighborhood of 1 in $\mathcal{T}[\mathcal{F}]$.*

Proof. Without loss of generality one may assume that for every $x \in A$, U_x is open, so $U = \bigcup_{x \in A} xU_x$ is open. Define $M : S \to \mathcal{F}$ by

$$M(x) = \begin{cases} A & \text{if } x = 1 \\ x^{-1}U & \text{if } x \in U \setminus \{1\} \\ S & \text{otherwise.} \end{cases}$$

Then $[M]_1 \subseteq U \cup \{1\}$. $\qquad\square$

Corollary 4.10. *If S is left cancellative and $\bigcap \mathcal{F} = \emptyset$, then $\mathcal{T}[\mathcal{F}]$ is a T_1-topology.*

Proof. Let $1 \neq a \in S$. Since S is left cancellative and $\bigcap \mathcal{F} = \emptyset$, there is $M : S \to \mathcal{F}$ such that $a \notin xM(x)$ for all $x \in S$. Then $a \notin [M]_1$. $\qquad\square$

Of special importance is the case where $\mathcal{F}$ is an ultrafilter.

Theorem 4.11. *For every nonprincipal ultrafilter $\mathcal{F}$ on S, $\mathcal{T}[\mathcal{F}]$ is strongly extremally disconnected.*

Proof. Let U be an open nonclosed subset of $(S, \mathcal{T}[\mathcal{F}])$ and let $C = S \setminus U$. Consider two cases.

Case 1: there is $a \in C$ such that $U \in a\mathcal{F}$. Choose $M : S \to \mathcal{F}$ such that $aM(a) \subseteq U$ and for every $x \in U$, $xM(x) \subseteq U$. Then $[M]_a \subseteq U \cup \{a\}$. Hence, $U \cup \{a\}$ is open.

Case 2: for every $x \in C$, $C \in x\mathcal{F}$. Choose $M : S \to \mathcal{F}$ such that for every $x \in C$, $xM(x) \subseteq C$. Then $[M]_b \subseteq C$ for every $b \in C$. Consequently, C is open, and so U is closed. Hence, this case is impossible. $\qquad\square$

4.3 Strongly Discrete Filters

Definition 4.12. Let S be a semigroup with identity. Given a mapping $M : S \to \mathcal{P}(S)$, an *M-product* is any product of elements of S of the form $x_0 \cdots x_n$ where $x_{i+1} \in M(x_0 \cdots x_i)$ for each $i < n$, and an *M-decomposition* of an element $x \in S$ is a decomposition $x = x_0 \cdots x_n$, where $x_0 \cdots x_n$ is an M-product. An M-decomposition $x = x_0 \cdots x_n$ is *trivial* if $n = 0$ (and then $x_0 = x$). A mapping $M : S \to \mathcal{P}(S)$ is *basic* if every $x \in S$ has a unique M-decomposition $x = x_0 \cdots x_n$ with $x_0 = 1$.

Lemma 4.13. *Let $M : S \to \mathcal{P}(S)$ be a basic mapping. Then*

(1) *there is no nontrivial M-decomposition of 1,*

(2) *the subsets $xM(x)$, where $x \in S$, are pairwise disjoint, and*

(3) *for each $x \in S$, the mapping $M(x) \ni y \mapsto xy \in xM(x)$ is injective.*

Proof. (1) It follows from the definition of a basic mapping that there is no nontrivial M-decomposition $1 = x_0 \cdots x_n$ with $x_0 = 1$. So suppose that $1 = x_0 \cdots x_n$ is an M-decomposition and $x_0 \neq 1$. Then there is an M-decomposition $x_0 = y_0 \cdots y_m$ with $y_0 = 1$ and $m > 0$. But then $1 = y_0 \cdots y_m x_1 \cdots x_n$ is a nontrivial M-decomposition with $y_0 = 1$, which is a contradiction.

(2) Suppose that $xu = yv = z$ for some $x, y, z \in S$, $u \in M(x)$ and $v \in M(y)$. Let $x = x_0 \cdots x_n$ and $y = y_0 \cdots y_m$ be M-decompositions with $x_0 = y_0 = 1$. Then $z = x_0 \cdots x_n u$ and $z = y_0 \cdots y_m v$ are M-decompositions with $x_0 = y_0 = 1$. It follows that $u = v$, $n = m$ and $x_i = y_i$ for each $i \leq n$, and so $x = y$.

(3) Suppose that $xu = xv$ for some $x \in S$ and $u, v \in M(x)$. Let $x = x_0 \cdots x_n$ be M-decomposition with $x_0 = 1$. Then $x_0 \cdots x_n u$ and $x_0 \cdots x_n v$ are M-products and $x_0 \cdots x_n u = x_0 \cdots x_n v$. Hence $u = v$. $\qquad\square$

Lemma 4.14. *Let $M : S \to \mathcal{P}(S)$ and suppose that the subsets $xM(x)$, where $x \in S$, are pairwise disjoint and for each $x \in S$, $M(x) \ni y \mapsto xy \in xM(x)$ is injective. Let $x_0 \cdots x_n$ and $y_0 \cdots y_m$ be M-products and let $x_0 \cdots x_n = y_0 \cdots y_m$. Then either*

(1) $x_0 \cdots x_{n-m} = y_0$ *and* $x_{n-m+j} = y_j$, $1 \leq j \leq m$, *if* $n \geq m$, *or*

(2) $x_0 = y_0 \cdots y_{m-n}$ *and* $x_i = y_{m-n+i}$, $1 \leq i \leq n$, *if* $n < m$.

Proof. We proceed by induction on $\min\{n, m\}$. It is trivial for $\min\{n, m\} = 0$. Let $\min\{n, m\} > 0$. We claim that $x_0 \cdots x_{n-1} = y_0 \cdots y_{m-1}$ and then $x_n = y_m$, since

$$M(x_0 \cdots x_{n-1}) \ni y \mapsto x_0 \cdots x_{n-1} y \in x_0 \cdots x_{n-1} M(x_0 \cdots x_{n-1})$$

is injective. Indeed, if $x_0 \cdots x_{n-1} \neq y_0 \cdots y_{m-1}$, then $x_0 \cdots x_n \neq y_0 \cdots y_m$, since

$$x_0 \cdots x_{n-1} M(x_0 \cdots x_{n-1}) \cap y_0 \cdots y_{m-1} M(y_0 \cdots y_{m-1}) = \emptyset.$$

We then apply the inductive assumption. $\qquad\square$

Recall that a subset D of a space X is called *strongly discrete* if for every $x \in D$, there is a neighborhood U_x of $x \in X$ such that the subsets U_x, where $x \in D$, are pairwise disjoint.

The *Fréchet filter* on an infinite set X consists of all subsets $A \subseteq X$ such that $|X \setminus A| < |X|$.

Definition 4.15. Let S be a semigroup with identity and let $\mathcal{F}$ be a filter on S. We say that $\mathcal{F}$ is *strongly discrete* if

(1) $\mathcal{F}$ contains the Fréchet filter, and

(2) there is $M : S \to \mathcal{F}$ such that the subsets $xM(x)$, where $x \in S$, are pairwise disjoint, and for every $x \in S$, $M(x) \ni y \mapsto xy \in xM(x)$ is injective.

Theorem 4.16. *For every strongly discrete filter $\mathcal{F}$ on S, there is a basic mapping $M : S \to \mathcal{F}$.*

Proof. Let $\kappa = |S|$ and enumerate S as $\{s_\alpha : \alpha < \kappa\}$ with $s_0 = 1$. Choose $M_0 : S \to \mathcal{F}$ such that

(i) the subsets $xM_0(x)$, where $x \in S$, are pairwise disjoint, and for every $x \in S$, $M_0(x) \ni y \mapsto xy \in xM_0(x)$ is injective, and

(ii) $s_\alpha \notin s_\beta M_0(s_\beta)$ for all $\alpha \le \beta < \kappa$.

It follows from (ii) that $1 \notin xM_0(x)$ for all $x \in S$, which in turn implies that there is no nontrivial M_0-decomposition of 1.

Construct inductively a λ-sequence $(a_\gamma)_{\gamma<\lambda}$ with maximally possible $\lambda \le \kappa$ by putting $a_0 = 1$ and taking a_γ, for $\gamma > 0$, to be the first element in the sequence $\{s_\alpha : \alpha < \kappa\}$ such that $a_\gamma \notin X_\gamma$ where $X_\gamma = \bigcup_{\delta<\gamma}[M_0]a_\delta$.

We define $M : S \to \mathcal{F}$ by

$$M(x) = \begin{cases} M_0(1) \cup \{a_\gamma : 0 < \gamma < \lambda\} & \text{if } x = 1 \\ M_0(x) & \text{otherwise.} \end{cases}$$

Clearly, for every $x \in S$, $M(x) \ni y \mapsto xy \in xM(x)$ is injective. Let us check that the subsets $xM(x)$, where $x \in S$, are pairwise disjoint. Assume the contrary. Then by the definition of M and (i), $a_\gamma \in xM_0(x)$ for some $1 \ne x \in S$ and $0 < \gamma < \lambda$. We have also that $a_\gamma = s_\alpha$ and $x = s_\beta$ for some $\alpha, \beta < \kappa$, so $s_\alpha \in s_\beta M_0(s_\beta)$. It follows from this and (ii) that $\alpha > \beta$. But then by the choice of $a_\gamma = s_\alpha$, one has $s_\beta \in X_\gamma$. This implies that also every M_0-product $x_0 \cdots x_n$ with $x_0 = s_\beta$ belongs to X_γ. Since $s_\alpha \in s_\beta M_0(s_\beta)$, we obtain that $s_\alpha = a_\gamma \in X_\gamma$, which is a contradiction.

Now we claim that M is basic.

Indeed, for every $x \in S$, there is $\gamma < \lambda$ such that $x \in [M_0]a_\gamma$, so there is an M_0-decomposition $x = x_0 \cdots x_n$ with $x_0 = a_\gamma$. If $\gamma = 0$, this is M-decomposition with $x_0 = 1$. Otherwise so is $x = 1x_0 \cdots x_n$.

To see that every x has only one M-decomposition $x = x_0 \cdots x_n$ with $x_0 = 1$, it suffices, by Lemma 4.14, to check that if $1 = x_0 \cdots x_n$ is an M-decomposition with $x_0 = 1$, then $n = 0$. Assume on the contrary that $n > 0$. Then $1 = x_1 \cdots x_n$ is a nontrivial M_0-decomposition of 1, which is a contradiction. □

Definition 4.17. Let D be a discrete subset of a space X and let x be an accumulation point of D. We say that D is *locally maximal* with respect to x if for every discrete subset $E \subset X$ such that $D \cap U \subseteq E \cap U$ for some neighborhood U of x, there is a neighborhood V of x such that $D \cap V = E \cap V$.

The next theorem gives us the main properties of topologies determined by strongly discrete filters.

Theorem 4.18. *Let $\mathcal{F}$ be a strongly discrete filter on S and let $S = (S, \mathcal{T}[\mathcal{F}])$. Then*

(1) *S is zero-dimensional and Hausdorff,*

(2) *there is $D \in \mathcal{F}$ such that*

    (i) *D is a strongly discrete subset of S with exactly one accumulation point, and*

    (ii) *D is a locally maximal discrete subset of S with respect to 1.*

Furthermore, if in addition $\mathcal{F}$ is an ultrafilter, then

(3) *S is strongly extremally disconnected.*

Proof. (1) To see that $\mathcal{T}[\mathcal{F}]$ is a T_1-topology, let a, b be distinct elements of S. Pick $N : S \to \mathcal{F}$ such that $b \notin x N(x)$ for all $x \in S$. Then $b \notin [N]_a$.

Now to show that $\mathcal{T}[\mathcal{F}]$ is zero-dimensional, let $a \in S$. By Theorem 4.16, there is a basic mapping $M : S \to \mathcal{F}$, and let $N : S \to \mathcal{F}$ be any mapping such that $N(x) \subseteq M(x)$ for all $x \in S$. We show that $[N]_a$ is closed. To this end, pick $K : S \to \mathcal{F}$ such that for every $x \in S$, $K(x) \subseteq M(x)$ and $a \notin x K(x)$. We claim that for every $b \in S \setminus [N]_a$, one has $[K]_b \cap [N]_a = \emptyset$.

Indeed, assume the contrary. Then $x_0 \cdots x_n = y_0 \cdots y_m$ for some $n, m < \omega$ and $x_i, y_j \in S$ such that $x_0 = b$, $x_{i+1} \in K(x_0 \cdots x_i)$ for $i < n$, $y_0 = a$, and $y_{j+1} \in N(y_0 \cdots y_j)$ for $j < m$. Since $K(x), N(x) \subseteq M(x)$ for all $x \in S$, we obtain by Lemma 4.14, that either $x_0 \cdots x_{n-m} = y_0$ (if $n \geq m$) or $x_0 = y_0 \cdots y_{m-n}$ (if $n < m$). The first possibilty gives us that $a \in [K]_b$, a contradiction with $a \notin x K(x)$ for all $x \in S$. And the second gives us that $b \in [N]_a$, again a contradiction.

(2) Pick a basic mapping $M : S \to \mathcal{F}$ and let $D = M(1)$.

(i) We claim that the subsets $[M]_a$, where $a \in D$, are pairwise disjoint. Indeed, let $x_1, y_1 \in D$ and let $[M]_{x_1} \cap [M]_{y_1} \neq \emptyset$. Then $x_1 \cdots x_n = y_1 \cdots y_m$ for some $n, m \geq 1$ and x_i, y_j such that $x_{i+1} \in M(x_1 \cdots x_i)$ and $y_{j+1} \in M(y_1 \cdots y_j)$. But then $x_0 x_1 \cdots x_n = y_0 y_1 \cdots y_m$ where $x_0 = y_0 = 1$. Since $x_1, y_1 \in M(1)$, it follows from this and Lemma 4.14 that $n = m$ and $x_i = y_i$ for each $i \leq n$, in particular, $x_1 = y_1$.

To see that $D \cup \{1\}$ is closed, let $1 \neq x_1 \in [M]_1$ and let $[M]_{x_1} \cap D \neq \emptyset$. Then $x_1 \cdots x_n = y_1$ for some $y_1 \in D$, $n \geq 1$ and x_i such that $x_{i+1} \in M(x_1 \cdots x_i)$. But then $x_0 x_1 \cdots x_n = y_0 y_1$ where $x_0 = y_0 = 1$. It follows from this that $n = 1$ and $x_1 = y_1$.

(ii) Let E be a discrete subset of S such that $D \cap U \subseteq E \cap U$ for some neighborhood U of 1. One may suppose that U is open. For every $x \in D \cap U$, choose a neighborhood V_x of 1 such that $xV_x \subseteq U$ and $xV_x \cap E = \{x\}$. Put $V = \left(\bigcup_{x \in D \cap U} xV_x\right) \cup \{1\}$. By Corollary 4.9, V is a neighborhood of 1, and by the construction, $D \cap V = E \cap V$.

(3) is immediate from Theorem 4.11.   □

The next proposition shows how naturally strongly discrete filters arise on left topological groups.

Proposition 4.19. *Let $(S, \mathcal{T})$ be a regular left topological group such that the intersection of $< |S|$ open sets is open and let D be a discrete subset of $(S, \mathcal{T})$ with exactly one accumulation point 1. Then any filter on S containing D and converging to 1 is strongly discrete.*

Proof. Let $\mathcal{F}$ be a filter on S containing D and converging to 1 and let $\kappa = |S|$. Enumerate S as $\{s_\alpha : \alpha < \kappa\}$. Construct inductively a sequence $(M(s_\alpha))_{\alpha < \kappa}$ such that $M(s_\alpha) \in \mathcal{F}$, $M(s_\alpha) \subseteq D$ and

$$s_\alpha M(s_\alpha) \cap \left(\bigcup_{\beta < \alpha} s_\beta M(s_\beta) \right) = \emptyset.$$

This can be done because

$$\left(\bigcup_{\beta < \alpha} s_\beta M(s_\beta) \right)' = \{s_\beta : \beta < \alpha\}.$$

(For every subset Y of a space X, Y' denotes the set of accumulation points of $Y \subseteq X$.)   □

We now show that strongly discrete filters exist in profusion on any infinite cancellative semigroup.

Proposition 4.20. *Let S be a cancellative semigroup with identity and let $|S| = \kappa \geq \omega$. Enumerate S without repetitions as $\{s_\alpha : \alpha < \kappa\}$. Then there is a one-to-one κ-sequence $(x_\alpha)_{\alpha < \kappa}$ in S such that the subsets $s_\alpha X_\alpha$, where $X_\alpha = \{x_\beta : \alpha \leq \beta < \kappa\}$, are pairwise disjoint.*

Proof. Pick as x_0 any element of S. Now let $0 < \gamma < \kappa$ and assume that we have constructed a one-to-one sequence $(x_\alpha)_{\alpha < \gamma}$ such that the subsets $s_\alpha X_{\alpha,\gamma}$, where $X_{\alpha,\gamma} = \{x_\beta : \alpha \leq \beta < \gamma\}$, are pairwise disjoint. Choose $x_\gamma \in S$ satisfying the condition $s_\alpha x_\gamma \neq s_\beta x_\delta$, where $\alpha \leq \gamma$ and $\beta \leq \delta < \gamma$ (this can be done because S is left cancellative). The condition $s_\alpha x_\gamma \neq s_\beta x_\gamma$, where $\alpha, \beta \leq \gamma$ and $\alpha \neq \beta$, is satisfied automatically (because S is right cancellative). Hence, the subsets $s_\alpha X_{\alpha,\gamma+1}$, where $\alpha < \gamma + 1$, are also pairwise disjoint.   □

Let $\mathcal{F}$ and $\mathcal{G}$ be filters on sets X and Y, respectively. We say that $\mathcal{F}$ and $\mathcal{G}$ are *isomorphic* if there is a bijection $f : X \to Y$ such that $f(\mathcal{F}) = \mathcal{G}$.

Definition 4.21. Let $\mathcal{F}$ be a filter on an infinite set X containing the Fréchet filter. We say that $\mathcal{F}$ is *locally Fréchet* if there is $A \in \mathcal{F}$ such that $\mathcal{F}|_A$ is the Fréchet filter on A. A locally Fréchet filter is *proper* if it is not Fréchet.

Note that all proper locally Fréchet filters on X are isomorphic, and if $\mathcal{F}$ is not locally Fréchet, then for every $A \in \mathcal{F}$, $\mathcal{F}$ is isomorphic to $\mathcal{F}|_A$. To see the latter, pick $B \in \mathcal{F}$ such that $B \subset A$ and $|A \setminus B| = |X|$ and take any bijection $g : X \setminus B \to A \setminus B$. Define $f : X \to A$ by

$$f(x) = \begin{cases} x & \text{if } x \in B \\ g(x) & \text{otherwise.} \end{cases}$$

Then f is a bijection and $f(\mathcal{F}) = \mathcal{F}|_A$.

Now we obtain from Proposition 4.20 the following.

Corollary 4.22. *Let S be a cancellative semigroup with identity and let $|S| = \kappa \geq \omega$. For every filter $\mathcal{F}$ on κ properly containing the Fréchet filter, there is a strongly discrete filter $\mathcal{G}$ on S isomorphic to $\mathcal{F}$.*

Proof. Let $(x_\alpha)_{\alpha < \kappa}$ be a sequence in S guaranteed by Proposition 4.20, let $A = \{x_\alpha : \alpha < \kappa\}$, and let $\mathcal{H}$ be the filter on S such that $A \in \mathcal{H}$ and $\mathcal{H}|_A$ is the Fréchet filter. Then $\mathcal{H}$ is proper locally Fréchet and strongly discrete. Now let $\mathcal{F}$ be a filter on κ properly containing the Fréchet filter. If $\mathcal{F}$ is locally Fréchet, it is isomorphic to $\mathcal{H}$. Otherwise take a filter $\mathcal{G}$ on S such that $A \in \mathcal{G}$ and $\mathcal{G}|_A$ is isomorphic to $\mathcal{F}$. Then $\mathcal{G}$ is isomorphic to $\mathcal{F}$ and is strongly discrete. $\qquad\qquad\square$

We conclude the section with a topological classification of topologies determined by strongly discrete filters.

Definition 4.23. Given a space X with exactly one accumulation point $1 \in X$, define an increasing sequence $(X_n)_{n=1}^\infty$ of extensions of $X = X_1$ as follows. For each $n \geq 1$, put

$$X_{n+1} = X_n' \cup \bigcup_{y \in X_n \setminus X_n'} X(y),$$

where $X(y)$ is a copy of X with $y = 1$ (we suppose that $X(y) \cap X_n = \{y\}$ and that $X(y) \cap X(z) = \emptyset$ if $y \neq z$), and topologize X_{n+1} by declaring a subset $U \subseteq X_{n+1}$ to be open if and only if $U \cap X_n$ is open in X_n and for every $y \in X_n \setminus X_n'$, $U \cap X(y)$ is open in $X(y)$. Finally, we define

$$X_\omega = \lim_{n \to \infty} X_n,$$

that is, $X_\omega = \bigcup_{n=1}^\infty X_n$ and a subset $U \subseteq X$ is open if and only if $U \cap X_n$ is open in X_n for every n.

Theorem 4.24. *Let $\mathcal{F}$ be a strongly discrete filter on S and let X be a space with exactly one accumulation point $1 \in X$ such that the filter on X with a base consisting of subsets $U \setminus \{1\}$, where U runs over neighborhoods of 1, is isomorphic to $\mathcal{F}$ if $\mathcal{F}$ is not locally Fréchet and to the Fréchet filter otherwise. Then $(S, \mathcal{T}[\mathcal{F}])$ is homeomorphic to X_ω.*

Proof. Suppose first that $\mathcal{F}$ is not locally Fréchet. By Theorem 4.16, there is a basic $M : S \to \mathcal{F}$. Define an increasing sequence $(Y_n)_{n=1}^\infty$ of subspaces of $(S, \mathcal{T}[\mathcal{F}])$) by putting $Y_1 = M(1)$ and

$$Y_{n+1} = Y_n' \cup \bigcup_{y \in Y_n \setminus Y_n'} (M(y) \cup \{y\}).$$

Then

$$(S, \mathcal{T}[\mathcal{F}]) = \lim_{n \to \infty} Y_n.$$

For every $y \in S$, $M(y) \cup \{y\}$ is homeomorphic to X. It follows that for every n, Y_n is homeomorphic to X_n, and consequently, $(S, \mathcal{T}[\mathcal{F}])$ is homeomorphic to X_ω.

Now suppose that $\mathcal{F}$ is locally Fréchet, so there is $B \in \mathcal{F}$ such that $\mathcal{F}|_B$ is the Fréchet filter. Choose a basic $M : S \to \mathcal{F}$ such that $M(x) \subseteq B$ for all $x \in S$. Then for every $N : S \to \mathcal{F}$ such that $N(x) \subseteq M(x)$ and for every $a \in S$, $[N]_a$ is homeomorphic to X_ω, in particular, $[M]_1$ is homeomorphic to X_ω. We shall show that there are $N : S \to \mathcal{F}$ and $A \subset S$ with $1 \in A$ such that $N(x) \subseteq M(x)$, $|A| = |S| = \kappa$, and the subsets $[N]_a$, where $a \in A$, form a partition of $[M]_1$. This implies that both $[M]_1$ and $(S, \mathcal{T}[\mathcal{F}])$ are homeomorphic to the sum of κ copies of X_ω, and consequently, $(S, \mathcal{T}[\mathcal{F}])$ is homeomorphic to X_ω.

For each $x \in M(1)$, pick $y_x \in M(x)$, and let $C = \{xy_x : x \in M(1)\}$. Then $|C| = \kappa$, the subsets $[M]_a \subseteq [M]_1$, where $a \in C$, are pairwise disjoint, and $U = [M]_1 \setminus \bigcup_{a \in C}[M]_a$ is an open neighborhood of 1 (see Theorem 4.18 (2)). Define $N : S \to \mathcal{F}$ by

$$N(x) = \begin{cases} (x^{-1}U) \cap M(x) & \text{if } x \in U \\ M(x) & \text{otherwise.} \end{cases}$$

The mapping N, in turn, determines $D \subset S$ such that the subsets $[N]_a$, where $a \in D$, form a partition of S. Put $A = D \cap [M]_1$. $\qquad\qquad\qquad\qquad\square$

Remark 4.25. Let κ be an infinite cardinal, let X be a space with $|X| = \kappa$ and exactly one accumulation point $1_X \in X$ such that the filter on X with a base consisting of subsets $U \setminus \{1_X\}$, where U runs over neighborhoods of 1_X, is isomorphic to the Fréchet filter on κ, and let Y be a space with $|Y| = \kappa$ and exactly one accumulation point $1_Y \in Y$ such that the filter on Y with a base consisting of subsets $U \setminus \{1_Y\}$, where U runs over neighborhoods of 1_Y, is isomorphic to the proper locally Fréchet filter on κ. It is not hard to see that the space Y_ω is homeomorphic to the sum of κ copies of X_ω, and so just to X_ω. Hence Theorem 4.24 can be stated also as follows:

Let $\mathcal{F}$ be a strongly discrete filter on S and let X be a space with exactly one accumulation point $1 \in X$ such that the filter on X with a base consisting of subsets $U \setminus \{1\}$, where U runs over neighborhoods of 1, is isomorphic to $\mathcal{F}$. Then $(S, \mathcal{T}[\mathcal{F}])$ is homeomorphic to X_ω.

Corollary 4.26. *Let $\mathcal{F}$ and $\mathcal{G}$ be strongly discrete filters on S. Then the topologies $\mathcal{T}[\mathcal{F}]$ and $\mathcal{T}[\mathcal{G}]$ are homeomorphic if and only if the filters $\mathcal{F}$ and $\mathcal{G}$ are isomorphic.*

Proof. Let $f : (S, \mathcal{T}[\mathcal{F}]) \to (S, \mathcal{T}[\mathcal{G}])$ be a homeomorphism. Without loss of generality one may suppose that $f(1) = 1$. We claim that $f(\mathcal{F}) = \mathcal{G}$. Let $D_{\mathcal{F}} \in \mathcal{F}$ and $D_{\mathcal{G}} \in \mathcal{G}$ be sets satisfying Theorem 4.18 (2). It suffices to show that for every $A \in \mathcal{F}$ with $A \subseteq D_{\mathcal{F}}$, $f(A) \in \mathcal{G}$. Assume the contrary. Then $B = D_{\mathcal{G}} \setminus f(A)$ is a discrete subset of $(S, \mathcal{T}[\mathcal{G}])$ with cl $B = B \cup \{1\}$. Consequently, $C = A \cup f^{-1}(B)$ is a discrete subset of $(S, \mathcal{T}[\mathcal{F}])$ such that $A \subset C$ and for every neighborhood U of 1, $(C \setminus A) \cap U \neq \emptyset$. But this contradicts Theorem 4.18 (2).

Conversely, let $\mathcal{F}$ and $\mathcal{G}$ be isomorphic. Then $\mathcal{T}[\mathcal{F}]$ and $\mathcal{T}[\mathcal{G}]$ are homeomorphic by Theorem 4.24. $\qquad \square$

4.4 Invariant Topologies

Definition 4.27. We say that a topology on a group is *invariant* if all translations and the inversion are continuous.

In this section we show that

Theorem 4.28. *Every infinite group admits a nondiscrete zero-dimensional Hausdorff invariant topology.*

By Theorem 4.18 (1), in order to prove Theorem 4.28, it suffices to show that for every infinite group G, there is a strongly discrete filter $\mathcal{F}$ on G such that the topology $\mathcal{T}[\mathcal{F}]$ is invariant.

We first derive a simple sufficient condition for the topology $\mathcal{T}[\mathcal{F}]$ to be invariant.

Lemma 4.29. *Let $\mathcal{T}$ be a left invariant topology on G and let $\mathcal{N}$ be the neighborhood filter of 1 in $\mathcal{T}$. Then $\mathcal{T}$ is invariant if and only if $\mathcal{N}^{-1} = \mathcal{N}$ and $x\mathcal{N}x^{-1} = \mathcal{N}$ for all $x \in G$.*

Proof. Necessity is obvious. Sufficiency follows from the fact that for every $U \in \mathcal{N}$ and $x, y \in G$, one has $(xU)y = xy(y^{-1}Uy)$ and $(xU)^{-1} = x^{-1}(xU^{-1}x^{-1})$. $\qquad \square$

Lemma 4.30. *Let $\mathcal{F}$ be a filter on G such that $\mathcal{F}^{-1} = \mathcal{F}$ and $x\mathcal{F}x^{-1} = \mathcal{F}$ for all $x \in G$. Then the topology $\mathcal{T}[\mathcal{F}]$ is invariant.*

Proof. By Lemma 4.29 and Theorem 4.8, it suffices to show that

(a) for every $M : G \to \mathcal{F}$ and $a \in G$, there is $N : G \to \mathcal{F}$ such that $a[N]_1 a^{-1} \subseteq [M]_1$, and

(b) for every $M : G \to \mathcal{F}$, there is $N : G \to \mathcal{F}$ such that $[N]_1^{-1} \subseteq [M]_1$.

To see (a), define $N : G \to \mathcal{F}$ by

$$N(x) = a^{-1}M(axa^{-1})a.$$

Let $x \in [N]_1$ and let $x = x_0 \cdots x_n$ be N-decomposition with $x_0 = 1$. Then

$$axa^{-1} = ax_0 \cdots x_n a^{-1} = ax_0 a^{-1} \cdots ax_n a^{-1}$$

and for each $i < n$, one has

$$
\begin{aligned}
ax_{i+1}a^{-1} &\in aN(x_0 \cdots x_i)a^{-1} \\
&= M(ax_0 \cdots x_i a^{-1}) \\
&= M(ax_0 a^{-1} \cdots ax_i a^{-1}).
\end{aligned}
$$

To see (b), define $N : G \to \mathcal{F}$ by

$$N(x) = x^{-1}(M(x^{-1}))^{-1}x.$$

Let $x \in [N]_1$ and let $x = x_0 \cdots x_n$ be N-decomposition with $x_0 = 1$. Then

$$x^{-1} = (x_0 \cdots x_n)^{-1} = x_0^{-1} \cdot x_0 x_1^{-1} x_0^{-1} \cdots (x_0 \cdots x_{n-1})x_n^{-1}(x_0 \cdots x_{n-1})^{-1}$$

and for each $i < n$, one has

$$
\begin{aligned}
(x_0 \cdots x_i)&x_{i+1}^{-1}(x_0 \cdots x_i)^{-1} \\
&\in (x_0 \cdots x_i)(N(x_0 \cdots x_i))^{-1}(x_0 \cdots x_i)^{-1} \\
&= M((x_0 \cdots x_i)^{-1}) \\
&= M(x_0^{-1} \cdot x_0 x_1^{-1} x_0^{-1} \cdots (x_0 \cdots x_{i-1})x_i^{-1}(x_0 \cdots x_{i-1})^{-1}). \qquad \square
\end{aligned}
$$

In the case where $\mathcal{F}$ is a strongly discrete filter, the condition in Lemma 4.30 is also necessary for the topology $\mathcal{T}[\mathcal{F}]$ to be invariant.

Proposition 4.31. *Let $\mathcal{F}$ be a strongly discrete filter on G. Then the topology $\mathcal{T}[\mathcal{F}]$ is invariant if and only if $\mathcal{F}^{-1} = \mathcal{F}$ and $x\mathcal{F}x^{-1} = \mathcal{F}$ for all $x \in G$.*

Proof. Sufficiency is immediate from Lemma 4.30. Necessity follows from Theorem 4.18 (2).   □

Thus, by Theorem 4.18 (1) and Lemma 4.30, in order to prove Theorem 4.28, it suffices to show that

Theorem 4.32. *For every infinite group G, there is a strongly discrete filter $\mathcal{F}$ on G such that $\mathcal{F}^{-1} = \mathcal{F}$ and $x\mathcal{F}x^{-1} = \mathcal{F}$ for all $x \in G$.*

Proof. We will prove the assertion:
For every infinite subgroup H of G and for every subset $F \subset G$ with $|F| < |G|$, there is a filter $\mathcal{F}$ on G and a mapping $M : H \to \mathcal{F}$ such that

(1) $\mathcal{F}^{-1} = \mathcal{F}$ and $x\mathcal{F}x^{-1} = \mathcal{F}$ for all $x \in H$,

(2) $\mathcal{F}|_{M(1)}$ contains the Fréchet filter on $M(1)$, $|M(1)| = |H|$ and $M(x) \subseteq M(1)$ for all $x \in H$,

(3) the subsets $xM(x)$, where $x \in H$, are pairwise disjoint and disjoint from F.

Here, $\mathcal{F}|_{M(1)}$ is the trace of $\mathcal{F}$ on $M(1)$.
We proceed by induction on $\lambda = |H|$.
Let $\lambda = \omega$. Enumerate H as $\{x_n : n < \omega\}$. It suffices to construct a one-to-one sequence $(y_n)_{n<\omega}$ in G such that the subsets $x_n Z_n$, where

$$Z_n = \{x_i y_j^{\pm 1} x_i^{-1} : i \le j, n \le j < \omega\},$$

are pairwise disjoint and disjoint from F. Then one could define the filter $\mathcal{F}$ on G by taking as a base the subsets of the form

$$\bigcup_{i<\omega} \{x_i y_j^{\pm 1} x_i^{-1} : n(i) \le j < \omega\},$$

where $n : \omega \to \omega$, and the mapping $M : H \to \mathcal{F}$ by $M(x_n) = Z_n$ (see Lemma 1.16).
Pick as y_0 any element of G such that $x_0 y_0 x_0^{-1} \notin F$. Fix $0 < k < \omega$ and assume that we have constructed an one-to-one sequence $(y_n)_{n<k}$ such that the subsets $x_n Z_{n,k}$, where $n < k$ and

$$Z_{n,k} = \{x_i y_j^{\pm 1} x_i^{-1} : i \le j, n \le j < k\},$$

are pairwise disjoint and disjoint from F. We claim that there exists $x_k \in G$ such that the subsets $x_n Z_{n,k+1}$, where $n < k + 1$, are also pairwise disjoint and disjoint from F. Assume the contrary. Then there exist a finite set T_1 of terms of the form $x_n x_i y^{\pm 1} x_i^{-1}$, where $n, i \le k$, and a finite set T_2 of pairs $(x_n x_i y^{\pm 1} x_i^{-1}, x_m x_j y^{\pm 1} x_j^{-1})$ of terms, where $n, m, i, j \le k$ and $n \ne m$, such that for every $y \in G \setminus \{y_n : n < k\}$, either $f(y) \in F$ for some $f \in T_1$ or $f(y) \ne g(y)$ for some $(f, g) \in T_2$. But this contradicts the following lemma.

Lemma 4.33. *Every finite system of inequalities of the form $ayb \neq y^\varepsilon$, where $a, b \in G$, $\varepsilon = \pm 1$ and $ab \neq 1$, has $|G|$-many solutions in G.*

Proof. Assume the contrary. Then there is a subset $F \subset G$ with $|F| < |G|$ such that for every $x \in G \setminus F$, there exists an inequality $ayb \neq y^\varepsilon$ of the system with $axb = x^\varepsilon$. By Corollary 1.42, there is a 3-sequence $(x_n)_{n<3}$ in G and an inequality $ayb \neq y^\varepsilon$ of the system such that $axb = x^\varepsilon$ for every $x \in \mathrm{FP}((x_n)_{n<3})$. Consider two cases.

Case 1: $\varepsilon = 1$. Then on the one hand

$$ax_1 x_2 b = x_1 x_2$$

and on the other hand

$$ax_1 x_2 b = ax_1 b \cdot b^{-1} a^{-1} \cdot ax_2 b = x_1 (ab)^{-1} x_2,$$

so

$$x_1 (ab)^{-1} x_2 = x_1 x_2$$

which contradicts $ab \neq 1$.

Case 2: $\varepsilon = -1$. Then for any $x \in \mathrm{FP}((x_n)_{1 \leq n < 3})$, on the one hand

$$ax_0 xb = (x_0 x)^{-1} = x^{-1} x_0^{-1}$$

and on the other hand

$$ax_0 xb = ax_0 b \cdot b^{-1} a^{-1} \cdot axb = x_0^{-1} (ab)^{-1} x^{-1},$$

so

$$x_0^{-1} (ab)^{-1} x^{-1} = x^{-1} x_0^{-1}$$

and consequently

$$x_0^{-1} (ab)^{-1} x^{-1} x_0 = x^{-1}.$$

Hence

$$(x_0^{-1}) x (abx_0) = x$$

and since $x_0^{-1} abx_0 \neq 1$, we can apply Case 1.   □

Fix $\kappa > \omega$, assume that the assertion has been proved for all infinite $\lambda < \kappa$, and prove it for $\lambda = \kappa$. Choose an increasing sequence $(H_\alpha)_{\alpha < \kappa}$ of infinite subgroups of H with $|H_\alpha| < \kappa$ and $\bigcup_{\alpha < \kappa} H_\alpha = H$. For every $\alpha < \kappa$, construct a filter $\mathcal{F}_\alpha$ on G and a mapping $M_\alpha : H_\alpha \to \mathcal{F}_\alpha$ such that

(1) $\mathcal{F}_\alpha^{-1} = \mathcal{F}_\alpha$ and $x\mathcal{F}_\alpha x^{-1} = \mathcal{F}_\alpha$ for all $x \in H_\alpha$,

(2) $\mathcal{F}|_{M_\alpha(1)}$ contains the Fréchet filter on $M_\alpha(1)$, $|M_\alpha(1)| = |H_\alpha|$ and $M\alpha(x) \subseteq M_\alpha(1)$ for all $x \in H_\alpha$, and

(3) the subsets $xM_\alpha(x)$, where $x \in H_\alpha$, are pairwise disjoint and disjoint from F and all $yM_\gamma(y)$, where $\gamma < \alpha$ and $y \in H_\gamma$.

Define the filter $\mathcal{F}$ on G by taking as a base the subsets of the form

$$\bigcup_{\gamma \leq \alpha < \kappa} A_\alpha,$$

where $\gamma < \kappa$ and $A_\mu \in \mathcal{F}_\mu$, and the mapping $M : II \to \mathcal{F}$ by

$$M(x) = \bigcup_{\gamma \leq \alpha < \kappa} M_\alpha(x),$$

where $\gamma = \min\{\alpha : x \in H_\alpha\}$. Then $\mathcal{F}$ is as required.   □

Remark 4.34. In fact, the proof of Theorem 4.32 shows more: there are $2^{2^{|G|}}$ nondiscrete zero-dimensional Hausdorff invariant topologies $\mathcal{T}_\alpha$ on G such that for any distinct $\alpha, \gamma < 2^{2^{|G|}}$, the topology $\mathcal{T}_\alpha \vee \mathcal{T}_\gamma$ is discrete. (The latter means that there are neighborhoods U_α, U_γ of 1 in $\mathcal{T}_\alpha, \mathcal{T}_\gamma$ respectively with $U_\alpha \cap U_\gamma = \{1\}$.) To see this, let $H = G$ and for every uniform ultrafilter p on λ, define the filter $\mathcal{F}(p)$ on G by taking as a base the subsets of the form

$$\bigcup_{\alpha \in P} A_\alpha,$$

where $P \in p$ and $A_\alpha \in \mathcal{F}_\alpha$. It is clear that $(\mathcal{F}(p))^{-1} = \mathcal{F}(p)$, $x\mathcal{F}(p)x^{-1} = \mathcal{F}(p)$ for all $x \in G$, and for any distinct p and q, there exist $A(p) \in \mathcal{F}(p)$ and $A(q) \in \mathcal{F}(q)$ with $A(p) \cap A(q) = \emptyset$. That $\mathcal{T}(\mathcal{F}(p)) \vee \mathcal{T}(\mathcal{F}(q))$ is discrete follows from Theorem 4.18 (2) and Corollary 4.9.

References

The results about the topology $\mathcal{T}[\mathcal{F}]$ and strongly discrete filters in the case where S is a group were proved in [103]. That every infinite group admits a nondiscrete zero-dimensional Hausdorff invariant topology was proved in [100].

Chapter 5
Topological Groups with Extremal Properties

In this chapter, assuming MA, we construct important topological groups with extremal properties. We also show that some of them cannot be constructed in ZFC.

Recall that starting from this chapter, all topological groups are assumed to be Hausdorff.

5.1 Extremally Disconnected Topological Groups

Theorem 5.1. *For every $n < \omega$, let G_n be a nontrivial finite group written additively, and let $G = \bigoplus_{n<\omega} G_n$. Let p be a nonprincipal ultrafilter on ω, let $\mathcal{T}$ be the group topology on G with a neighborhood base at 0 consisting of subgroups*

$$H_A = \{x \in G : \mathrm{supp}(x) \subset A\}$$

where $A \in p$, let $\mathcal{F}$ be the filter on G with a base consisting of subsets

$$D_A = \{x \in G : \mathrm{supp}(x) \subset A \text{ and } |\mathrm{supp}(x)| = 1\}$$

where $A \in p$, and let $D = D_\omega$. Then the following statements are equivalent:

(1) *p is Ramsey,*

(2) *$\mathcal{T} = \mathcal{T}[\mathcal{F}]$,*

(3) *D is a locally maximal discrete subset of $(G, \mathcal{T})$ with respect to 0.*

Furthermore, if $G = \bigoplus_\omega \mathbb{Z}_2$, statements (1)–(3) are equivalent to each of the following:

(4) *$\mathcal{T}$ is extremally disconnected,*

(5) *every nonempty nowhere dense subset of $(G, \mathcal{T})$ has an isolated point.*

Proof. (1) $\Rightarrow$ (2) Since $\mathcal{F}$ converges to 0 in $\mathcal{T}$, $\mathcal{T} \subseteq \mathcal{T}[\mathcal{F}]$. To show that $\mathcal{T}[\mathcal{F}] \subseteq \mathcal{T}$, let $M : G \to \mathcal{F}$. For each $n < \omega$, choose $B_n \in p$ such that for every $x \in G$ with $\max \mathrm{supp}(x) = n$, one has $D_{B_n} \subseteq M(x)$. Choose B_0 in addition so that also $D_{B_0} \subseteq M(0)$. Define the coloring $\chi : [\omega]^2 \to \{0, 1\}$ by putting for every $\{m, k\} \in [\omega]^2$ with $m < k$,

$$\chi(\{m, k\}) = \begin{cases} 1 & \text{if } k \in B_m \\ 0 & \text{otherwise.} \end{cases}$$

Since p is Ramsey, there exists $A \in p$ such that $A \subseteq B_0$ and $[A]^2$ is monochrome. Observe that $\chi([A]^2) = \{1\}$. We claim that $H_A \subseteq [M]_0$. To see this, let $0 \neq x \in H_A$ and let $\operatorname{supp}(x) = \{n_1, \ldots, n_k\}$, where $n_1 < \cdots < n_k$. We have that $n_1 \in B_0$ and $n_{i+1} \in B_{n_i}$ for each $i = 1, \ldots, k-1$. Write x as $x = x_1 + \cdots + x_k$, where $\operatorname{supp}(x_i) = \{n_i\}$. Then $x_1 \in D_{B_0} \subseteq M(0)$ and

$$x_{i+1} \in D_{B_{n_i}} \subseteq M(x_1 + \cdots + x_i).$$

Hence $x \subset [M]_0$.

(2) $\Rightarrow$ (3) Let E be a discrete subset of $(G, \mathcal{T})$ such that $D \cap U \subseteq E \cap U$ for some neighborhood U of 0. One may suppose that U is open. For every $x \in D \cap U$, choose a neighborhood V_x of 0 such that $x + V_x \subseteq U$ and $(x + V_x) \cap E = \{x\}$. Let

$$V = \bigcup_{x \in D \cap U} (x + V_x) \cup \{0\}.$$

By Corollary 4.9, V is a neighborhood of 0 and by the construction, $D \cap V = E \cap V$.

(3) $\Rightarrow$ (1) Let $\{A_n : n < \omega\}$ be any partition of ω such that $A_n \notin p$ for all $n < \omega$. Define $E \subseteq G$ by

$$E = \{x \in G : \operatorname{supp}(x) \subseteq A_n \text{ for some } n < \omega \text{ and } |\operatorname{supp}(x)| = 2\}.$$

Notice that every point from E is isolated in $D \cup E$. Since $A_n \notin p$ for all $n < \omega$, it follows that also every point from D is isolated in $D \cup E$. Consequently $D \cup E$ is discrete. Since D is a locally maximal discrete subset with respect to 0, there is a neighborhood U of 0 such that $U \cap E = \emptyset$. Choose $A \in p$ such that $H_A \subseteq U$. Then $|A \cap A_n| \leq 1$ for all $n < \omega$.

Now let $G = \bigoplus_\omega \mathbb{Z}_2$. Then $\mathcal{F}$ is an ultrafilter. (2) $\Rightarrow$ (4) and (2) $\Rightarrow$ (5) follow from Theorem 4.11 and Proposition 3.15.

To show (4) $\Rightarrow$ (3), let E be a discrete subset of $G = (G, \mathcal{T})$ such that $D \cap U \subseteq E \cap U$ for some neighborhood U of 0. For every $x \in E$, choose an open neighborhood U_x of 0 such that the subsets $x + U_x$ are pairwise disjoint. Put

$$U_D = \bigcup_{x \in D} (x + U_x) \quad \text{and} \quad U_{E \setminus D} = \bigcup_{x \in E \setminus D} (x + U_x).$$

Then U_D and $U_{E \setminus D}$ are disjoint open subsets and, obviously, $0 \in \operatorname{cl} U_D$. But then, since G is extremally disconnected, $0 \notin \operatorname{cl} U_{E \setminus D}$. Hence, $V = U \setminus U_{E \setminus D}$ is a neighborhood of 0 and $D \cap V = E \cap V$.

To show (5) $\Rightarrow$ (1), suppose that p is not Ramsey. Then there is a partition $\{A_n : n < \omega\}$ of ω such that $A_n \notin p$ for all $n < \omega$ and for every $A \in p$, $|A \cap A_n| \geq 2$ for some $n < \omega$. Define the subset F of $(G, \mathcal{T})$ by

$$F = \{x \in G : |\operatorname{supp}(x) \cap A_n| \leq 1 \text{ for all } n < \omega\}.$$

We claim that F is perfect and nowhere dense.

Clearly, F is closed. To see that F is dense in itself, let $x \in F$. Pick $n_0 < \omega$ such that

$$\max \mathrm{supp}(x) < \min A_n$$

for all $n \geq n_0$ and define $A \in p$ by

$$A = \bigcup_{n_0 \leq n < \omega} A_n.$$

Then $x + D_A \subset F$ and x is an accumulation point of $x + D_A$.

Finally, to see that F is nowhere dense, let $x \in G$ and let U be a neighborhood of 0 in $\mathcal{T}$. Pick $A \in p$ such that

$$\max \mathrm{supp}(x) < \min A$$

and $H_A \subseteq U$. Then $|A \cap A_n| \geq 2$ for some n. Consequently, there is $y \in H_A$ such that $\mathrm{supp}(y)$ is a 2-element subset of A_n, and so $x + y \notin F$. Since F is closed, this shows that F is nowhere dense.   □

Corollary 5.2. *Assume* $\mathfrak{p} = \mathfrak{c}$. *Then there exists a countable nondiscrete extremally disconnected topological group.*

A group of exponent 2 is called *Boolean*. Note that if G is a Boolean group, then for every $x, y \in G$, one has $xy = (xy)^{-1} = y^{-1}x^{-1} = yx$, so G is Abelian. It then follows that G is isomorphic to $\bigoplus_\kappa \mathbb{Z}_2$ for some cardinal κ.

For every group G, let

$$B(G) = \{x \in G : x^2 = 1\}.$$

Note that if G is Abelian, then $B(G)$ is the largest Boolean subgroup of G.

Lemma 5.3. *Let G be a topological group. If $B(G)$ is a neighborhood of* 1, *then G contains an open Boolean subgroup.*

Proof. Let $B = B(G)$. Since B is a neighborhood of 1, there is a neighborhood W of 1 such that $W^2 \subseteq B$. Then for every $x, y \in W$, $xy \in B$ and consequently $xy = (xy)^{-1} = y^{-1}x^{-1} = yx$. It then follows that for every $x_1, \ldots, x_n \in W$, $(x_1 \cdots x_n)^2 = x_1^2 \cdots x_n^2 = 1$ and so $x_1 \cdots x_n \in B$. Hence, $\langle W \rangle$ is an open Boolean subgroup of G.   □

Theorem 5.4. *Every extremally disconnected topological group contains an open Boolean subgroup.*

Proof. Let G be an extremally disconnected topological group and let $\iota : G \ni x \mapsto x^{-1} \in G$. Then by Theorem 3.13, $\{x \in G : \iota(x) = x\}$ is clopen. Since $\{x \in G : \iota(x) = x\} = B(G)$, G contains an open Boolean subgroup by Lemma 5.3.   □

5.2 Maximal Topological Groups

We say that a topological group (or a group topology) is *maximal* if the underlying space is maximal.

Theorem 5.5. *Assume* $\mathfrak{p} = \mathfrak{c}$. *Then there exists a maximal topological group.*

Proof. Let $G = \bigoplus_\omega \mathbb{Z}_2$. Enumerate the subsets of G as $\{Z_\alpha : \alpha < \mathfrak{c}\}$ with $Z_0 = G$. For every $\alpha < \mathfrak{c}$, we shall construct a sequence $(x_{\alpha,n})_{n<\omega}$ in $G \setminus \{0\}$ such that

(i) $\max \operatorname{supp}(x_{\alpha,n}) < \min \operatorname{supp}(x_{\alpha,n+1})$ for all $n < \omega$,

(ii) for every $\gamma < \alpha$, $\{x_{\alpha,n} : n < \omega\} \setminus \operatorname{FS}((x_{\gamma,n})_{n<\omega})$ is finite, and

(iii) either $\operatorname{FS}((x_{\alpha,n})_{n<\omega}) \subseteq Z_\alpha$ or $\operatorname{FS}((x_{\alpha,n})_{n<\omega}) \subseteq G \setminus Z_\alpha$.

It follows from (i) and (ii) that

(iv) whenever $k < \omega$, $\gamma_0 < \cdots < \gamma_k < \alpha$ and $n_0, \ldots, n_{k-1} \in \mathbb{N}$, there exists $n_k \in \mathbb{N}$ such that $\operatorname{FS}((x_{\gamma_k,n})_{n_k \le n < \omega}) \subseteq \operatorname{FS}((x_{\gamma_i,n})_{n_i \le n < \omega})$ for all $i < k$.

As $(x_{0,n})_{n<\omega}$ pick any sequence in $G \setminus \{0\}$ satisfying (i) ((ii) and (iii) are satisfied trivially).

Fix $\alpha < \mathfrak{c}$ and suppose that we have constructed sequences $(x_{\gamma,n})_{n<\omega}$, $\gamma < \alpha$, satisfying (i)–(iii).

Since $\mathfrak{p} = \mathfrak{c}$ and $\alpha < \mathfrak{c}$, there is $Y_\alpha \subseteq G \setminus \{0\}$ such that $Y_\alpha \setminus \operatorname{FS}((x_{\gamma,n})_{m \le n < \omega})$ is finite for all $\gamma < \alpha$ and $m < \omega$. Pick a sequence $(y_{\alpha,n})_{n<\omega}$ in Y_α such that

$$\max \operatorname{supp}(y_{\alpha,n}) < \min \operatorname{supp}(y_{\alpha,n+1})$$

for all $n < \omega$. By Theorem 1.41, there is a sum subsystem $(x_{\alpha,n})_{n<\omega}$ of $(y_{\alpha,n})_{n<\omega}$ satisfying condition (iii). Then conditions (i) and (ii) are satisfied as well.

Having constructed the sequences $(x_{\alpha,n})_{n<\omega}$, $\alpha < \mathfrak{c}$, define the group topology on G by taking as a neighborhood base at 0 the subgroups $\operatorname{FS}((x_{\alpha,n})_{m \le n < \omega}) \cup \{0\}$, where $\alpha < \mathfrak{c}$ and $m < \omega$. That this topology is maximal follows from (iii). $\square$

A family $\mathcal{A}$ of sets is called a Δ-*system* if there is a fixed set R, called the *root* of $\mathcal{A}$, such that $A \cap B = R$ whenever A and B are distinct members of $\mathcal{A}$.

Lemma 5.6 (Δ-system Lemma). *Every uncountable family of finite sets contains an uncountable Δ-system.*

Proof. It suffices to show that for every positive integer n and for every uncountable family $\mathcal{A}$ of n-sets, there is an uncountable Δ-system $\mathcal{B} \subseteq \mathcal{A}$. We proceed by induction on n. If $n = 1$, then $\mathcal{A}$ itself is a Δ-system with root $\emptyset$. Now fix $n > 1$ and suppose that the statement holds for all positive integers less than n. Let $X = \bigcup \mathcal{A}$ and for every $x \in X$, let $\mathcal{A}_x = \{A \in \mathcal{A} : x \in A\}$. Consider two cases.

Case 1: for every $x \in X$, $\mathcal{A}_x$ is countable. By transfinite recursion on $\alpha < \omega_1$, pick $A_\alpha \in \mathcal{A}$ so that $A_\alpha \cap A_\gamma = \emptyset$ for all $\gamma < \alpha$. This can be done because $\bigcup \{ \mathcal{A}_x : x \in \bigcup_{\gamma < \alpha} A_\gamma \}$ is countable. Let $\mathcal{B} = \{ A_\alpha : \alpha < \omega_1 \}$. Then $\mathcal{B}$ is a Δ-system with root $\emptyset$.

Case 2: there is $x \in X$ with uncountable $\mathcal{A}_x$. Let $\mathcal{A}' = \{ A \setminus \{x\} : A \in \mathcal{A}_x \}$. By the inductive hypothesis, there is an uncountable Δ-system $\mathcal{B}' \subseteq \mathcal{A}'$. Let R denote the root of $\mathcal{B}'$. Define $\mathcal{B} \subseteq \mathcal{A}_x$ by $\mathcal{B} = \{ A \cup \{x\} : A \in \mathcal{B}' \}$. Then $\mathcal{B}$ is a Δ-system with root $R \cup \{x\}$. $\square$

Theorem 5.7. *Every maximal topological group contains a countable open Boolean subgroup.*

Proof. Let G be a maximal topological group. Then G is extremally disconnected. Consequently by Theorem 5.4, one may suppose that $G = \bigoplus_\kappa \mathbb{Z}_2$ for some infinite cardinal κ.

For every $a \in \mathbb{N}$, let

$$\theta_2(a) = \max\{n < \omega : 2^n | a\}.$$

Define the partition $\{A_i : i < 2\}$ of $G \setminus \{0\}$ by

$$A_i = \{x \in G \setminus \{0\} : \theta_2(|\mathrm{supp}(x)|) \equiv i \pmod 2\}.$$

Since G is maximal, there exists a neighborhood U of 0 such that either $U \setminus \{0\} \subseteq A_0$ or $U \setminus \{0\} \subseteq A_1$. Choose a neighborhood V of 0 such that

$$V + V + V + V \subseteq U.$$

We claim that V is countable.

Indeed, assume the contrary. Then applying the Δ-system Lemma gives us an uncountable $X \subseteq V$ and a finite $F \subseteq \kappa$ such that, whenever x and y are different elements of X, one has $\mathrm{supp}(x) \cap \mathrm{supp}(y) = F$ and $|\mathrm{supp}(x)| = |\mathrm{supp}(y)|$. Pick different $x_0, x_1, x_2, x_3 \in X$ and let $x = x_0 + x_1$ and $y = x_0 + x_1 + x_2 + x_3$. Then $x, y \in U$ and $|\mathrm{supp}(y)| = 2 \cdot |\mathrm{supp}(x)|$, so x, y belong to different sets A_0, A_1, a contradiction. $\square$

5.3 Nodec Topological Groups

Theorem 5.8. *Assume $\mathfrak{p} = \mathfrak{c}$. Then every nondiscrete group topology on an Abelian group of character $< \mathfrak{c}$ can be refined to a nondiscrete nodec group topology.*

Before proving Theorem 5.8 we establish several auxiliary statements.

Lemma 5.9. *Let Y be a topological space, let X be a nowhere dense subset of Y, and let F be a finite set of continuous open mappings $f : Y \to Y$. Then for every nonempty open set $U \subseteq Y$, there is a nonempty open set $V \subseteq U$ such that $f(V) \cap X = \emptyset$ for all $f \in F$.*

Proof. Let $f \in F$. Since f is open and X is nowhere dense, there is a nonempty open $W_f \subseteq f(U)$ such that $W_f \cap X = \emptyset$. Then, since f is continuous, there is a nonempty open $V_f \subseteq U$ such that $f(V_f) \subseteq W_f$, and so $f(V_f) \cap X = \emptyset$.

Now let $F = \{f_1, \ldots, f_n\}$. Construct inductively a decreasing sequence $(V_i)_{i=1}^n$ of nonempty open subsets of U such that $f_i(V_i) \cap X = \emptyset$ for each $i = 1, \ldots, n$. Then the set $V = V_n$ is as required. $\qquad\square$

Lemma 5.10. *Let G be a nondiscrete metrizable Abelian topological group of prime period and let X be a nowhere dense subset of G such that $0 \notin X$. Then there is a nondiscrete subgroup H of G such that $H \cap X = \emptyset$.*

Proof. Let $\{U_n : n \in \mathbb{N}\}$ be a neighborhood base at $0 \in G$. It suffices to construct a sequence $(a_n)_{n=1}^\infty$ in G such that $0 \neq a_n \in U_n$ and $\langle a_1, \ldots, a_n \rangle \cap X = \emptyset$ for every $n < \omega$. Then the group

$$H = \langle a_n : n \in \mathbb{N} \rangle = \bigcup_{n=1}^\infty \langle a_1, \ldots, a_n \rangle$$

would be as required.

Without loss of generality one may suppose that $X \cup \{0\}$ is closed. Let p be the period of G. Since p is prime, the mapping $x \mapsto mx$ is a homeomorphism for each $m = 1, \ldots, p-1$. Then by Lemma 5.9, there is $a_1 \in U_1 \setminus \{0\}$ such that $\langle a_1 \rangle \cap X = \emptyset$.

Now fix $n < \omega$ and suppose that we have constructed a sequence $(a_i)_{i=1}^n$ in G such that $\langle a_1, \ldots, a_n \rangle \cap X = \emptyset$. Then there is a neighborhood V of 0 such that

$$(\langle a_1, \ldots, a_n \rangle \setminus \{0\} + V) \cap X = \emptyset.$$

Choose a neighborhood W of 0 such that

$$\underbrace{W + \cdots + W}_{p-1} \subseteq V.$$

By Lemma 5.9, there is $a_{n+1} \in (W \cap U_{n+1}) \setminus \{0\}$ such that $\langle a_{n+1} \rangle \cap X = \emptyset$. Since

$$\langle a_1, \ldots, a_{n+1} \rangle \setminus \langle a_1, \ldots, a_n \rangle = (\langle a_1, \ldots, a_n \rangle \setminus \{0\}) + \langle a_{n+1} \rangle$$

and $\langle a_{n+1} \rangle \subseteq V$, it follows that $\langle a_1, \ldots, a_{n+1} \rangle \cap X = \emptyset$. $\qquad\square$

Lemma 5.11. *Let $(G, \mathcal{T})$ be an Abelian topological group with no open subgroup of finite period. Then $\mathcal{T}$ can be refined to a nondiscrete group topology $\mathcal{T}'$ of the same character that $\mathcal{T}$ and such that for every $m \in \mathbb{N}$, the mapping $x \mapsto mx$ is open in $\mathcal{T}'$.*

Proof. Define the group topology $\mathcal{T}'$ on G by taking as a neighborhood base at 0 the subsets of the form mU, where U runs over a neighborhood base at 0 in $\mathcal{T}$ and $m \in \mathbb{N}$.   □

Lemma 5.12. *Let $(G, \mathcal{T})$ be a nondiscrete metrizable Abelian topological group such that for every $m \in \mathbb{N}$, the mapping $x \mapsto mx$ is open, and let X be a nowhere dense subset of $(G, \mathcal{T})$ such that $0 \notin X$. Then the topology $\mathcal{T}$ can be refined to a nondiscrete metrizable group topology $\mathcal{T}'$ such that $0 \notin \mathrm{cl}_{\mathcal{T}'} X$.*

Proof. Let $\{U_n : n \in \mathbb{N}\}$ be a neighborhood base at 0 in $\mathcal{T}$. We shall construct a sequence $(a_n)_{n=1}^{\infty}$ in G such that $0 \neq a_n \in U_n$ and

$$\left(\sum_{i=1}^{n} B_i^n \right) \cap X = \emptyset$$

for all $n \in \mathbb{N}$, where $B_i^n = \{0, \pm a_i, \ldots, \pm a_n\}$.

Without loss of generality one may suppose that $X \cup \{0\}$ is closed. Clearly, there is $a_1 \in U_1 \setminus \{0\}$ such that $\pm a_1 \notin X$.

Fix $n \in \mathbb{N}$ and suppose that we have constructed a sequence $(a_i)_{i=1}^{n}$ such that

$$\left(\sum_{i=1}^{n} B_i^n \right) \cap X = \emptyset.$$

Then there is a neighborhood V of 0 such that

$$\left(\left(\sum_{i=1}^{n} B_i^n \right) \setminus \{0\} + V \right) \cap X = \emptyset.$$

Choose a neighborhood W of 0 such that

$$\underbrace{W + \cdots + W}_{n+1} \subseteq V.$$

By Lemma 5.9, there is $a_{n+1} \in W \cap U_{n+1} \setminus \{0\}$ such that

$$\{\pm a_{n+1}, \ldots, \pm(n+1)a_{n+1}\} \cap X = \emptyset.$$

Then

$$\left(\sum_{i=1}^{n+1} B_i^{n+1} \right) \cap X = \emptyset.$$

Now let $\mathcal{T}_1 = \mathcal{T}((a_n)_{n=1}^{\infty})$ and let

$$U = \bigcup_{n=1}^{\infty} \sum_{i=1}^{n} B_i^n$$

By Theorem 1.19, U is a neighborhood of 0 in $\mathcal{T}_1$, and by the construction, $U \cap X = \emptyset$. Since $a_n \in U_n$, $(a_n)_{n=1}^{\infty}$ converges to 0 in $\mathcal{T}$, and consequently, $\mathcal{T} \subseteq \mathcal{T}_1$. By Lemma 1.31, $\mathcal{T}_1$ can be weakened to a metrizable group topology $\mathcal{T}'$ in which U remains a neighborhood of 0.   □

Now we are ready to prove Theorem 5.8.

Proof of Theorem 5.8. Let $(G, \mathcal{T})$ be a nondiscrete Abelian topological group of character $< \mathfrak{c}$. Without loss of generality one may assume that G is countable. By Theorem 1.28, one may assume also that $\mathcal{T}$ is metrizable. Enumerate the subsets of $G \setminus \{0\}$ into a sequence $(X_\xi)_{\xi < \mathfrak{c}}$. We shall construct an increasing sequence $(\mathcal{T}_\alpha)_{\alpha < \mathfrak{c}}$ of nondiscrete group topologies on G such that $\mathcal{T} \subseteq \mathcal{T}_0$ and the following conditions are satisfied:

(i) for every nonlimit ordinal $\alpha < \mathfrak{c}$, $\mathcal{T}_\alpha$ is metrizable,

(ii) for every limit ordinal $\alpha < \mathfrak{c}$, $\mathcal{T}_\alpha = \bigcup_{\gamma < \alpha} \mathcal{T}_\gamma$, and

(iii) for every $\xi < \mathfrak{c}$, there is $\alpha(\xi) < \mathfrak{c}$ such that, whenever $\alpha(\xi) \leq \gamma < \mathfrak{c}$ and X_ξ is nowhere dense in $\mathcal{T}_\gamma$, one has $0 \notin \mathrm{cl}_{\mathcal{T}_\gamma} X_\xi$.

For every $m \in \mathbb{N}$, let

$$G[m] = \{x \in G : mx = 0\}.$$

Consider two cases.

Case 1: there is a prime p such that the subgroup $G[p]$ is nondiscrete in $\mathcal{T}$. Define the group topology $\mathcal{T}_0$ on G by taking as a neighborhood base at 0 the subsets of the form $U \cap G[p]$, where U runs over a neighborhood base at 0 in $\mathcal{T}$.

Fix $\alpha < \mathfrak{c}$ and suppose that for every $\gamma < \alpha$, the topology $\mathcal{T}_\gamma$ has already been defined. If α is a limit ordinal, define $\mathcal{T}_\alpha$ by (ii). Let $\alpha = \delta + 1$ for some $\delta < \mathfrak{c}$. By Lemma 1.31, there is a nondiscrete metrizable group topology $\mathcal{T}'_\alpha$ on G such that $\mathcal{T}_\delta \subseteq \mathcal{T}'_\alpha$. Let $X_{\xi(\alpha)}$ be the first set in the sequence $(X_\xi)_{\xi < \mathfrak{c}}$ which is nowhere dense in $\mathcal{T}'_\alpha$ and such that $0 \in \mathrm{cl}_{\mathcal{T}'_\alpha} X_{\xi(\alpha)}$. By Lemma 5.10, there is a nondiscrete subgroup H_α in $(G, \mathcal{T}'_\alpha)$ such that $H_\alpha \subseteq G[p]$ and $H_\alpha \cap X_{\xi(\alpha)} = \emptyset$. Define the group topology $\mathcal{T}_\alpha$ on G by taking as a neighborhood base at 0 the subsets of the form $U \cap H_\alpha$, where U runs over a neighborhood base at 0 in $\mathcal{T}'_\alpha$.

Case 2: for every prime p, the subgroup $G[p]$ of $(G, \mathcal{T})$ is discrete. Then for every $m \in \mathbb{N}$, the subgroup $G[m]$ is discrete, and so every subgroup of $(G, \mathcal{T})$ of finite period is discrete. Indeed, otherwise for every neighborhood U of 0, there is a prime $p_U | m$ such that $U \cap G[p_U] \neq \emptyset$, and since there are only finitely many primes dividing m, there is a prime $p | m$ such that $U \cap G[p] \neq \emptyset$ for every neighborhood U of 0.

By Lemma 5.11, there is a nondiscrete first countable group topology $\mathcal{T}_0$ on G such that $\mathcal{T} \subseteq \mathcal{T}_0$ and the mappings $x \mapsto mx$, where $m \in \mathbb{N}$, are open in $\mathcal{T}_0$.

Fix $\alpha < c$ and suppose that for every $\gamma < \alpha$, the topology $\mathcal{T}_\gamma$ has already been defined and the mappings $x \mapsto mx$, where $m \in \mathbb{N}$, are open in $\mathcal{T}_\gamma$. If α is a limit ordinal, define $\mathcal{T}_\alpha$ by (ii). Then the mappings $x \mapsto mx$, where $m \in \mathbb{N}$, are open in $\mathcal{T}_\alpha$ as well. Let $\alpha = \delta + 1$ for some $\delta < c$. By Theorem 1.28 and Lemma 5.17, there is a nondiscrete metrizable group topology $\mathcal{T}'_\alpha$ on G such that $\mathcal{T}_\delta \subseteq \mathcal{T}'_\alpha$ and the mappings $x \mapsto mx$, where $m \in \mathbb{N}$, are open in $\mathcal{T}'_\alpha$. Let $X_{\xi(\alpha)}$ be the first set in the sequence $(X_\xi)_{\xi<c}$ which is nowhere dense in $\mathcal{T}'_\alpha$ and such that $0 \in \text{cl}_{\mathcal{T}'_\alpha} X_{\xi(\alpha)}$. By Lemma 5.12, there is a nondiscrete metrizable group topology $\mathcal{T}''_\alpha$ on G such that $\mathcal{T}'_\alpha \subseteq \mathcal{T}''_\alpha$ and $0 \notin \text{cl}_{\mathcal{T}''_\alpha} X$. Then refine $\mathcal{T}''_\alpha$ to a nondiscrete metrizable group topology $\mathcal{T}_\alpha$ in which the mappings $x \mapsto mx$, where $m \in \mathbb{N}$, are open.

Clearly, the sequence $(\mathcal{T}_\alpha)_{\alpha<c}$ so constructed satisfies conditions (i) and (ii). To check (iii), let $\xi < c$ and suppose that for every $\eta < \xi$, (iii) holds. Put $\delta = \sup_{\eta<\xi} \alpha(\eta)$ (if $\xi = 0$, $\delta = 0$). If for every γ from the interval $[\delta, c)$, X_ξ is not a nowhere dense set in $\mathcal{T}_\gamma$ with $0 \in \text{cl}_{\mathcal{T}_\gamma} X_\xi$, then one can take δ as $\alpha(\xi)$. If for some γ from $[\delta, c)$, X_ξ is a nowhere dense set in $\mathcal{T}_\gamma$ with $0 \in \text{cl}_{\mathcal{T}_\gamma} X_\xi$, then X_ξ is the first such set and so one can take $\gamma + 1$ as $\alpha(\xi)$.

Now let $\mathcal{T}' = \bigcup_{\alpha<c} \mathcal{T}_\alpha$. We claim that $\mathcal{T}'$ is nodec.

Assume the contrary. Then there is $\xi < c$ such that X_ξ is a nowhere dense set in $\mathcal{T}$ with $0 \in \text{cl}_{\mathcal{T}} X_\xi$. Construct inductively an increasing sequence $(\gamma_n)_{n<\omega}$ of nonlimit ordinals in $[\alpha(\xi), c)$ such that for every $n < \omega$ and for every nonempty $U \in \mathcal{T}_{\gamma_n}$, there exists a nonempty $V \in \mathcal{T}_{\gamma_{n+1}}$ such that $V \cap X_\xi = \emptyset$. Let $\gamma = \sup_{n<\omega} \alpha_n$. Then $\alpha(\xi) < \gamma < c$ and X_ξ is a nowhere dense set in $\mathcal{T}_\gamma$ with $0 \in \text{cl}_{\mathcal{T}_\gamma} X_\xi$. But this contradicts (iii). $\square$

5.4   *P*-point Theorems

In this section we prove two theorems. The first of them is concerned with extremally disconnected topological groups containing countable nonclosed discrete subsets.

Lemma 5.13. *Let X be an extremally disconnected space, let D be a strongly discrete subset of X, and let $x \in \text{cl} \, D \setminus D$. Then there is exactly one ultrafilter on X containing D and converging to x.*

Proof. Assume on the contrary that there are two different ultrafilters on G containing D and converging to x. Then there are disjoint subsets A and B of D with $x \in (\text{cl } A) \cap (\text{cl } B)$. For every $y \in D$, there is a neighborhood U_y of y such that the subsets U_y, where $y \in D$, are pairwise disjoint. Let $U_A = \bigcup_{y \in A} U_y$ and $U_B = \bigcup_{y \in B} U_y$. Then U_A and U_B are disjoint open subsets of X with $x \in (\text{cl } U_A) \cap (\text{cl } U_B)$, which is a contradiction. $\square$

Theorem 5.14. *Let G be an extremally disconnected topological group and let p be a nonprincipal converging ultrafilter on G. Suppose that p contains a countable* *discrete subset of G. Then there is a mapping $f : G \to \omega$ such that $f(p)$ is a P-point.*

Proof. Without loss of generality one may assume that G is Boolean and p converges to 0. Let D be a countable discrete subset of G such that $D \in p$. Then by Lemma 3.34, D is strongly discrete. Enumerate D without repetitions as $\{x_n : n < \omega\}$. Construct a decreasing sequence $(U_n)_{n<\omega}$ of open neighborhoods of $0 \in G$ such that

(i) for every $n < \omega$, $x_n \notin U_n$, and

(ii) the subsets $x_n + U_n$, where $n < \omega$, are pairwise disjoint.

It follows from (i) that $D \cap \bigcap_{n<\omega} U_n = \emptyset$. Define a mapping $f : G \to \omega$ by the condition that

$$f(x) = n \text{ if } x \in U_n \setminus U_{n+1}.$$

Clearly, the ultrafilter $f(p)$ is nonprincipal.

To show that $f(p)$ is a P-point, let $\{A_m : m < \omega\}$ be a partition of ω with $A_m \notin f(p)$. For every $m < \omega$, put $D_m = f^{-1}(A_m) \cap D$. Then $\{D_m : m < \omega\}$ is a partition of D with $D_m \notin p$. So for every $n < \omega$, there is a unique $m(n) < \omega$ such that $x_n \in D_{m(n)}$. Consider the subset $x_n + (D_{m(n)} \cap U_n)$. Since $D_{m(n)} \notin p$,

$$x \notin \mathrm{cl}\,(x_n + (D_{m(n)} \cap U_n))$$

by Lemma 5.13. Consequently, there are disjoint open subsets P_n and Q_n of $x_n + U_n$ such that $x_n \in P_n$ and $x_n + (D_{m(n)} \cap U_n) \subset Q_n$. Let $P = \bigcup_{n<\omega} P_n$, $Q = \bigcup_{n<\omega} Q_n$ and

$$F = \bigcup_{n<\omega} (x_n + (D_{m(n)} \cap U_n)).$$

Then P and Q are disjoint open subsets of G, $F \subset Q$, and $0 \in \mathrm{cl}\,P$. It follows that $0 \notin \mathrm{cl}\,Q$ and so $0 \notin \mathrm{cl}\,F$. Choose neighborhoods V and W of 0 such that $V \cap F = \emptyset$ and $W + W \subseteq V$. Let $E = D \cap W$ and $A = f(E)$. We claim that for every $m < \omega$, $A \cap A_m$ is finite.

Indeed, assume the contrary. Pick any $x_n \in E \cap D_m$. Then there exists $x_k \in E \cap D_m \cap U_n$. But then, on the one hand,

$$x_n + x_k \in E + E \subseteq W + W \subseteq V$$

and, on the other hand,

$$x_n + x_k \in x_n + (D_m \cap U_n) \subseteq F \subseteq G \setminus V,$$

a contradiction.   □

It follows from Theorem 5.14 and Theorem 2.38 that

Corollary 5.15. *It is consistent with ZFC that there is no extremally disconnected topological group containing a countable discrete nonclosed subset.*

Our second theorem deals with countable topological Boolean groups containing no nonclosed discrete subset, and even more generally, containing no subset with exactly one accumulation point.

A nonempty subset of a topological space is called a *P-set* if the intersection of any countable family of its neighborhoods is again its neighborhood. Note that every isolated point of a *P*-set is a *P*-point. In particular, every point of a finite *P*-set is a *P*-point.

Definition 5.16. We say that a set of nonprincipal ultrafilters on ω is a *P-set* if it is a *P*-set in ω^*.

Lemma 5.17. *Let $\mathscr{F}$ be a filter on ω with $\bigcap \mathscr{F} = \emptyset$. Then the following statements are equivalent:*

(1) *$\overline{\mathscr{F}}$ is a P-set,*

(2) *for every decreasing sequence $(A_n)_{n<\omega}$ of members of $\mathscr{F}$, there is $A \in \mathscr{F}$ such that $A \cap (A_n \setminus A_{n+1})$ is finite for all $n < \omega$,*

(3) *for every mapping $f : \omega \to \omega$, there is $A \subseteq \omega$ such that either $A \cap B \neq \emptyset$ for all $B \in \mathscr{F}$ and $f(A)$ is finite or $A \in \mathscr{F}$ and $f^{-1}(n) \cap A$ is finite for all $n < \omega$.*

Proof. (1) $\Rightarrow$ (2) Pick $A \in \mathscr{F}$ such that $A^* \subseteq \bigcap_{n<\omega} A_n^*$. Then $A \setminus A_n$ is finite for all $n < \omega$. Consequently, $A \cap (A_n \setminus A_{n+1})$ is finite for all $n < \omega$.

(2) $\Rightarrow$ (3) For every $n < \omega$, let $A_n = \bigcup_{n \leq m < \omega} f^{-1}(m)$. If $A_n \in \mathscr{F}$ for all $n < \omega$, then there is $A \in \mathscr{F}$ such that $A \cap (A_n \setminus A_{n+1}) = A \cap f^{-1}(n)$ is finite for all n. If $A_n \notin \mathscr{F}$ for some n, then $A_n \notin p$ for some $p \in \overline{\mathscr{F}}$, so $A = \omega \setminus A_n \in p$ and $f(A) \subseteq \{0, \ldots, n-1\}$.

(3) $\Rightarrow$ (1) Let $(U_n)_{n<\omega}$ be any sequence of neighborhoods of $\overline{\mathscr{F}} \subseteq \omega^*$. Without loss of generality one may suppose that there is a decreasing sequence $(A_n)_{n<\omega}$ of members of $\mathscr{F}$ such that $A_0 = \omega$, $\bigcap_{n<\omega} A_n = \emptyset$ and $U_n = A_n^*$ for all $n < \omega$. Define $f : \omega \to \omega$ by

$$f(x) = n \text{ if } x \in A_n \setminus A_{n+1}.$$

For every $p \in \overline{\mathscr{F}}$ and $C \in p$, $f(C)$ is infinite. Consequently, there is $A \in \mathscr{F}$ such that $f^{-1}(n) \cap A$ is finite for all $n < \omega$. It follows that $A^* \subseteq \bigcap_{n<\omega} U_n$. $\square$

Definition 5.18. Define the functions $\theta, \phi : (\bigoplus_\omega \mathbb{Z}_2) \setminus \{0\} \to \omega$ by

$$\theta(x) = \min \operatorname{supp}(x) \quad \text{and} \quad \phi(x) = \max \operatorname{supp}(x).$$

Theorem 5.19. *Let $\mathcal{T}$ be a nondiscrete group topology on $\bigoplus_\omega \mathbb{Z}_2$ finer than that induced by the product topology on $\prod_\omega \mathbb{Z}_2$ and let $\mathcal{F}$ denote the neighborhood filter of 0 in $\mathcal{T}$. Suppose that $(\bigoplus_\omega \mathbb{Z}_2, \mathcal{T})$ contains no subset with exactly one accumulation point. Then $\overline{\phi(\mathcal{F})}$ is a P-set. In particular, if $\overline{\phi(\mathcal{F})}$ is finite, then each of its points is a P-point.*

Proof. To show that $\overline{\phi(\mathcal{F})}$ is a P-set, we use characterization (3) from Lemma 5.17. Let $f : \omega \to \omega$. Denote $S = \overline{\mathcal{F}} \setminus \{0\}$ and for each $p \in S$, pick $A_p \in p$ such that either $f(\phi(x)) \geq \theta(x)$ for all $x \in A_p$ or $f(\phi(x)) < \theta(x)$ for all $x \in A_p$. Consider two cases.

   Case 1: there is $p \in S$ such that $f(\phi(x)) < \theta(x)$ for all $x \in A_p$. Pick an accumulation point $a \neq 0$ of A_p. Then there is $q \in S$ and $Q \in q$ such that $a + Q \subseteq A_p$. Choose Q in addition so that for every $x \in Q$, one has $\phi(a) < \theta(x)$. Then for every $x \in Q$, we have that

$$f(\phi(x)) = f(\phi(a + x)) < \theta(a + x) = \theta(a).$$

Hence $f(\phi(Q))$ is finite.

   Case 2: for every $p \in S$, $f(\phi(x)) \geq \theta(x)$ for all $x \in A_p$. Define the neighborhood V of 0 in $\mathcal{T}$ by

$$V = \bigcup_{p \in S} A_p \cup \{0\}$$

and choose a neighborhood W of 0 such that $W + W \subseteq V$. We claim that for every $n < \omega$, $f^{-1}(n) \cap \phi(W)$ is finite.

   Indeed, assume the contrary. Then there exist $m < \omega$ and a sequence $(x_n)_{n<\omega}$ in W such that $f(\phi(x_n)) = m$ and $\phi(x_n) < \phi(x_{n+1})$ for all $n < \omega$. It follows that there exist $n < k < \omega$ such that $x_n(i) = x_k(i)$ for all $i \leq m$. Let $x = x_n + x_k$. Then $x \in V$ and so $f(\phi(x)) \geq \theta(x)$. But $\phi(x) = \phi(x_k)$ and $\theta(x) > m$. Hence $f(\phi(x_\gamma)) > m$, which is a contradiction.   □

   We say that a topological group $(G, \mathcal{T})$ is *maximally nondiscrete* if $\mathcal{T}$ is maximal among nondiscrete group topologies on G. By Zorn's Lemma, every nondiscrete group topology can be refined to a maximally nondiscrete group topology. Clearly, every maximal topological group is maximally nondiscrete.

Lemma 5.20. *Let $(G, \mathcal{T}_0)$ be a totally bounded topological group and let $\mathcal{T}_1$ be a maximally nondiscrete group topology on G. Then $\mathcal{T}_0 \subseteq \mathcal{T}_1$.*

Proof. Pick any nonprincipal ultrafilter $\mathcal{U}$ on G converging to 1 in $\mathcal{T}_1$. Considering $(G, \mathcal{T}_0)$ as a subgroup of a compact group shows that the filter $\mathcal{U}\mathcal{U}^{-1}$ (with a base of subsets AA^{-1}, where $A \in \mathcal{U}$) converges to 1 in $\mathcal{T}_0$. Clearly $\mathcal{U}\mathcal{U}^{-1}$ converges to 1 in $\mathcal{T}_1$ as well. Since $\mathcal{T}_1$ is maximally nondiscrete, it follows that $\mathcal{T}_1 = \mathcal{T}(\mathcal{U}\mathcal{U}^{-1})$. Hence $\mathcal{T}_0 \subseteq \mathcal{T}_1$.   □

Now as a consequence we obtain from from Theorem 5.19 that

Corollary 5.21. *The existence of a maximal topological group implies the existence of a P-point in ω^*.*

Proof. Let $(G, \mathcal{T})$ be a maximal topological group. By Theorem 5.7, one may suppose that $G = \bigoplus_\omega \mathbb{Z}_2$. Let $\mathcal{F}$ be the nonprincipal ultrafilter on G converging to 0 in $\mathcal{T}$ and let $\mathcal{T}_0$ be the topology on G induced by the product topology on $\prod_\omega \mathbb{Z}_2$. By Lemma 5.20, $\mathcal{T}_0 \subseteq \mathcal{T}$. Hence by Theorem 5.19, $\phi(\mathcal{F})$ is a P-point. $\square$

Combining Corollary 5.21 and Theorem 2.38 gives us that

Corollary 5.22. *It is consistent with ZFC that there is no maximal topological group.*

As another consequence we obtain from Theorem 5.19 the following result.

Corollary 5.23. *Let p be a nonprincipal ultrafilter on ω not being a P-point and let $\mathcal{T}^p$ be the group topology on $G = \bigoplus_\omega \mathbb{Z}_2$ with a neighborhood base at 0 consisting of subgroups*
$$H_A = \{x \in G : \operatorname{supp}(x) \subset A\}$$
where $A \in p$. Then for every nondiscrete group topology $\mathcal{T}$ on G finer than $\mathcal{T}^p$, $(G, \mathcal{T})$ has a discrete subset with exactly one accumulation point.

Proof. Let $\mathcal{F}$ be the neighborhood filter of 0 in $\mathcal{T}$. It is clear that $\phi(\mathcal{F}) = p$. Now assume on the contrary that $(G, \mathcal{T})$ has a subset with exactly one accumulation point. Then by Theorem 5.19, $\phi(\mathcal{F}) = p$ is a P-point, which is a contradiction. $\square$

Lemma 5.24. *There exists a nonprincipal ultrafilter on ω not being a P-point.*

Proof. Let $\{A_n : n < \omega\}$ be a partition of ω into infinite subsets. Define the filter $\mathcal{F}$ on ω by taking as a base the subsets of the form
$$\bigcup_{n \le i < \omega} (A_i \setminus F_i)$$
where $n < \omega$ and F_i is a finite subset of A_i for each i. It then follows that every ultrafilter on ω containing $\mathcal{F}$ is not a P-point. $\square$

Combining Corollary 5.23 and Lemma 5.24 gives us that

Corollary 5.25. *There exists a maximally nondiscrete group topology $\mathcal{T}$ on $G = \bigoplus_\omega \mathbb{Z}_2$ such that $(G, \mathcal{T})$ has a discrete subset with exactly one accumulation point.*

References

The question whether there exists in ZFC a nondiscrete extremally disconnected topological group was raised by A. Arhangel'skiĭ [3]. The first consistent example of such a group was constructed by S. Sirota [71], the implication (1) $\Rightarrow$ (4) in Theorem 5.1 for $\bigoplus_{\omega} \mathbb{Z}_2$. The equivalence (1) $\Leftrightarrow$ (2) and the fact that $\mathcal{T}[\mathcal{F}]$ is extremally disconnected for $\bigoplus_{\omega} \mathbb{Z}_2$ are due to A. Louveau [45]. Theorem 5.1, except for statement (5), is from [103]. Theorem 5.5 is a result of V. Malykhin [46]. Theorem 5.4 and Theorem 5.7 are also due to him [46, 47]. Theorem 5.8 was proved in [88]. Theorems 5.14 and 5.19 are from [97]. Corollary 5.21 is due to I. Protasov [56]. Corollary 5.25 is a partial case of a result from [65].

Chapter 6

The Semigroup βS

In this chapter we extend the operation of a discrete semigroup S to βS making βS a right topological semigroup with S contained in its topological center. We show that every compact Hausdorff right topological semigroup has an idempotent and, as a consequence, a smallest two sided ideal which is a completely simple semigroup. The structure of a completely simple semigroup is given by the Rees-Suschkewitsch Theorem. As a combinatorial application of the semigroup βS we prove Hindman's Theorem. We conclude by characterizing ultrafilters from the smallest ideal of βS.

6.1 Extending the Operation to βS

Theorem 6.1. *The operation of a discrete groupoid S extends uniquely to βS so that*

(1) *for each $a \in S$, the left translation $\lambda_a : \beta S \ni x \mapsto ax \in \beta S$ is continuous, and*

(2) *for each $q \in \beta S$, the right translation $\rho_q : \beta S \ni x \mapsto xq \in \beta S$ is continuous.*

If S is a semigroup, the extended operation is associative.

Proof. For each $a \in S$, the left translation $l_a : S \ni x \mapsto ax \in S$ extends uniquely to a continuous mapping $\overline{l_a} : \beta S \to \beta S$. For each $a \in S$ and $q \in \beta S$, define $a \cdot q \in \beta S$ by

$$a \cdot q = \overline{l_a}(q).$$

Next, for each $q \in \beta S$, the mapping $r_q : S \ni x \mapsto x \cdot q \in \beta S$ extends uniquely to a continuous mapping $\overline{r_q} : \beta S \to \beta S$. For each $p \in S^*$ and $q \in \beta S$, define $p \cdot q \in \beta S$ by

$$p \cdot q = \overline{r_q}(p).$$

Under the extended operation $\lambda_a = \overline{l_a}$ and $\rho_q = \overline{r_q}$, consequently, conditions (1) and (2) are satisfied. Furthermore, whenever the operation of S extends to βS so that conditions (1) and (2) are satisfied, one has $\lambda_a = \overline{l_a}$ and $\rho_q = \overline{r_q}$, hence such an extension is unique.

Finally, suppose that S is a semigroup and let $p, q, r \in \beta S$. Then

$$
\begin{aligned}
(pq)r &= (\lim_{x \to p} xq)r & \text{because } \rho_q \text{ is continuous} \\
&= (\lim_{x \to p} \lim_{y \to q} xy)r & \text{because } \lambda_x \text{ is continuous} \\
&= \lim_{x \to p} \lim_{y \to q} (xy)r & \text{because } \rho_r \text{ is continuous} \\
&= \lim_{x \to p} \lim_{y \to q} \lim_{z \to r} (xy)z & \text{because } \lambda_{xy} \text{ is continuous} \\
&= \lim_{x \to p} \lim_{y \to q} \lim_{z \to r} xyz
\end{aligned}
$$

and

$$
\begin{aligned}
p(qr) &= \lim_{x \to p} (x(qr)) & \text{because } \rho_{qr} \text{ is continuous} \\
&= \lim_{x \to p} (x \lim_{y \to q} yr) & \text{because } \rho_r \text{ is continuous} \\
&= \lim_{x \to p} (x \lim_{y \to q} \lim_{z \to r} yz) & \text{because } \lambda_y \text{ is continuous} \\
&= \lim_{x \to p} \lim_{y \to q} \lim_{z \to r} x(yz) & \text{because } \lambda_x \text{ is continuous} \\
&= \lim_{x \to p} \lim_{y \to q} \lim_{z \to r} xyz,
\end{aligned}
$$

so $(pq)r = p(qr)$. $\qquad\qquad\qquad\qquad\qquad\qquad\qquad\qquad\qquad\qquad$ $\square$

Definition 6.2. A semigroup T endowed with a topology is a *right topological semigroup* if for each $p \in T$, the right translation

$$
\rho_p : T \ni x \mapsto xp \in T
$$

is continuous. The *topological center* of a right topological semigroup T, denoted $\Lambda(T)$, consists of all $a \in T$ such that the left translation

$$
\lambda_a : T \ni x \mapsto ax \in T
$$

is continuous.

Theorem 6.1 tells us that the operation of a discrete semigroup S extends uniquely to βS so that βS is a right topological semigroup with $S \subseteq \Lambda(\beta S)$.

Lemma 6.3. *Let R and T be Hausdorff right topological semigroups, let S be a dense subsemigroup of R such that $S \subseteq \Lambda(R)$, and let $\varphi : R \to T$ be a continuous mapping such that $\varphi(S) \subseteq \Lambda(T)$. If $\varphi|_S$ is a homomorphism, so is φ.*

Proof. First let $x \in S$ and $q \in R$. Then

$$
\begin{aligned}
\varphi(xq) &= \varphi(\lim_{S \ni y \to q} xy) && \text{because } \lambda_x \text{ is continuous} \\
&= \lim_{y \to q} \varphi(xy) && \text{because } \varphi \text{ is continuous} \\
&= \lim_{y \to q} \varphi(x)\varphi(y) && \text{because } \varphi|_S \text{ is a homomorphism} \\
&= \varphi(x) \lim_{y \to q} \varphi(y) && \text{because } \lambda_{\varphi(x)} \text{ is continuous} \\
&= \varphi(x)\varphi(q).
\end{aligned}
$$

Now let $p, q \in R$. Then

$$
\begin{aligned}
\varphi(pq) &= \varphi(\lim_{S \ni x \to p} xq) && \text{because } \rho_q \text{ is continuous} \\
&= \lim_{x \to p} \varphi(xq) && \text{because } \varphi \text{ is continuous} \\
&= \lim_{x \to p} \varphi(x)\varphi(q) \\
&= (\lim_{x \to p} \varphi(x))\varphi(q) && \text{because } \rho_{\varphi(q)} \text{ is continuous} \\
&= \varphi(p)\varphi(q). && \qquad\qquad \square
\end{aligned}
$$

Corollary 6.4. *Let S be a discrete semigroup and let $\varphi : S \to T$ be any homomorphism of S into a compact Hausdorff right topological semigroup T such that $\varphi(S) \subseteq \Lambda(T)$. Then the continuous extension $\overline{\varphi} : \beta S \to T$ of φ is a homomorphism.*

Definition 6.5. Given a semigroup S endowed with a topology, a *semigroup compactification* of S is a pair (φ, T) where T is a compact right topological semigroup and $\varphi : S \to T$ is a continuous homomorphism such that $\varphi(S)$ is dense in T and $\varphi(S) \subseteq \Lambda(T)$.

Corollary 6.4 tells us that the Stone–Čech compactification βS of a discrete semigroup S is the largest semigroup compactification of S.

From now on, for any semigroup S, βS denotes the Stone–Čech compactification of the discrete semigroup S.

The next lemma describes the operation of βS in terms of ultrafilters. That is, given ultrafilters $p, q \in \beta S$, one characterizes the subsets of S which are members of the ultrafilter pq.

Recall that given a semigroup S, $A \subseteq S$ and $s \in S$,

$$
s^{-1}A = \{x \in S : sx \in A\} = \lambda_s^{-1}(A).
$$

Lemma 6.6. *Let S be a semigroup, $A \subseteq S$, $s \in S$ and $p, q \in \beta S$. Then*

(1) *$A \in sq$ if and only if $s^{-1}A \in q$, and*

(2) *$A \in pq$ if and only if $\{x \in S : x^{-1}A \in q\} \in p$.*

Proof. (1) Let $A \in sq$. Then $\overline{A}$ is a neighborhood of sq. Since λ_s is continuous, there is $Q \in q$ such that $s\overline{Q} \subseteq \overline{A}$. It follows that $sQ \subseteq A$ and so $s^{-1}A \in q$.

Conversely, let $s^{-1}A \in q$. Assume on the contrary that $A \notin sq$. Consequently, $S \setminus A \in sq$. Then by the already established necessity, $s^{-1}(S \setminus A) \in q$. But

$$(s^{-1}A) \cap (s^{-1}(S \setminus A)) = \emptyset,$$

a contradiction.

(2) Let $A \in pq$. Since ρ_q is continuous, there is $P \in p$ such that $\overline{P}q \subseteq \overline{A}$. Then for every $x \in P$, $A \in xq$ and so by (1), $x^{-1}A \in q$. Hence, $\{x \in S : x^{-1}A \in q\} \in p$.

Conversely, let $\{x \in S : x^{-1}A \in q\} \in p$. Assume on the contrary that $A \notin pq$. Consequently, $S \setminus A \in pq$. Then by the already established necessity, $\{x \in S : x^{-1}(S \setminus A) \in q\} \in p$. But $(x^{-1}A) \cap (x^{-1}(S \setminus A)) = \emptyset$ for each $x \in S$. It follows that

$$\{x \in S : x^{-1}A \in q\} \cap \{x \in S : x^{-1}(S \setminus A) \in q\} = \emptyset,$$

a contradiction. □

Corollary 6.7. *Let S be a semigroup, $a \in S$ and $p, q \in \beta S$. Then*

(1) *the ultrafilter aq has a base consisting of subsets of the form aQ where $Q \in q$, and*

(2) *the ultrafilter pq has a base consisting of subsets of the form $\bigcup_{x \in P} xQ_x$ where $P \in p$ and $Q_x \in q$.*

We conclude this section by showing that if S is a cancellative semigroup, then S^* is a two-sided ideal of βS and the translations of βS are injective on S.

A nonempty subset I of a semigroup S is a *left ideal* (*two-sided ideal* or just *ideal*) if $SI \subseteq I$ (both $SI \subseteq I$ and $IS \subseteq I$).

Lemma 6.8. *Let S be a cancellative (left cancellative, right cancellative) semigroup. Then S^* is an ideal (left ideal, right ideal) of βS.*

Proof. Suppose that S is left cancellative. Let $p \in \beta S$ and $q \in S^*$. To see that $pq \in S^*$, let $A \in pq$. Then there exist $x \in S$ and $B \in q$ such that $xB \subseteq A$. It follows that A is infinite.

Now suppose that S is right cancellative. Let $p \in S^*$ and $q \in \beta S$. Assume on the contrary that $pq = a \in S$. Then there exist $A \in p$ and, for each $x \in A$, $B_x \in q$ such that $xB_x = \{a\}$. Pick distinct $x, y \in A$ and any $z \in B_x \cap B_y$. Then $xz = yz$, a contradiction. □

Lemma 6.9. *Let S be a cancellative semigroup, let a and b be distinct elements of S, and let $p \in \beta S$. Then $ap \neq bp$ and $pa \neq pb$.*

Proof. Define $f : S \to S$ by putting $f(ax) = bx$ for every $x \in S$ and $f(y) = a^2$ for every $y \in S \setminus aS$. Then f has no fixed points and $\overline{f}(ap) = bp$. Hence, $ap \neq bp$ by Corollary 2.17. The proof that $pa \neq pb$ is similar. $\square$

As usual, for every semigroup S, we use S^1 to denote the semigroup with identity obtained from S by adjoining one if necessary.

Lemma 6.10. *If S is a cancellative semigroup, so is S^1.*

Proof. We first show that if e is an idempotent in S, then $e = 1$. To see this, let $x \in S$. Multiplying the equality $e = ee$ by x from the left gives us $xe = xee$. Then cancelling the latter equality by e from the right we obtain $x = xe$. Similarly $x = ex$.

Now assume on the contrary that S^1 is not cancellative. Then some translation in S^1 is not injective. One may suppose that this is a left translation. It follows that there exist $a, b \in S$ such that $ab = a$. But then $ab^2 = ab$, and so $b^2 = b$. Hence $b = 1 \in S$, a contradiction. $\square$

Combining Lemma 6.9 and Lemma 6.10, we obtain the following.

Corollary 6.11. *Let S be a cancellative semigroup, let $a \in S$, and let $p \in \beta S$. If $ap = p$, then $a = 1$.*

6.2   Compact Right Topological Semigroups

Recall that an element p of a semigroup is an *idempotent* if $pp = p$.

Theorem 6.12. *Every compact Hausdorff right topological semigroup has an idempotent.*

Proof. Let S be a compact Hausdorff right topological semigroup. Consider the set $\mathcal{P}$ of all closed subsemigroups of S partially ordered by the inclusion. Since $S \in \mathcal{P}$, $\mathcal{P} \neq \emptyset$. For every chain $\mathcal{C}$ in $\mathcal{P}$, $\bigcap \mathcal{C} \in \mathcal{P}$. Hence by Zorn's Lemma, $\mathcal{C}$ has a minimal element A. Pick $x \in A$. We shall show that $xx = x$. (It will follow that $A = \{x\}$, but we do not need this.)

We start by showing that $Ax = A$. Let $B = Ax$. Clearly B is nonempty. Since $B = \rho_x(A)$, B is closed. Also $BB \subseteq AxAx \subseteq AAAx \subseteq Ax = B$, so B is a subsemigroup. Thus $B \in \mathcal{P}$. But then $B = A$, since $B = Ax \subseteq AA \subseteq A$ and A is minimal.

Now let $C = \{y \in A : yx = x\}$. Since $x \in A = Ax$, C is nonempty. And since $C = \rho_x^{-1}(x)$, C is closed. Given any $y, z \in C$, $yz \in AA \subseteq A$ and $yzx = yx = x$, so C is a subsemigroup. Thus $C \in \mathcal{P}$. But then $C = A$, since $C \subseteq A$ and A is minimal. Hence $x \in C$ and so $xx = x$ as required.   □

A left ideal L of a semigroup S is *minimal* if S has no left ideal strictly contained in L.

Corollary 6.13. *Let S be a compact Hausdorff right topological semigroup. Then S contains a minimal left ideal. Every minimal left ideal of S is closed and has an idempotent.*

Proof. If L is any left ideal of S and $x \in L$, then $Sx = \rho_x(S)$ is a closed left ideal contained in L. It follows that every minimal left ideal of S is closed and, by Theorem 6.12, has an idempotent. Thus we need only to show that S contains a minimal left ideal. To this end, consider the set $\mathcal{P}$ of all closed left ideals of S, partially ordered by the inclusion. Applying Zorn's Lemma gives us a minimal element $L \in \mathcal{P}$. Since L is minimal among closed left ideals and every left ideal contains a closed left ideal, it follows that L is a minimal left ideal.   □

A semigroup can have many minimal left (right) ideals. However, it can have at most one minimal two-sided ideal. Indeed, if K is a minimal ideal, then for any ideal I, KI is an ideal and $KI \subseteq K \cap I$, so $K \subseteq I$.

Definition 6.14. For every semigroup S, let $K(S)$ denote the smallest ideal of S, if exists.

Lemma 6.15. *Let S be a semigroup and assume that there is a minimal left ideal L of S. Then*

(1) *for every $a \in S$, La is a minimal left ideal of S,*

(2) *every left ideal of S contains a minimal left ideal,*

(3) *different minimal left ideals of S are disjoint and their union is $K(S)$.*

Proof. (1) If M is a left ideal of S contained in La, then $L' = \{x \in L : xa \in M\}$ is a left ideal of S contained in L, so $L' = L$ and consequently $M = La$.

(2) Let N be a left ideal of S. Pick $a \in N$. Then $La \subseteq N$ and by (1), La is a minimal left ideal.

(3) If M and N are minimal left ideals and $L = M \cap N \neq \emptyset$, then L is a left ideal, so $M = L = N$.

Now let $K = LS = \bigcup_{a \in S} La$. Clearly K is an ideal. If I is any ideal of S, then $IL \subseteq I \cap L \subseteq L$, so $IL = I \cap L = L$ and consequently $L \subseteq I$. Hence, $K = LS \subseteq IS \subseteq I$.   □

A semigroup is *simple* (*left simple*) if it has no proper ideal (left ideal). It is easy to see that a smallest ideal (a minimal left ideal) is a simple (left simple) semigroup.

A semigroup S is *completely simple* if it is simple and there is a minimal left ideal of S which has an idempotent.

Corollary 6.16. *Every compact Hausdorff right topological semigroup has a smallest ideal which is a completely simple semigroup.*

Definition 6.17. Given a semigroup S, let $E(S)$ denote the set of idempotents of S.

Lemma 6.18. *Let S be a left simple semigroup with $E(S) \neq \emptyset$ and let $e \in E(S)$. Then*

(1) *e is a right identity of S, and*

(2) *eS is a group.*

Proof. (1) Let $s \in S$. Since $Se = S$, there is $t \in S$ such that $te = s$. Then $se = tee = te = s$.

(2) Clearly, eS is a semigroup and e is a left identity of eS. By (1), e is also a right identity, so e is identity of eS. Now let $x \in eS$. We have to find $y \in eS$ such that $yx = xy = e$. Since $Sx = S$, there is $s \in S$ such that $sx = e$. Let $y = es$. Then $yx = esx = ee = e$. To see that $xy = e$, note that there is $t \in S$ such that $ty = e$, since $Sy = S$. It then follows that $t = te = tyx = ex = x$.   □

A semigroup satisfying the identity $xy = x$ ($xy = y$) is called a *left* (*right*) *zero semigroup*. By Lemma 6.18 (1), if S is a left simple semigroup and $E(S) \neq \emptyset$, then $E(S)$ is a left zero semigroup.

Proposition 6.19. *Let S be a left simple semigroup with $E(S) \neq \emptyset$. Let $e \in E(S)$ and define $f : E(S) \times eS \to S$ by $f(x, a) = xa$. Then f is an isomorphism.*

Proof. To see that f is a homomorphism, let $(u_1, a_1), (u_2, a_2) \in E(S) \times eS$. Then

$$f((u_1, a_1)(u_2, a_2)) = f(u_1 u_2, a_1 a_2) = u_1 u_2 a_1 a_2$$
$$= u_1 a_1 a_2 = u_1 a_1 u_2 a_2 = f(u_1, a_1) f(u_2, a_2).$$

To see that f is surjective, let $x \in S$. Since S is left simple, there is $y \in S$ such that $yx = e$. Then $xyxy = xey = xy$, so $xy \in E(S)$. It follows that

$$f(xy, ex) = xyex = xyx = xe = x.$$

To see that f is injective, let $(u, a) \in E(S) \times eS$ and let $x = f(u, a)$. We claim that $a = ex$ and $u = xy$ where y is the inverse of $ex \in eS$. Indeed, $a = ea = eua = ex$ and $u = ue = uexy = uxy = xy$, since $ux = uua = ua = x$.   □

Lemma 6.20. *Let S be a semigroup and let $e \in E(S)$. If Se is a minimal left ideal of S, then eS is a minimal right ideal.*

Proof. Let $a \in S$. We have to show that $e \in eaS$. Since $ae \in Se$ and Se is a minimal left ideal, there is $b \in S$ such that $aeb = u \in E(Se)$. Then $e = eu = eaeb \in eaS$. $\square$

Lemma 6.21. *Let S be a semigroup and let R and L be minimal right and left ideals of S. Then $R \cap L = RL$ is a maximal subgroup of S.*

Proof. Let $G = RL$. Then for every $a \in G$, $aG = aRL = RL = G$ and $Ga = RLa = RL = G$, so G is a group. Clearly $G \subseteq R \cap L$. To see the converse inclusion, let $b \in R \cap L$ and let e denote the identity of G. Since $b \in L$ and $e \in E(L)$, it follows that $b = be$, and consequently $b \in RL$. Finally, if G' is any subgroup of S with the identity e, then $G' = eG' \subseteq R$ and $G' = G'e \subseteq L$, so $G' \subseteq R \cap L = G$. $\square$

Definition 6.22. Let G be a group, let I, Λ be nonempty sets, and let $P = (p_{\lambda i})$ be a $\Lambda \times I$ matrix with entries in G. The *Rees matrix semigroup over the group G with $\Lambda \times I$ sandwich matrix P*, denoted $\mathcal{M}(G; I, \Lambda; P)$, is the set $I \times G \times \Lambda$ with the operation defined by
$$(i, a, \lambda)(j, b, \mu) = (i, a p_{\lambda j} b, \mu).$$

It is straightforward to check that $\mathcal{M}(G; I, \Lambda; P)$ is a completely simple semigroup, the minimal right ideals being the subsets $\{i\} \times G \times \Lambda$, where $i \in I$, the minimal left ideals the subsets $I \times G \times \{\lambda\}$, where $\lambda \in \Lambda$, and the maximal groups the subsets $\{i\} \times G \times \{\lambda\}$, where $i \in I$ and $\lambda \in \Lambda$. The next theorem tells us that every completely simple semigroup is isomorphic to some Rees matrix semigroup.

Theorem 6.23. *Let S be a completely simple semigroup. Pick $e \in E(S)$ such that Se is a minimal left ideal. Let $I = E(Se)$, $\Lambda = E(eS)$, $G = eSe$, and $p_{\lambda i} = \lambda i$, and define $f : \mathcal{M}(G; I, \Lambda; P) \to S$ by $f(i, a, \lambda) = i a \lambda$. Then f is an isomorphism.*

Proof. Let (i, a, λ) and (j, b, μ) be arbitrary elements of $\mathcal{M}(G; I, \Lambda; P)$. Then
$$f((i, a, \lambda)(j, b, \mu)) = f(i, a\lambda jb, \mu) = i a \lambda j b \mu = f(i, a, \lambda) f(j, b, \mu),$$
so f is a homomorphism. Since
$$E(Se)eSeE(eS) = E(Se)eSeeSeE(eS) = SeeS = SeS,$$
f is surjective. To see that f is injective, let $i a \lambda = j b \mu$. Then
$$a = eae = ei a \lambda e = ej b \mu e = ebe = b,$$
$$i = ie = i a a^{-1} = i a \lambda e a^{-1} = j a \mu e a^{-1} = j a a^{-1} = je = j,$$
$$\lambda = e\lambda = a^{-1} a \lambda = a^{-1} ei a \lambda = a^{-1} ej a \mu = a^{-1} a \mu = e\mu = \mu.   \square$$

If S is a completely simple semigroup and $\mathcal{M}(G; I, \Lambda; P)$ is a Rees matrix semigroup isomorphic to S, then every maximal subgroup of S is isomorphic to G. We call G the *structure group* of S.

The next proposition and lemma are useful in identifying the smallest ideal of subsemigroups and homomorphic images.

Proposition 6.24. *Let S be a semigroup and let T be a subsemigroup S. Suppose that both S and T have a smallest ideal which is a completely simple semigroup. If $K(S) \cap T \neq \emptyset$, then $K(T) = K(S) \cap T$.*

Proof. Clearly, $K(S) \cap T$ is an ideal of T, so $K(T) \subseteq K(S) \cap T$. For the reverse inclusion, let $x \in K(S) \cap T$. Then Sx is a minimal left ideal of S, $x \in Sx$, and Tx is a left ideal of T. Since $K(T)$ is completely simple, there is an idempotent $e \in K(T) \cap Tx$. Then $Se = Sx$, so $x \in Se$. It follows that $x = xe \in Te \subseteq K(T)$.   □

Lemma 6.25. *Let S and T be semigroups and let $f : S \to T$ be a surjective homomorphism. If S has a smallest ideal, so does T and $K(T) = f(K(S))$.*

Proof. By surjectivity, $f(K(S))$ is a two-sided ideal of T. If K' is any ideal of T, then $f^{-1}(K')$ is an ideal of S, so contains $K(S)$ by minimality. Thus $f(K(S)) \subseteq K'$, whence $f(K(S))$ is the smallest ideal of T.   □

We conclude this section with discussing standard preorderings on the idempotents of a semigroup.

Definition 6.26. Let S be a semigroup with $E(S) \neq \emptyset$. Define the relations $\leq_L, \leq_R$, and $\leq$ on $E(S)$ by

(a) $e \leq_L f$ if and only if $ef = e$,

(b) $e \leq_R f$ if and only if $fe = e$, and

(a) $e \leq f$ if and only if $e \leq_L f$ and $e \leq_R f$, that is, $ef = fe = e$.

Note that $e \leq_L f$ if and only if $Se \subseteq Sf$.

Indeed, if $ef = e$, then $Se = Sef \subseteq Sf$. Conversely, if $Se \subseteq Sf$, then $e = ee = sf$ for some $s \in S$, and consequently $ef = sff = sf = e$.

Similarly, $e \leq_R f$ if and only if $eS \subseteq fS$.

It is easy to see that the relations $\leq_L$, $\leq_R$, and $\leq$ are reflexive and transitive, and $\leq$, in addition, is antisymmetric. Thus, $\leq_L$ and $\leq_R$ are preorderings on $E(S)$ and $\leq$ is an ordering.

Given a preordering $\preceq$ on a set E, an element $e \in E$ is *minimal* (*maximal*) if for every $f \in E$, $f \preceq e$ implies $e \preceq f$ ($e \preceq f$ implies $f \preceq e$).

Lemma 6.27. *Let S be a semigroup and let $e \in E(S)$. Then the following statements are equivalent:*

(a) e is minimal with respect to $\leq_L$,

(b) e is minimal with respect to $\leq_R$, and

(c) e is minimal with respect to $\leq$.

Proof. It suffices to show the equivalence of (a) and (c).

(a) $\Rightarrow$ (c) Assume that e is minimal with respect to $\leq_L$ and let $f \leq e$. Since e is $\leq_L$-minimal, it follows from $f \leq_L e$ that $e \leq_L f$, that is, $e = ef$. But $ef = f$, since $f \leq_R e$. Hence, $e = f$.

(c) $\Rightarrow$ (a) Assume that e is minimal with respect to $\leq$ and let $f \leq_L e$. Denote $g = ef$. Then $gg = efef = eff = ef = g \in E(S)$. Also, $ge = efe = ef = g$ and $eg = eef = ef = g$, so $g \leq e$. Consequently $g = e$. Thus $ef = e$ and so $e \leq_L f$. Hence, e is $\leq_L$-minimal. $\qquad\square$

We say that an idempotent e of a semigroup S is *minimal* if e is minimal with respect to any of the preorderings $\leq_L$, $\leq_R$, or $\leq$ on $E(S)$.

Thus, to say that a semigroup S contains a minimal left ideal which has an idempotent is the same as saying that S has a minimal idempotent. So a completely simple semigroup may be defined as a simple one having a minimal idempotent. It is clear also that if a semigroup S has a minimal idempotent, then the minimal idempotents of S are precisely the idempotents of $K(S)$.

An idempotent e of a semigroup S is *right (left) maximal* if e is maximal in $E(S)$ with respect to $\leq_R$ ($\leq_L$).

Theorem 6.28. *Let S be a compact Hausdorff right topological semigroup. Then for every idempotent $e \in S$, there is a right maximal idempotent $f \in S$ with $e \leq_R f$.*

Proof. Let $P = \{x \in E(S) : e \leq_R x\}$. Clearly $P \neq \emptyset$. It suffices to show that any chain C in P has an upper bound. Then by Zorn's Lemma, C would have a maximal element.

For each $p \in C$, let $T_p = \{x \in S : xp = p\}$. If $x, y \in T_p$, then $xyp = xp = p$, so $xy \in T_p$. Also $T_p = \rho_p^{-1}(p)$. Thus, T_p is a closed subsemigroup of S. Furthermore, if $p \leq_R q$ and $x \in T_q$, then $xp = xqp = qp = p$, so $x \in T_p$ and, consequently, $T_p \supseteq T_q$. It follows that $T = \bigcap_{p \in C} T_p$ is a closed subsemigroup of S. Now by Theorem 6.12, there is an idempotent $q \in T$. Since $\bigcap_{p \in C} T_p = \{x \in S : xp = p$ for all $p \in C\}$, q is an upper bound of C. $\qquad\square$

6.3   Hindman's Theorem

In this section we prove Hindman's Theorem using the semigroup βS.

Lemma 6.29. *Let S be a semigroup and let $p \in \beta S$. Then p is an idempotent if and only if for every $A \in p$, one has $\{x \in S : x^{-1}A \in p\} \in p$.*

Proof. It is immediate from Lemma 6.6. $\qquad\square$

Proposition 6.30. *Let S be a semigroup and let p be an idempotent in S^*. Then there is a left invariant topology $\mathcal{T}$ on S^1 in which for each $s \in S^1$, $\{sA \cup \{s\} : A \in p\}$ is a neighborhood base at s. Furthermore, if S is cancellative, then $\mathcal{T}$ is Hausdorff.*

Proof. Let $\mathcal{N}$ be the filter on S^1 with a base consisting of subsets $A \cup \{1\}$ where $A \in p$. Then for every $U \in \mathcal{N}$, $1 \in U$ and and by Lemma 6.29, $\{x \in S : x^{-1}U \in \mathcal{N}\} \in \mathcal{N}$. Hence by Theorem 4.3, there is a left invariant topology $\mathcal{T}$ on S^1 in which for each $s \in S^1$, $s\mathcal{N}$ is a neighborhood base at s.

Now suppose that S is cancellative. Then

(i) ap is a nonprincipal ultrafilter for every $a \in S^1$, and

(ii) $ap \neq bp$ for all distinct $a, b \in S^1$.

It follows that $\mathcal{T}$ is Hausdorff. $\square$

Lemma 6.31. *Let S be a left topological semigroup with identity and let U be an open subset of S such that $1 \in \mathrm{cl}\, U$. Then there is a sequence $(x_n)_{n=1}^{\infty}$ in S such that $\mathrm{FP}((x_n)_{n=1}^{\infty}) \subseteq U$.*

Proof. Pick $x_1 \in U$. Fix $n \in \mathbb{N}$ and suppose we have chosen a sequence $(x_i)_{i=1}^n$ such that $\mathrm{FP}((x_i)_{i=1}^n) \subseteq U$. Let $F = \mathrm{FP}((x_i)_{i=1}^n)$. Then there is a neighborhood V of 1 such that $FV \subseteq U$. Pick $x_{n+1} \in U \cap V$. $\square$

Theorem 6.32. *Let S be a semigroup, let p be an idempotent in βS, and let $A \in p$. Then there is a sequence $(x_n)_{n=1}^{\infty}$ in S such that $\mathrm{FP}((x_n)_{n=1}^{\infty}) \subseteq A$.*

Proof. The statement is obvious if $p \in S$. Let $p \in S^*$. By Proposition 6.30, there is a left invariant topology $\mathcal{T}$ on S^1 in which for each $s \in S^1$, $\{sA \cup \{s\} : A \in p\}$ is a neighborhood base at s. Let $U = \mathrm{int}_{\mathcal{T}}\, A$. Note that $U = A \cap \{x \in S : x^{-1}A \in p\}$. We have that U is an open subset of $(S^1, \mathcal{T})$, $1 \in \mathrm{cl}_{\mathcal{T}}U$, and $U \subseteq A$. Then apply Lemma 6.31. $\square$

Corollary 6.33. *Let S be a semigroup, let $r \in \mathbb{N}$, and let $S = \bigcup_{i=1}^r A_i$. Then there exist $i \in \{1, 2, \ldots, r\}$ and a sequence $(x_n)_{n=1}^{\infty}$ in S such that $\mathrm{FS}((x_n)_{n=1}^{\infty}) \subseteq A_i$.*

Proof. Pick by Theorem 6.12 an idempotent $p \in \beta S$ and pick $i \in \{1, 2, \ldots, r\}$ such that $A_i \in p$. Then apply Theorem 6.32. $\square$

As a special case of Corollary 6.33 we obtain Hindman's Theorem.

Corollary 6.34 (Hindman's Theorem). *Let $r \in \mathbb{N}$ and let $\mathbb{N} = \bigcup_{i=1}^r A_i$. Then there exist $i \in \{1, 2, \ldots, r\}$ and a sequence $(x_n)_{n=1}^{\infty}$ in $\mathbb{N}$ such that $\mathrm{FS}((x_n)_{n=1}^{\infty}) \subseteq A_i$.*

Corollary 6.34 is strong enough to derive all other versions of Hindman's Theorem including Theorem 1.41.

Definition 6.35. For every $x \in \mathbb{N}$, define $\mathrm{supp}_2(x) \in \mathcal{P}_f(\omega)$ by

$$x = \sum_{i \in \mathrm{supp}_2(x)} 2^i$$

and let

$$\theta_2(x) = \min \mathrm{supp}_2(x) \quad \text{and} \quad \phi_2(x) = \max \mathrm{supp}_2(x).$$

In other words, $\mathrm{supp}_2(x)$ is the set of indexes of nonzero digits in the binary expansion of x, and $\theta_2(x)$ and $\phi_2(x)$ are the indexes of the first and the last nonzero digit, respectively. Equivalently,

$$\theta_2(x) = \max\{i < \omega : 2^i | x\} \quad \text{and} \quad \phi_2(x) = \max\{i < \omega : 2^i \leq x\}.$$

$\theta_2(x)$ can be defined also as $n < \omega$ such that

$$x \equiv 2^n \pmod{2^{n+1}}.$$

Note that we have already used the function $\theta_2(x)$ in the proof of Theorem 5.7.

Lemma 6.36. *Let $(x_n)_{n=1}^\infty$ be a sequence in $\mathbb{N}$. Then there is a sum subsystem $(y_n)_{n=1}^\infty$ of $(x_n)_{n=1}^\infty$ such that $\phi_2(y_n) < \theta_2(y_{n+1})$ for every $n \in \mathbb{N}$.*

Proof. It suffices to show that for every $m, k < \omega$, there is $F \in \mathcal{P}_f(\mathbb{N})$ such that $\min F > m$ and $2^k | \sum_{n \in F} x_n$.

We proceed by induction on k. If $k = 0$, put $F = \{m + 1\}$. Now assume the statement holds for some k. Then there are $F_1, F_2 \in \mathcal{P}_f(\mathbb{N})$ such that $m < \min F_1$, $\max F_1 < \min F_2$, and $2^k | \sum_{n \in F_i} x_n$ for each $i \in \{1, 2\}$. If $2^{k+1} | \sum_{n \in F_j} x_n$ for some $j \in \{1, 2\}$, put $F = F_j$. Otherwise $2^{k+1} | \sum_{n \in F_1} x_n + \sum_{n \in F_2} x_n$, so put $F = F_1 \cup F_2$. □

Corollary 6.37. *Let $r \in \mathbb{N}$ and let $\mathcal{P}_f(\mathbb{N}) = \bigcup_{i=1}^r A_i$. Then there exist $i \in \{1, 2, \ldots, r\}$ and a sequence $(G_n)_{n=1}^\infty$ in $\mathcal{P}_f(\mathbb{N})$ such that $\mathrm{FU}((G_n)_{n=1}^\infty) \subseteq A_i$ and for each $n \in \mathbb{N}$, $\max G_n < \min G_{n+1}$.*

Proof. Consider the bijection $\mathbb{N} \ni x \mapsto \mathrm{supp}_2(x) \in \mathcal{P}_f(\omega)$ and apply Corollary 6.34 and Lemma 6.36. □

Corollary 6.37 gives us, in turn, the following.

Corollary 6.38. *Let S be a semigroup, let $(x_n)_{n=1}^\infty$ be a sequence in S, let $r \in \mathbb{N}$, and let $S = \bigcup_{i=1}^r A_i$. Then there exist $i \in \{1, 2, \ldots, r\}$ and a product subsystem $(y_n)_{n=1}^\infty$ of $(x_n)_{n=1}^\infty$ such that $\mathrm{FP}((y_n)_{n=1}^\infty) \subseteq A_i$.*

The next proposition tells us that the relationship between idempotents and finite products is even more intimate than indicated by Theorem 6.32.

Proposition 6.39. *Let S be a semigroup and let $(x_n)_{n=1}^{\infty}$ be a sequence in S. Then $\bigcap_{m=1}^{\infty} \overline{\mathrm{FP}((x_n)_{n=m}^{\infty})}$ is a closed subsemigroup of βS. Consequently, there is an idempotent $p \in \beta S$ such that $\mathrm{FP}((x_n)_{n=m}^{\infty}) \in p$ for every $m \in \mathbb{N}$.*

Before proving Proposition 6.39 we establish the following simple general fact.

Lemma 6.40. *Let S be a semigroup and let $\mathcal{F}$ be a filter on S. Suppose that for every $A \in \mathcal{F}$, $\{x \in S : x^{-1}A \in \mathcal{F}\} \in \mathcal{F}$. Then $\overline{\mathcal{F}}$ is a closed subsemigroup of βS.*

Proof. Clearly, $\overline{\mathcal{F}}$ is a closed subset of βS. To see that it is a subsemigroup, let $p, q \in \overline{\mathcal{F}}$ and let $A \in \mathcal{F}$. We have to show that $A \in pq$. Let $B = \{x \in S : x^{-1}A \in \mathcal{F}\}$, and for every $x \in B$, let $C_x = x^{-1}A$. Then $B \in \mathcal{F} \subseteq p$, $C_x \in \mathcal{F} \subseteq q$, and $\bigcup_{x \in B} xC_x \subseteq A$. Hence, $A \in pq$. □

Proof of Proposition 6.39. Let $m \in \mathbb{N}$ and let $x \in \mathrm{FP}((x_n)_{n=m}^{\infty})$. Pick $F \in \mathcal{P}_f(\mathbb{N})$ with $\min F \geq m$ such that $x = \prod_{n \in F} x_n$. Let $k = \max F + 1$. Then

$$x \cdot \mathrm{FP}((x_n)_{n=k}^{\infty}) \subseteq \mathrm{FP}((x_n)_{n=m}^{\infty}).$$

Consequently, by Lemma 6.40, $\bigcap_{m=1}^{\infty} \overline{\mathrm{FP}((x_n)_{n=m}^{\infty})}$ is a closed subsemigroup of βS. For the second part, apply Theorem 6.12. □

6.4   Ultrafilters from $K(\beta S)$

In this section we characterize ultrafilters from $K(\beta S)$.

Definition 6.41. Let S be a semigroup.

(a) A subset $A \subseteq S$ is *syndetic* if there is a finite $F \subseteq S$ such that $F^{-1}A = S$.

(b) Let $\mathcal{F}$ and $\mathcal{G}$ be filters on S. A subset $A \subseteq S$ is *$(\mathcal{F}, \mathcal{G})$-syndetic* if for every $V \in \mathcal{F}$, there is a finite $F \subseteq V$ such that $F^{-1}A \in \mathcal{G}$.

If $\mathcal{F} = \mathcal{G}$, we say $\mathcal{F}$*-syndetic* instead of $(\mathcal{F}, \mathcal{G})$-syndetic. Note that if $\mathcal{F} \subseteq \mathcal{G}$ and A is either $\mathcal{F}$-syndetic or $\mathcal{G}$-syndetic, then A is also $(\mathcal{F}, \mathcal{G})$-syndetic.

Lemma 6.42. *Let T be a closed subsemigroup of βS, let L be a minimal left ideal of T, let $\mathcal{F}$ and $\mathcal{G}$ be the filters on S such that $\overline{\mathcal{F}} = T$ and $\overline{\mathcal{G}} = L$, and let $A \subseteq S$. Then the following statements are equivalent:*

(1) $\overline{A} \cap L \neq \emptyset$,

(2) *A is $\mathcal{G}$-syndetic, and*

(3) *A is $(\mathcal{F}, \mathcal{G})$-syndetic.*

Proof. (1) $\Rightarrow$ (2) Pick $p \in \overline{A} \cap L$. Then for every $q \in L$, one has $p \in Lq$. Now to show that A is $\mathcal{G}$-syndetic, let $V \in \mathcal{G}$. For every $q \in L$, there is $r \in L$ such that $p = rq$, consequently, there is $x \in V$ such that $xq \in \overline{A}$, and so $q \in x^{-1}\overline{A}$. Thus, the sets of the form $x^{-1}A$, where $x \in V$, cover the compact L. Hence, there is a finite $F \subseteq V$ such that $L \subseteq \overline{F^{-1}A}$, and so $F^{-1}A \in \mathcal{G}$.

(2) $\Rightarrow$ (3) is obvious.

(3) $\Rightarrow$ (1) Pick $q \in L$. For every $V \in \mathcal{F}$, there is a finite $F \subseteq V$ such that $F^{-1}A \in \mathcal{G}$, and so $F^{-1}A \in q$. It follows that for every $V \in \mathcal{F}$, there is $x_V \in V$ such that $A \in x_V q$. Pick $r \in T \cap c\ell_{\beta S}\{x_V : V \in \mathcal{G}\}$. Then $A \in rq$ and $rq \in L$. Hence, $\overline{A} \cap L \neq \emptyset$.   □

The next theorem characterizes ultrafilters from $K(T)$.

Theorem 6.43. *Let T be a closed subsemigroup of βS, let $\mathcal{F}$ be the filter on S such that $\overline{\mathcal{F}} = T$, and let $p \in T$. Then $p \in K(T)$ if and only if for every $A \in p$, $\{x \in S : x^{-1}A \in p\}$ is $\mathcal{F}$-syndetic.*

Proof. Suppose that $p \in K(T)$. Let $L = Tp$ and let $\mathcal{G}$ be the filter on S such that $\overline{\mathcal{G}} = L$. Now let $A \in p$, let $B = \{x \in S : x^{-1}A \in p\}$, and let $V \in \mathcal{F}$. By Lemma 6.42, A is $\mathcal{G}$-syndetic, so there is $F \subseteq V$ such that $F^{-1}A \in \mathcal{G}$. Since $L = Tp$, there is $W \in \mathcal{F}$ such that $Wp \subseteq \overline{F^{-1}A}$. We claim that $W \subseteq F^{-1}B$. Indeed, let $y \in W$. Then there is $x \in F$ such that $yp \in \overline{x^{-1}A}$. It follows that $(xy)^{-1}A \in p$. Hence, $xy \in B$, and then $y \in x^{-1}B$.

Conversely, suppose that $p \notin K(T)$. Pick $q \in K(T)$. Then $p \notin Tqp$, since $Tqp \subseteq K(T)$. It follows that there is $A \in p$ such that $\overline{A} \cap Tqp = \emptyset$. Now let $B = \{x \in S : x^{-1}A \in p\}$. We claim that B is not $\mathcal{F}$-syndetic. To show this, pick a minimal left ideal L of T contained in Tp and let $\mathcal{G}$ be the filter on S such that $\overline{\mathcal{G}} = L$. Assume on the contrary that B is $\mathcal{F}$-syndetic. Then B is also $(\mathcal{F}, \mathcal{G})$-syndetic. Hence by Lemma 6.42, $\overline{B} \cap L \neq \emptyset$. Consequently, $B \in rq$ for some $r \in T$, and so $\{x \in S : x^{-1}A \in p\} \in rq$. But then $A \in rqp$, which is a contradiction.   □

As a consequence we obtain the following characterization of ultrafilters from $K(\beta S)$.

Corollary 6.44. *Let $p \in \beta S$. Then $p \in K(\beta S)$ if and only if for every $A \in p$, $\{x \in S : x^{-1}A \in p\}$ is syndetic.*

References

The fact that the operation of a discrete semigroup S can be extended to βS was implicitly established by M. Day [15] using a multiplication of the second conjugate of a Banach algebra, in this case $l_1(S)$, first introduced by R. Arens [2] for arbitrary

Banach algebras. P. Civin and B. Yood [8] explicitly stated that if S is a discrete group, then the above operation produced an operation on βS, viewed as a subspace of that second dual. R. Ellis [21] carried out the extension in βS viewed as a space of ultrafilters, again assuming that S is a group.

Theorem 6.12 was proved by K. Numakura [52] for topological semigroups and by R. Ellis [20] in the general case. Theorem 6.23 is a special case of the Rees-Suschkewitsch Theorem proved by A. Suschkewitsch [73] for finite semigroups and by D. Rees [63] in the general case. Theorem 6.28 is due to W. Ruppert [67]. For more information about compact right topological semigroups, including references, see [67]. An introduction to this topic can be found also in [39].

Corollary 6.34 is Hindman's Theorem [35] known also as the Finite Sums Theorem. The proof based on the semigroup $\beta \mathbb{N}$ is due to F. Galvin and S. Glazer. For other combinatorial applications of βS see [37].

The exposition of Section 6.4 is based on the treatment in [70].

Chapter 7
Ultrafilter Semigroups

Given a T_1 left topological group $(G, \mathcal{T})$, the set $\mathrm{Ult}(\mathcal{T})$ of all nonprincipal ultrafilters on G converging to 1 in $\mathcal{T}$ is a closed subsemigroup of βG called the ultrafilter semigroup of $\mathcal{T}$. In this chapter we study the relationship between algebraic properties of $\mathrm{Ult}(\mathcal{T})$ and topological properties of $(G, \mathcal{T})$. Not every closed subsemigroup of G^* is the ultrafilter semigroup of a left invariant topology on G. However, every finite sub-semigroup is. A special attention is paid to the question when a closed subsemigroup of G^* is the ultrafilter semigroup of a regular left invariant topology. We conclude by showing how to construct homomorphisms of $\mathrm{Ult}(\mathcal{T})$.

7.1 The Semigroup $\mathrm{Ult}(\mathcal{T})$

Lemma 7.1. *Let $(S, \mathcal{T})$ be a left topological semigroup with identity and let $\mathcal{N}$ be the neighborhood filter of 1. Then*

(1) $\overline{\mathcal{N}}$ *is a closed subsemigroup of βS,*

(2) *for every open subset U of $(S, \mathcal{T})$, $\overline{U} \cdot \overline{\mathcal{N}} \subseteq \overline{U}$, and*

(3) *if $\mathcal{T}$ satisfies the T_1 separation axiom, then*

$$\overline{\mathcal{N}} \setminus \{1\} = \{p \in S^* : p \text{ converges to } 1 \text{ in } \mathcal{T}\}$$

is a closed subsemigroup.

Proof. (1) follows from Theorem 4.3 and Lemma 6.40.

(2) Let $p \in \overline{U}$ and $q \in \overline{\mathcal{N}}$. Since U is open, for every $x \in U$, there is $V_x \in \mathcal{N}$ such that $x V_x \subseteq U$. Then $\bigcup_{x \in U} x V_x \subseteq U$. Since $U \in p$ and $V_x \in q$, it follows that $U \in pq$, so $pq \in \overline{U}$.

(3) Obviously, $\overline{\mathcal{N}} \setminus \{1\}$ is a closed subset of βS. Since $\mathcal{T}$ is a T_1-topology, $\bigcap \mathcal{N} = \{1\}$. It follows that $\overline{\mathcal{N}} \setminus \{1\} \subseteq S^*$, and so $\overline{\mathcal{N}} \setminus \{1\} = \{p \in S^* : p \text{ converges to } 1 \text{ in } \mathcal{T}\}$.

To see that $\overline{\mathcal{N}} \setminus \{1\}$ is a subsemigroup, let $p, q \in \overline{\mathcal{N}} \setminus \{1\}$. We have to show that for every $U \in \mathcal{N}$, $pq \in \overline{U} \setminus \{1\}$. Clearly, one may suppose that U is open. Then $U \setminus \{1\}$ is also open, because $\mathcal{T}$ is a T_1-topology. Since $p \in \overline{U} \setminus \{1\}$, we obtain by (2) that $pq \in \overline{U \setminus \{1\}} = \overline{U} \setminus \{1\}$. $\square$

Definition 7.2. Let $(S, \mathcal{T})$ be a T_1 left topological semigroup with identity. Define $\mathrm{Ult}(\mathcal{T}) \subseteq \beta S$ by

$$\mathrm{Ult}(\mathcal{T}) = \{p \in S^* : p \text{ converges to } 1 \text{ in } \mathcal{T}\}.$$

By Lemma 7.1, $\mathrm{Ult}(\mathcal{T})$ is a closed subsemigroup of βS (if nonempty). We refer to $\mathrm{Ult}(\mathcal{T})$ as to the *ultrafilter semigroup* of $\mathcal{T}$ (or $(S, \mathcal{T})$).

Recall that a filter on a space is called *open* (*closed*) if it has a base consisting of open (closed) sets.

Lemma 7.3. *Let $(G, \mathcal{T})$ be a T_1 left topological group, let $\mathcal{F}$ be a filter on G, and let $Q = \mathrm{Ult}(\mathcal{T})$.*

(1) *If $\mathcal{F}$ is open, then $\overline{\mathcal{F}} Q \subseteq \overline{\mathcal{F}}$.*

(2) *If $\mathcal{F}$ is closed, then $pQ \cap \overline{\mathcal{F}} = \emptyset$ for every $p \in \beta G \setminus \overline{\mathcal{F}}$.*

In the case where Q is finite, the converses of the statements (1)–(2) also hold.

Proof. (1) For every open $U \in \mathcal{F}$, $\overline{\mathcal{F}} Q \subseteq \overline{U} Q$, and by Lemma 7.1, $\overline{U} Q \subseteq \overline{U}$, so $\overline{\mathcal{F}} Q \subseteq \overline{U}$. It follows that $\overline{\mathcal{F}} Q \subseteq \overline{\mathcal{F}}$.

Conversely, suppose that $\overline{\mathcal{F}} Q \subseteq \overline{\mathcal{F}}$ and Q is finite. To show that $\mathcal{F}$ is open, let $U \in \mathcal{F}$. For every $p \in \overline{\mathcal{F}}$ and $q \in Q$, pick $A_{p,q} \in p$ such that $A_{p,q} \subseteq U$ and $\overline{A_{p,q}q} \subseteq \overline{U}$. Let

$$V = \bigcup_{p \in \overline{\mathcal{F}}} \bigcap_{q \in Q} A_{p,q}.$$

Then $V \in \mathcal{F}$, $V \subseteq U$, and $\overline{V} Q \subseteq \overline{U}$. It follows that $V \subseteq \mathrm{int}\, U$, and so $\mathrm{int}\, U \in \mathcal{F}$. To see this, for every $x \in V$ and $q \in Q$, pick $B_{x,q} \in q$ such that $xB_{x,q} \subseteq U$, and let

$$W_x = \{x\} \cup \bigcup_{q \in Q} B_{x,q}.$$

Then W_x is a neighborhood of 1 and $xW_x \subseteq U$.

(2) Since $\mathcal{F}$ is closed, for every $p \in \beta S \setminus \overline{\mathcal{F}}$, there is an open $U \in p$ such that $\overline{U} \cap Q = \emptyset$. By Lemma 7.1, $\overline{U} Q \subseteq \overline{U}$. It follows that $pQ \cap Q = \emptyset$.

Conversely, suppose that $pQ \cap \overline{\mathcal{F}} = \emptyset$ for every $p \in \beta G \setminus \overline{\mathcal{F}}$ and that Q is finite. To show that $\mathcal{F}$ is closed, assume the contrary. Then there is $U \in \mathcal{F}$ such that for every $V \in \mathcal{F}$, $(\mathrm{cl}\, V) \setminus U \neq \emptyset$. Since Q is finite, it follows that there is $q \in Q$ and, for every $V \in \mathcal{F}$, $x_V \in G \setminus U$ such that $V \in x_V q$. Let p be an ultrafilter on G extending the family of subsets

$$X_V = \{x_W : V \supseteq W \in \mathcal{F}\}$$

where $V \in \mathcal{F}$. Then $p \in \overline{G \setminus U}$ and $V \in pq$ for every $V \in \mathcal{F}$. Consequently, $p \in \beta G \setminus \overline{\mathcal{F}}$ and $pq \in \overline{\mathcal{F}}$, a contradiction. $\square$

Definition 7.4. For every filter $\mathcal{F}$ on a space X, denote by $\mathrm{int}\, \mathcal{F}$ (respectively $\mathrm{cl}\, \mathcal{F}$) the largest open (closed) filter on X, possibly improper, containing (contained in) $\mathcal{F}$.

Corollary 7.5. *Let $(G, \mathcal{T})$ be a T_1 left topological group, let $\mathcal{F}$ be a filter on $(G, \mathcal{T})$, and let $Q = \mathrm{Ult}(\mathcal{T})$. Suppose that Q is finite. Then*

(1) $\overline{\text{int } \mathcal{F}} = \{p \in \overline{\mathcal{F}} : pQ \subseteq \overline{\mathcal{F}}\}$, and

(2) $\overline{\text{cl } \mathcal{F}} = \overline{\mathcal{F}} \cup \{p \in \beta G : pQ \cap \overline{\mathcal{F}} \neq \emptyset\}$.

Proof. (1) By Lemma 7.3, $\overline{\text{int } \mathcal{F}} \cdot Q \subseteq \overline{\text{int } \mathcal{F}} \subseteq \overline{\mathcal{F}}$, so $\overline{\text{int } \mathcal{F}} \subseteq \{p \in \overline{\mathcal{F}} : pQ \subseteq \overline{\mathcal{F}}\}$.
To see the converse inclusion, note that $\{p \in \overline{\mathcal{F}} : pQ \subseteq \overline{\mathcal{F}}\}$ is closed. Let $\mathcal{G}$ denote
the filter on G such that $\overline{\mathcal{G}} = \{p \in \overline{\mathcal{F}} : pQ \subseteq \overline{\mathcal{F}}\}$. For every $p \in \overline{\mathcal{G}}$ and $q \in Q$,
one has $pqQ \subseteq pQ \subset \overline{\mathcal{F}}$, so $\overline{\mathcal{G}}Q \subset \overline{\mathcal{G}}$. Now, since Q is finite, applying Lemma 7.3
gives us that $\mathcal{G}$ is open. Hence $\overline{\mathcal{G}} \subseteq \overline{\text{int } \mathcal{F}}$.

(2) If $pQ \cap \overline{\mathcal{F}} \neq \emptyset$ for some $p \in \beta G$, then $pQ \cap \overline{\text{cl } \mathcal{F}} \neq \emptyset$, and so $p \in \overline{\text{cl } \mathcal{F}}$
by Lemma 7.3. Consequently, $\overline{\mathcal{F}} \cup \{p \in \beta G : pQ \cap \overline{\mathcal{F}} \neq \emptyset\} \subseteq \overline{\text{cl } \mathcal{F}}$. To see the
converse inclusion, note that $\{p \in \beta G : pQ \cap \overline{\mathcal{F}} \neq \emptyset\}$ is closed, since Q is finite.
Let $\mathcal{G}$ denote the filter on G such that $\overline{\mathcal{G}} = \overline{\mathcal{F}} \cup \{p \in \beta G : pQ \cap \overline{\mathcal{F}} \neq \emptyset\}$. We
claim for every $p \in \beta G \setminus \overline{\mathcal{G}}$, $pQ \cap \overline{\mathcal{G}} \neq \emptyset$. Indeed, otherwise there exists $q \in Q$
such that $pqQ \cap \overline{\mathcal{F}} \neq \emptyset$. It then follows that $pQ \cap \overline{\mathcal{F}} \neq \emptyset$, and so $p \in \overline{\mathcal{G}}$, which is a
contradiction. Now, since Q is finite, applying Lemma 7.3 gives us that $\mathcal{G}$ is closed.
Hence $\overline{\text{cl } \mathcal{F}} \subseteq \overline{\mathcal{G}}$.   □

Recall that an ultrafilter p on a space is *dense* if for every $A \in p$, $\text{int cl } A \neq \emptyset$.

Lemma 7.6. *Let $(G, \mathcal{T})$ be a T_1 left topological group. Then the set of all dense
ultrafilters on $(G, \mathcal{T})$ converging to 1 is a closed ideal of* Ult($\mathcal{T}$).

Proof. Let $\mathcal{F}$ be the filter on $(G, \mathcal{T})$ with a base consisting of subsets of the form
$U \setminus Y$ where U is a neighborhood of 1 and Y is a nowhere dense subset of $(G, \mathcal{T})$.
Then by Proposition 3.9, $\overline{\mathcal{F}}$ is the set of all dense ultrafilters on $(G, \mathcal{T})$ converging to
1. Since $\mathcal{F}$ is open, $\overline{\mathcal{F}}$ is a right ideal of Ult($\mathcal{T}$), by Lemma 7.3. To see that $\overline{\mathcal{F}}$ is a
left ideal, let $p \in \text{Ult}(\mathcal{T})$, $q \in \overline{\mathcal{F}}$, and $A \in pq$. Then there is $x \in G$ and $B \in q$ such
that $xB \subseteq A$. Since q is dense, $\text{int cl } B \neq \emptyset$. But $x(\text{int cl } B) \subseteq \text{int}(x(\text{cl } B))$ and
$\text{int}(x(\text{cl } B)) \subseteq \text{int cl}(xB)$. It follows that $\text{int cl } A \neq \emptyset$. Hence $pq \in \overline{\mathcal{F}}$.   □

The next proposition shows how the ultrafilter semigroup of a left topological group
reflects topological properties of the group itself.

Proposition 7.7. *Let $(G, \mathcal{T})$ be a T_1 left topological group and let $Q = \text{Ult}(\mathcal{T})$.*

(1) *If Q has only one minimal right ideal, then $\mathcal{T}$ is extremally disconnected.*

(2) *If $\mathcal{T}$ is irresolvable, then $K(Q)$ is a left zero semigroup.*

(3) *If $\mathcal{T}$ is n-irresolvable, then a minimal right ideal of Q consists of $< n$ elements.*

(4) *If $p \in K(Q)$, then p is dense.*

(5) *If $Q = K(Q)$, then $\mathcal{T}$ is nodec.*

In the case where Q is finite, the converses of the statements (1)–(5) also hold.

Proof. (1) Suppose that $\mathcal{T}$ is not extremally disconnected. Then there are two disjoint open subsets U and V of $(G, \mathcal{T})$ such that $1 \in (\text{cl } U) \cap (\text{cl } V)$. But then by Lemma 7.1, $\overline{U} \cap Q$ and $\overline{V} \cap Q$ are two disjoint right ideals of Q.

Conversely, suppose that there are two disjoint right ideals R and J of Q and that Q is finite. Let $\mathcal{F}$ and $\mathcal{G}$ denote filters on G such that $\overline{\mathcal{F}} = R$ and $\overline{\mathcal{G}} = J$. Then, since Q is finite, both $\mathcal{F}$ and $\mathcal{G}$ are open by Lemma 7.3, so there are disjoint open $U \in \mathcal{F}$ and $V \in \mathcal{G}$. But then $1 \in (\text{cl } U) \cap (\text{cl } V)$. Hence $\mathcal{T}$ is not extremally disconnected.

(2) is a special case of (3).

(3) Suppose that $\mathcal{T}$ is n-irresolvable. Then by Theorem 3.31, there is an open filter $\mathcal{F}$ on $(G, \mathcal{T})$ converging to 1 with $|\overline{\mathcal{F}}| < n$. Then by Corollary 7.3, $\overline{\mathcal{F}} \setminus \{1\}$ is a right ideal of Q. Hence, a minimal right ideal of Q consists of $\leq |\overline{\mathcal{F}}|$ elements.

Conversely, let R be a minimal right ideal of Q and suppose that $|R| < n$ and Q is finite. Let $\mathcal{F}$ denote the filter on G such that $\overline{\mathcal{F}} = R$. Then by Lemma 7.3, $\mathcal{F}$ is open. Hence by Theorem 3.31, $\mathcal{T}$ is n-irresolvable.

(4) By Lemma 7.6, the set of all dense ultrafilters from Q is an ideal of Q. It follows that every ultrafilter from $K(Q)$ is dense.

Conversely, suppose that $p \notin K(Q)$ and Q is finite. Then by Corollary 7.5, $\overline{\text{cl } p} = \{p\} \cup \{q \in \beta G : p \in qQ\}$ and $\overline{\text{int cl } p} = \{q \in \overline{\text{cl } p} : qQ \subseteq \overline{\text{cl } p}\}$. Since $p \notin K(Q)$, $p \notin qQ$ for every $q \in K(Q)$. Consequently, $\overline{\text{cl } p} \cap K(Q) = \emptyset$. On the other hand, for every $q \in \overline{\text{cl } p}$, one has $qQ \cap K(Q) \neq \emptyset$. Indeed, this is certainly true if $q \in Q$. Otherwise $p \in qQ$. Then $pQ \subseteq qQ$ and therefore $qQ \cap K(Q) \neq \emptyset$. It follows that for every $q \in \overline{\text{cl } p}$, one has $qQ \not\subseteq \overline{\text{cl } p}$. Hence, $\text{int cl } p = \emptyset$ and so p is nowhere dense.

(5) If $Q = K(Q)$, then by (4) every ultrafilter from Q is dense. It follows that $\mathcal{T}$ is nodec. If $Q \neq K(Q)$ and Q is finite, then by (4) any ultrafilter from $Q \setminus K(Q)$ is nowhere dense, and so $\mathcal{T}$ is not nodec. $\square$

We now show that every finite semigroup in G^* is the ultrafilter semigroup of a left invariant topology.

Proposition 7.8. *Let S be a semigroup with identity, let Q be a finite semigroup in S^*, and let $\mathcal{F}$ be the filter on S such that $\overline{\mathcal{F}} = Q$. Then there is a left invariant topology $\mathcal{T}$ on S in which for each $s \in S$, $\{sA \cup \{s\} : A \in \mathcal{F}\}$ is a neighborhood base at s. If S is left cancellative, then $\mathcal{T}$ is a T_1-topology and $\text{Ult}(\mathcal{T}) = Q$.*

Proof. Let $\mathcal{N}$ be the filter on S such that $\overline{\mathcal{N}} = Q \cup \{1\}$. We claim that

(i) $\bigcap \overline{\mathcal{N}} = \{1\}$, and

(ii) for every $U \in \mathcal{N}$, $\{x \in S : x^{-1}U \in \mathcal{N}\} \in \mathcal{N}$.

Statement (i) is obvious. To show (ii), let $U \in \mathcal{N}$. For every $p, q \in Q$, pick $A_{p,q} \in p$ such that $A_{p,q} \subseteq U$ and $\overline{A_{p,q}} \cdot q \subseteq \overline{U}$. Put $V = \bigcup_{p \in Q} \bigcap_{q \in F} A_{p,q} \cup \{1\}$. Then $V \in \mathcal{N}$, $V \subseteq U$ and $VQ \subseteq \overline{U}$. Now let $x \in V$. For every $q \in Q$, pick $B_{x,q} \in q$ such that $xB_{x,q} \subseteq U$. Put $W_x = \bigcup_{q \in Q} B_{x,q} \cup \{1\}$. Then $W_x \in \mathcal{N}$ and $xW_x \subseteq U$.

It follows from (i)–(ii) and Theorem 4.3 that there is a left invariant topology $\mathcal{T}$ on S in which for each $a \in S$, $a\mathcal{N}$ is a neighborhood base at a.   □

The next example shows that not every closed subsemigroup of G^* is the ultrafilter semigroup of a left invariant topology.

Example 7.9. Let $(G, \mathcal{T}_0)$ be a regular left topological group and suppose that there is a one-to-one sequence $(x_n)_{n < \omega}$ in $G \setminus \{1\}$ converging to 1. Choose a family $\{I_n . n < \omega\}$ of pairwise disjoint infinite subsets of ω and a family $\{U_n : n < \omega\}$ of pairwise disjoint open subsets of $G \setminus \{1\}$ such that $x_n \in U_n \subseteq G \setminus \{1\}$. Let $X = \{x_n : n < \omega\}$ and $Y = \{x_n x_m : m \in I_n\} \cap U_n$, let $\mathcal{F}$ be the filter on G with a base consisting of subsets of the form $U \setminus Y$ where U runs over the neighborhoods of 1, and let $Q = \overline{\mathcal{F}} \setminus \{1\}$. We claim that Q is a closed subsemigroup of G^* and there is no left invariant topology $\mathcal{T}$ on G such that $\mathrm{Ult}(\mathcal{T}) = Q$.

To see that Q is a subsemigroup, let $p, q \in Q$ and let U be an open neighborhood of 1. It suffices to show that $U \setminus Y \in pq$. Consider two cases.

Case 1: $X \notin p$. Then $V = U \setminus (X \cup Y)$ is open and $V \in p$. Consequently by Lemma 7.1, $V \in pq$, and so $U \setminus Y \in pq$.

Case 2: $X \in p$. Assume on the contrary that $U \setminus Y \notin pq$. Then $Y \in pq$. Pick distinct $k, m < \omega$ such that $Y \in x_k q$ and $Y \in x_m q$. It then follows that $\{x_n : n \in I_k\} \in q$ and $\{x_n : n \in I_m\} \in q$, which is a contradiction.

To see that there is no left invariant topology $\mathcal{T}$ on G such that $\mathrm{Ult}(\mathcal{T}) = Q$, assume on the contrary that there is such $\mathcal{T}$, so $\mathcal{F}$ is the neighborhood filter of 1 in $\mathcal{T}$. Let $U = G \setminus Y$. Clearly $U \in \mathcal{F}$. Now let $V \in \mathcal{N}$. Then $V \cap X \neq \emptyset$. Pick $n < \omega$ such that $x_n \in V$. But then for every $W \in \mathcal{N}$, one has $(x_n W) \cap Y \neq \emptyset$. Hence $x_n W \not\subseteq U$, which contradicts Theorem 4.3.

We conclude this section by showing that

Lemma 7.10. *For every totally bounded topological group* $(G, \mathcal{T})$, $\mathrm{Ult}(\mathcal{T})$ *contains all the idempotents of* G^*.

Proof. Let $p \in G^*$ be an idempotent. Consider $(G, \mathcal{T})$ as a subgroup of a compact group K and let q be the ultrafilter on K extending p, that is, $p \subseteq q$. Note that q is an idempotent of βK_d, where K_d is the group K reendowed with the discrete topology. Since K is compact, q converges to some element $a \in K$. Then q converges to a^2 as well, because $q = q^2$. It follows that $a = 1$. Hence, p also converges to $1 \in G$.   □

7.2 Regularity

Definition 7.11. Let G be a group and let Q be a closed subsemigroup of βG. We say that Q is *left saturated* if $Q \subseteq G^*$ and for every $p \in \beta G \setminus (Q \cup \{1\})$, one has $pQ \cap Q = \emptyset$.

Lemma 7.12. *Let $(G, \mathcal{T})$ be a regular left topological group and let $Q = \mathrm{Ult}(\mathcal{T})$. Then Q is left saturated.*

Proof. Let $\mathcal{N}$ be the neighborhood filter of 1 in $\mathcal{T}$. Since $\mathcal{T}$ is regular, $\mathcal{N}$ is closed. Then by Lemma 7.3, for every $p \in \beta G \setminus \mathcal{N}$, $pQ \cap \overline{\mathcal{N}} = \emptyset$. It follows that for every $p \in \beta G \setminus (Q \cup \{1\})$, $pQ \cap Q = \emptyset$. Hence, Q is left saturated.   □

Theorem 7.13. *Let G be a group and let Q be a closed subsemigroup of G^*. Suppose that Q is left saturated and has a finite left ideal. Then there is a regular left invariant topology $\mathcal{T}$ on G such that $\mathrm{Ult}(\mathcal{T}) = Q$.*

Proof. Let $\mathcal{N}$ be the filter on G such that $\overline{\mathcal{N}} = Q \cup \{1\}$. We first show that

(i) $\bigcap \mathcal{N} = \{1\}$, and

(ii) for every $U \in \mathcal{N}$, $\{x \in G : x^{-1}U \in \mathcal{N}\} \in \mathcal{N}$.

Statement (i) is obvious. To see (ii), let L be a finite left ideal of Q and let $C = G \setminus U$. Since Q is left saturated, $\overline{C} \cdot Q \cap Q = \emptyset$. For every $q \in L$, choose $W_q \in \mathcal{N}$ such that $\overline{C} \cdot q \cap \overline{W_q} = \emptyset$, and put $W = \bigcap_{q \in L} W_q$. Then $W \in \mathcal{N}$, as L is finite, and $(\overline{C} \cdot L) \cap \overline{W} = \emptyset$. Next, since L is a left ideal of Q, $Q^1 \cdot L \subseteq L$. For every $q \in L$, choose $V_q \in \mathcal{N}$ such that $\overline{V_q} \cdot q \subseteq \overline{W}$, and put $V = \bigcap_{q \in L} V_q$. Then $V \in \mathcal{N}$ and $\overline{V} \cdot L \subseteq \overline{W}$. We claim that for all $x \in V$, $x^{-1}U \in \mathcal{N}$. Indeed, otherwise $xp \in \overline{C}$ for some $x \in V$ and $p \in Q$. Take any $q \in L$. Then, on the one hand,

$$xpq = xp \cdot q \in \overline{C} \cdot L \subseteq \beta G \setminus \overline{W},$$

and on the other hand,

$$xpq = x \cdot pq \in \overline{V} \cdot L \subseteq \overline{W},$$

which is a contradiction.

It follows from (i)–(ii) and Theorem 4.3 that there is a left invariant T_1-topology $\mathcal{T}$ on G for which $\mathcal{N}$ is the neighborhood filter of 1, and so $\mathrm{Ult}(\mathcal{T}) = Q$.

We now show that $\mathcal{T}$ is regular. Assume the contrary. Then there is a neighborhood U of 1 such that for every neighborhood V of 1, $(\mathrm{cl}\, V) \setminus U \neq \emptyset$. For every open neighborhood V of 1, choose $x_V \in (\mathrm{cl}\, V) \setminus U$ and $p_V \in Q$ such that $V \in x_V p_V$. Pick any $q \in L$. By Lemma 7.1, one has $V \in x_V p_V q$, and $p_V q \in L$ because L is a left ideal. Since L is finite, it follows that there exist $q \in L$ and $p \in G \setminus U$ such that for every neighborhood V of 1, $V \in pq$, and so $pq \in Q$. We have obtained that Q is not left saturated, which is a contradiction.   □

Definition 7.14. Given a group G and $p \in G^*$,

$$C(p) = \{x \in G^* : xp = p\} \quad \text{and} \quad C^1(p) = C(p) \cup \{1\}.$$

Note that $C(p)$ is a closed subset of G^* and $p \in C(p)$ if and only if p is an idempotent.

Lemma 7.15. $C(p)$ *is a closed left saturated subsemigroup of* βG *(if nonempty).*

Proof. To see that $C(p)$ is a subsemigroup, let $x, y \in C(p)$. Then $xy \in G^*$ by Lemma 6.8 and $xyp = xp = p$, so $xy \in C(p)$.

To see that $C(p)$ is left saturated, let $xq = r$ for some $x \in \beta G$ and $q, r \in C(p)$. Then $xqp = rp$, and so $xp = p$. If $x \in G$, this equality implies that $x = 1$ by Corollary 6.11. Hence $x \in C(p) \cup \{1\}$.   □

From Lemma 7.15 we obtain that

Corollary 7.16. $C(p) = \{x \in \beta G \setminus \{1\} : xp = p\}$ *and* $C^1(p) = \{x \in \beta G : xp = p\}$.

Now we deduce from Theorem 7.13 and Lemma 7.15 the following result.

Theorem 7.17. *Let G be a group and let $p \in G^*$ be an idempotent. Then*

(1) *there is a regular left invariant topology $\mathcal{T}$ on G such that* $\mathrm{Ult}(\mathcal{T}) = C(p)$,

(2) $\mathcal{T}$ *is the largest regular left invariant topology on G in which p converges to 1, and*

(3) $\mathcal{T}$ *is extremally disconnected.*

Proof. By Lemma 7.15, $C(p)$ is a closed left saturated subsemigroup of βG. Clearly, $\{p\}$ is a left ideal of $C(p)$. Then applying by Theorem 7.13 gives us (1).

To see (2), let $\mathcal{T}'$ be any regular left invariant topology on G in which p converges to 1, let $Q = \mathrm{Ult}(\mathcal{T}')$, and let $x \in C(p)$. We have that $xp = p$, $p \in Q$ and $x \in G^*$. Hence by Lemma 7.12, $x \in Q$.

To see (3), notice that, since $\{p\}$ is a minimal left ideal of $C(p)$, there is only one minimal right ideal of $C(p)$. Hence by Proposition 7.7, $\mathcal{T}$ is extremally disconnected.   □

Since a regular extremally disconnected space is zero-dimensional, we obtain from Theorem 7.17 that for every idempotent $p \in G^*$, there is a zero-dimensional Hausdorff left invariant topology $\mathcal{T}$ on G such that $\mathrm{Ult}(\mathcal{T}) = C(p)$. The next theorem tells us that this is true for every $p \in G^*$.

Theorem 7.18. *For every group G and for every $p \in G^*$, there is a zero-dimensional Hausdorff left invariant topology $\mathcal{T}$ on G such that* $\mathrm{Ult}(\mathcal{T}) = C(p)$.

The proof of Theorem 7.18 involves the following notion.

Definition 7.19. Let S be a semigroup, $A \subseteq S$, and $p \in \beta S$. Then

$$A'_p = \{x \in S : A \in xp\}.$$

The next lemma contains some simple properties of this notion.

Lemma 7.20. *Let S be a semigroup, $A, B \subseteq S$, and $p, q \in \beta S$. Then*

(1) $(A \cap B)'_p = A'_p \cap B'_p$,

(2) $A \in pq$ *if and only if* $A'_q \in p$, *and*

(3) $(A'_p)'_q = A'_{qp}$.

Proof. (1) Let $x \in G$. Then

$$x \in (A \cap B)'_p \Leftrightarrow A \cap B \in xp$$

$$\Leftrightarrow A \in xp \text{ and } B \in xp$$

$$\Leftrightarrow x \in A'_p \text{ and } x \in B'_p$$

$$\Leftrightarrow x \in A'_p \cap B'_p.$$

(2) is Lemma 6.6 (ii).

(3) Let $x \in G$. Then

$$x \in (A'_p)'_q \Leftrightarrow A'_p \in xq$$

$$\Leftrightarrow A \in xqp \quad \text{by (2)}$$

$$\Leftrightarrow x \in A'_{qp}. \qquad \square$$

Proof of Theorem 7.18. Let $\mathcal{B} = \{A'_p : A \in p\}$. Since $1 \in A'_p$ for every $A \in p$ and $A'_p \cap B'_p = (A \cap B)'_p$, $\mathcal{B}$ is a filter base on G. Now we show that $\mathcal{B}$ possesses the following properties:

(i) $\overline{\mathcal{B}} = C^1(p)$,

(ii) for every $U \in \mathcal{B}$ and for every $x \in U$, there is $V \in \mathcal{B}$ such that $xV \subseteq U$, and

(iii) for every $U \in \mathcal{B}$ and for every $x \in G \backslash U$, there is $V \in \mathcal{B}$ such that $xV \cap U = \emptyset$.

To see (i), let $q \in \beta G$. Then

$$q \in \overline{\mathcal{B}} \Leftrightarrow A'_p \in q \quad \text{for every } A \in p$$

$$\Leftrightarrow A \in qp \quad \text{for every } A \in p$$

$$\Leftrightarrow qp = p.$$

To see (ii) and (iii), it suffices to show that for every $U \in \mathcal{B}$ and for every $q \in \overline{\mathcal{B}}$, $U'_q = U$. To this end, pick $A \in p$ such that $U = A'_p$. Then

$$U'_q = (A'_p)'_q = A'_{qp} = A'_p = U.$$

It follows from (i) and (ii) that there is a left invariant T_1-topology $\mathcal{T}$ on G such that $\mathrm{Ult}(\mathcal{T}) = C(p)$, and by (iii), $\mathcal{T}$ is zero-dimensional. $\qquad \square$

We conclude this section by characterizing subsemigroups of a finite left saturated subsemigroup of βG which determine locally regular left invariant topologies.

A space is *locally regular* if every point has a neighborhood which is a regular subspace.

Proposition 7.21. *Let G be a group, let Q be a finite left saturated subsemigroup of G^*, let S be a subsemigroup of Q, and let $\mathcal{T}$ be a left invariant topology on G with $\mathrm{Ult}(\mathcal{T}) = S$. Then $\mathcal{T}$ is locally regular if and only if for every $q \in Q \setminus S$, $qS \cap S \neq \emptyset$ implies $Sq \cap S = \emptyset$.*

Proof. Suppose that $\mathcal{T}$ is locally regular and let $q \in Q \setminus S$ be such that $qS \cap S \neq \emptyset$. We have to show that $Sq \cap S = \emptyset$. Choose a regular open neighborhood X of $1 \in G$ in $\mathcal{T}$. It suffices to show that $Xq \cap \overline{X} = \emptyset$, as this implies $\overline{X}q \cap \overline{X} = \emptyset$ and then $Sq \cap S = \emptyset$.

Assume on the contrary that $xq \in \overline{X}$ for some $x \in X$. Since $q \notin S \cup \{1\}$, xq does not converge to x, so there is a neighborhood U of $x \in X$ such that $U \notin xq$. Since X is regular, U can be chosen to be closed. We have that $X \setminus U \in xq$ and $X \setminus U$ is open. Then by Lemma 7.1, $xqS \subseteq \overline{X \setminus U}$. It follows that $xqS \cap xS = \emptyset$, and consequently, $qS \cap S = \emptyset$, a contradiction.

Conversely, let $F = \{q \in Q \setminus S : qS \cap S \neq \emptyset\}$ and suppose that for each $q \in F$, $Sq \cap S = \emptyset$. Then for each $q \in F$, there is a neighborhood X_q of 1 in $\mathcal{T}$ such that $\overline{X_q}q \cap \overline{X_q} = \emptyset$. Put $X = \bigcap_{q \in F} X_q$. Since F is finite, X is a neighborhood of 1 in $\mathcal{T}$, and for each $q \in F$, one has $Xq \cap \overline{X} = \emptyset$. We claim that X is regular.

Assume the contrary. Then there is $x \in X$ and a neighborhood U of x such that for every neighborhood V of x, $\mathrm{cl}_X(V) \setminus U \neq \emptyset$. Since S is finite, it follows that there is $p \in S$, and for every neighborhood V of x, there is $y_V \in X \setminus U$ such that $V \in y_V p$. Let r be an ultrafilter on G extending the family of subsets

$$Y_V = \{y_W : W \text{ is a neighborhood of } x \text{ contained in } V\},$$

where V runs over neighborhoods of x. Then $r \in \overline{X \setminus U}$ and $rp \in xS$, so $x^{-1}rp \in S$. Put $q = x^{-1}r$. We have that (a) $qp \in S$ and $q \neq 1$, and (b) $r = xq$. Since Q is left saturated, it follows from (a) that $q \in F$. It is clear that $xq \in Xq$, and (b) gives us that $xq \in \overline{X}$. Hence $Xq \cap \overline{X} \neq \emptyset$, a contradiction. □

7.3 Homomorphisms

Constructing homomorphisms of ultrafilter semigroups is based on the following lemma.

Lemma 7.22. *Let S be a semigroup, let $\mathcal{F}$ be a filter on S, and let $X \in \mathcal{F}$. Let T be a compact Hausdorff right topological semigroup and let $f : X \to T$. Assume that*

(1) *for every $x \in X$, there is $U_x \in \mathcal{F}$ such that $f(xy) = f(x)f(y)$ for all $y \in U_x$, and*

(2) $f(X) \subseteq \Lambda(T)$.

Then for every $p \in \overline{X}$ and $q \in \overline{\mathcal{F}}$, one has $\overline{f}(pq) = \overline{f}(p)\overline{f}(q)$, where $\overline{f} : \overline{X} \to T$ is the continuous extension of f.

Proof. For every $x \in X$, one has

$$\overline{f}(xq) = \overline{f}(\lim_{U_x \ni y \to q} xy)$$

$$= \lim_{y \to q} f(xy)$$

$$= \lim_{y \to q} f(x)f(y) \quad \text{by (1)}$$

$$= f(x) \lim_{y \to q} f(y) \quad \text{by (2)}$$

$$= f(x)\overline{f}(q).$$

Then

$$\overline{f}(pq) = \overline{f}(\lim_{X \ni x \to p} xq)$$

$$= \lim_{x \to p} \overline{f}(xq)$$

$$= \lim_{x \to p} f(x)\overline{f}(q)$$

$$= (\lim_{x \to p} f(x))\overline{f}(q)$$

$$= \overline{f}(p)\overline{f}(q). \qquad \square$$

Definition 7.23. Let κ be an infinite cardinal and let $H = \bigoplus_\kappa \mathbb{Z}_2$. For every $\alpha < \kappa$, let $H_\alpha = \{x \in H : x(\gamma) = 0 \text{ for each } \gamma < \alpha\}$, and let $\mathcal{T}_0$ denote the group topology on H with a neighborhood base at 0 consisting of subgroups H_α, where $\alpha < \kappa$. Define the semigroup $\mathbb{H}_\kappa \subseteq \beta H$ by $\mathbb{H}_\kappa = \text{Ult}(\mathcal{T}_0)$. If $\kappa = \omega$, we write $\mathbb{H}$ instead of $\mathbb{H}_\omega$.

The next theorem tells us that the semigroup $\mathbb{H}_\kappa$ admits a continuous homomorphism onto any compact Hausdorff right topological semigroup containing a dense subset of cardinality $\leq \kappa$ within the topological center.

Theorem 7.24. *Let κ be an infinite cardinal and let T be a compact Hausdorff right topological semigroup. Assume that there is a dense subset $A \subseteq T$ such that $|A| \leq \kappa$ and $A \subseteq \Lambda(T)$. Then there is a continuous surjective homomorphism $\pi : \mathbb{H}_\kappa \to T$.*

Proof. For every $\alpha < \kappa$, let e_α denote the element of H with $\operatorname{supp}(e_\alpha) = \{\alpha\}$. Then every $x \in H \setminus \{0\}$ can be uniquely written in the form $x = e_{\alpha_1} + \cdots + e_{\alpha_n}$ where $n \in \mathbb{N}$ and $\alpha_1 < \cdots < \alpha_n < \kappa$. Pick a surjection $f_0 : \{e_\alpha : \alpha < \kappa\} \to A$ such that for each $a \in A$, one has $|f_0^{-1}(a)| = \kappa$. Extend f_0 to a mapping $f : H \to T$ by

$$f(x) = f_0(e_{\alpha_1}) \cdots f_0(e_{\alpha_n})$$

where $x = e_{\alpha_1} + \cdots + e_{\alpha_n}$ and $\alpha_1 < \cdots < \alpha_n$. (As $f(0)$ pick any element of T.) Define $\pi : \mathbb{H}_\kappa \to T$ by $\pi = \overline{f}|_{\mathbb{H}_\kappa}$.

Now let $x \in H \setminus \{0\}$ and let $t = \max \operatorname{supp}(x) + 1$. We claim that for every $y \in H_t \setminus \{0\}$, one has $f(x + y) = f(x)f(y)$.

Indeed, let $x = e_{\alpha_1} + \cdots + e_{\alpha_n}$ where $\alpha_1 < \cdots < \alpha_n$ and let $y = e_{\beta_1} + \cdots + e_{\beta_m}$ where $\beta_1 < \cdots < \beta_m$. Then $x + y = e_{\alpha_1} + \cdots + e_{\alpha_n} + e_{\beta_1} + \cdots + e_{\beta_m}$ and $\alpha_1 < \cdots < \alpha_n < \beta_1 < \cdots < \beta_m$. Consequently,

$$f(x + y) = f(e_{\alpha_1}) \cdots f(e_{\alpha_n}) f(e_{\beta_1}) \cdots f(e_{\beta_m}) = f(x)f(y).$$

It follows from this and Lemma 7.22 that π is a homomorphism.

To see that π is surjective, it suffices to show that $A \subseteq \pi(\mathbb{H}_\kappa)$, because $A \subseteq T$ is dense and π is continuous. Let $a \in A$. Pick $p \in U(H)$ such that $f_0^{-1}(a) \in p$. Then $p \in \mathbb{H}_\kappa$ and $\overline{f}(p) = a$. $\qquad\square$

Now we are going to show that for every cancellative semigroup S of cardinality κ, there is a zero-dimensional Hausdorff left invariant topology $\mathcal{T}$ on S^1 such that $\operatorname{Ult}(\mathcal{T})$ is topologically and algebraically isomorphic to $\mathbb{H}_\kappa$.

Lemma 7.25. *Let S be an infinite cancellative semigroup with identity and let $|S| = \kappa$. Then there are two κ-sequences $(x_\alpha)_{1 \leq \alpha < \kappa}$ and $(y_\alpha)_{\alpha < \kappa}$ in S with $y_0 = 1$ such that every element of S is uniquely representable in the form $y_{\alpha_0} x_{\alpha_1} \cdots x_{\alpha_n}$ where $n < \omega$ and $\alpha_0 < \alpha_1 < \cdots < \alpha_n < \kappa$.*

Proof. Enumerate S as $\{s_\alpha : \alpha < \kappa\}$. Put $y_0 = 1$. Fix $0 < \gamma < \kappa$ and suppose that we have constructed $(x_\alpha)_{1 \leq \alpha < \gamma}$ and $(y_\alpha)_{\alpha < \gamma}$ such that all products $y_{\alpha_0} x_{\alpha_1} \cdots x_{\alpha_n}$, where $n < \omega$ and $\alpha_0 < \alpha_1 < \cdots < \alpha_n < \gamma$, are different. Pick as y_γ the first element in the sequence $(s_\alpha)_{\alpha < \kappa}$ not belonging to the subset

$$S_\gamma = \{y_{\alpha_0} x_{\alpha_1} \cdots x_{\alpha_n} : n < \omega \text{ and } \alpha_0 < \alpha_1 < \cdots < \alpha_n < \gamma\}.$$

Then pick $x_\gamma \in S \setminus (S_\gamma^{-1} S_\gamma)$. (Here, $S_\gamma^{-1} S_\gamma = \bigcup_{x \in S_\gamma} x^{-1} S_\gamma$.) This can be done because $|S_\gamma^{-1} S_\gamma| \leq |S_\gamma|^2 < \kappa$. Then whenever $n < \omega$ and $\alpha_0 < \alpha_1 < \cdots < \alpha_n = \gamma$, one has $y_{\alpha_0} x_{\alpha_1} \cdots x_{\alpha_n} \notin S_\gamma$. Also if $y_{\alpha_0} x_{\alpha_1} \cdots x_{\alpha_n}$ and $y_{\beta_0} x_{\beta_1} \cdots x_{\beta_m}$ are different elements of S_γ, the elements $y_{\alpha_0} x_{\alpha_1} \cdots x_{\alpha_n} x_\gamma$ and $y_{\beta_0} x_{\beta_1} \cdots x_{\beta_m} x_\gamma$ are different as well. $\qquad\square$

Theorem 7.26. *Let S be an infinite cancellative semigroup with identity and let $|S| = \kappa$. Then there is a zero-dimensional Hausdorff left invariant topology $\mathcal{T}$ on S such that $\mathrm{Ult}(\mathcal{T}) \subseteq U(S)$ and $\mathrm{Ult}(\mathcal{T})$ is topologically and algebraically isomorphic to $\mathbb{H}_\kappa$.*

Proof. Let $(x_\alpha)_{1 \le \alpha < \kappa}$ and $(y_\alpha)_{\alpha < \kappa}$ be sequences guaranteed by Lemma 7.25. For every $\alpha \in [1, \kappa)$, define $B_\alpha \subseteq S$ by

$$B_\alpha = \{y_0 x_{\alpha_1} \cdots x_{\alpha_n} : n < \omega, \alpha \le \alpha_1 < \cdots < \alpha_n < \kappa\}.$$

Define also $\phi : S \to \kappa$ by

$$\phi(y_{\alpha_0} x_{\alpha_1} \cdots x_{\alpha_n}) = \alpha_n$$

where $n < \kappa$ and $\alpha_0 < \alpha_1 < \ldots < \alpha_n < \kappa$. It is easy to see that the subsets B_α possess the following properties:

(i) $(B_\alpha)_{1 \le \alpha < \kappa}$ is a decreasing sequence of subsets of S with $\bigcap_{1 \le \alpha < \kappa} B_\alpha = \{1\}$,

(ii) for every $\alpha \in [1, \kappa)$ and $x \in B_\alpha$, one has $x B_{\phi(x)+1} \subseteq B_\alpha$, and

(iii) for every $x \in S$, $\alpha \in [\phi(x) + 1, \kappa)$ and $y \in S \setminus (x B_\alpha)$, one has $(y B_{\phi(y)+1}) \cap (x B_\alpha) = \emptyset$.

It follows from (i)–(ii) and Corollary 4.4 that there is a left invariant T_1-topology $\mathcal{T}$ on S in which for each $x \in S$, $\{x B_\alpha : \phi(x) + 1 \le \alpha < \kappa\}$ is an open neighborhood base at x. Condition (iii) gives us that $\mathcal{T}$ is zero-dimensional. Since $\mathcal{T}$ is also a T_1-topology, it is Hausdorff. Obviously, $\mathrm{Ult}(\mathcal{T}) \subseteq U(S)$. To see that $\mathrm{Ult}(\mathcal{T})$ is topologically and algebraically isomorphic to $\mathbb{H}_\kappa$, let $X = B_1$. For every $\alpha < \kappa$, let e_α denote the element of H with $\mathrm{supp}\,(e_\alpha) = \{\alpha\}$. Define $f : X \to H$ by putting $f(1) = 0$ and

$$f(y_0 x_{\alpha_1} \cdots x_{\alpha_n}) = e_{\alpha_1} + \cdots + e_{\alpha_n}$$

where $n \in \mathbb{N}$ and $0 < \alpha_1 < \ldots < \alpha_n < \kappa$. Clearly f is bijective and $f(B_\alpha) = H_\alpha$ for every $\alpha \in [1, \kappa)$, so f homeomorphically maps X onto $(H, \mathcal{T}_0)$. Also f satisfies conditions of Lemma 7.22, since $f(xy) = f(x) + f(y)$ whenever $x \in X \setminus \{1\}$ and $y \in B_{\phi(x)+1}$. Let $\overline{f} : \mathrm{cl}_{\beta S} X \to \beta H$ denote the continuous extension of f and let $\varphi = \overline{f}|_{\mathrm{Ult}(\mathcal{T})}$. It then follows that $\varphi : \mathrm{Ult}(\mathcal{T}) \to \mathbb{H}_\kappa$ is an isomorphism. $\square$

We conclude this section by constructing a homomorphism of a semigroup generated by a strongly discrete ultrafilter.

Lemma 7.27. *Let $\mathcal{F}$ be a strongly discrete filter on S and let $M : S \to \mathcal{F}$ be a basic mapping. Let $x \in S \setminus \{1\}$ and let $x = x_0 \cdots x_n$ be an M-decomposition with $x_0 = 1$. Then there is a neighborhood U of 1 in $\mathcal{T}[\mathcal{F}]$ such that whenever $y \in U \setminus \{1\}$ and $y = y_0 \cdots y_m$ is an M-decomposition with $y_0 = 1$, $xy = x_0 x_1 \cdots x_n y_1 \cdots y_m$ is an M-decomposition.*

Proof. Define $N : S \to \mathcal{F}$ by $N(y) = M(xy) \cap M(y)$ and put $U = [N]_1$.

Let $y \in U \setminus \{1\}$. Then there is an N-decomposition $y = y_0 \cdots y_m$ with $y_0 = 1$. Clearly this is also an M-decomposition. We have that $xy = x_0 x_1 \cdots x_n y_1 \cdots y_m$. Since $x = x_0 \cdots x_n$ is an M-decomposition, $x_{i+1} \in M(x_0 \cdots x_i)$. Next, we have that $y_1 \in N(y_0) = N(1) \subseteq M(x1) = M(x_0 \cdots x_n)$ and, for $i > 1$,

$$y_{i+1} \in N(y_0 \cdots y_i) = N(y_1 \cdots y_i) \subseteq M(xy_1 \cdots y_i) = M(x_0 \cdots x_n y_1 \cdots y_i).$$

Hence, $xy = x_0 x_1 \cdots x_n y_1 \cdots y_m$ is an M-decomposition. □

Definition 7.28. Given a semigroup S and $p \in \beta S$, let C_p denote the smallest closed subsemigroup of βS containing p.

Theorem 7.29. *Let S be an infinite semigroup with identity and let p be a strongly discrete ultrafilter on S. Then there is a continuous homomorphism $\pi : \mathrm{Ult}(\mathcal{T}[p]) \to \beta \mathbb{N}$ such that $\pi(p) = 1$ and $\pi(C_p) = \beta \mathbb{N}$.*

Proof. Choose a basic mapping $M : S \to p$ such that $M(x) \subseteq M(1)$ for every $x \in S$ (see the proof of Theorem 4.16). For every $x \in M(1)$, put $f_0(x) = 1$. Extend $f_0 : M(1) \to \mathbb{N}$ to a mapping $f : (S, \mathcal{T}[p]) \to \mathbb{N} \subset \beta \mathbb{N}$ by

$$f(x) = f_0(x_1) + \cdots + f_0(x_n)$$

where $x = x_0 x_1 \cdots x_n$ is an M-decomposition with $x_0 = 1$. Define $\pi : \mathrm{Ult}(\mathcal{T}[p]) \to \beta \mathbb{N}$ by $\pi = \overline{f}|_{\mathrm{Ult}(\mathcal{T}[p])}$. Lemma 7.27 gives us that for every $x \in S \setminus \{1\}$, there is a neighborhood U of 1 in $\mathcal{T}[p]$ such that $f(xy) = f(x) + f(y)$ for all $y \in U \setminus \{1\}$. Then by Lemma 7.22, π is a homomorphism. Clearly, π is continuous and $\pi(p) = 1$. It follows that $\pi(C_p)$ is a closed subsemigroup of $\beta \mathbb{N}$ containing 1. Hence $\pi(C_p) = \beta \mathbb{N}$. □

References

Proposition 7.7 and Proposition 7.8 are from [99]. Example 7.9 is from [92] and Lemma 7.10 is from [98].

Theorem 7.13 is a result from [101]. Theorem 7.17 is due to T. Budak (Papazyan) [54]. Proposition 7.21 was proved in [99].

Theorem 7.24 is from [38], where it was also proved that for every infinite cancellative semigroup S of cardinality κ, S^* contains copies of $\mathbb{H}_\kappa$. Theorem 7.26 is from [108]. Theorem 7.29 is a result from [107].

Chapter 8
Finite Groups in βG

In this chapter we show that if G is a countable torsion free group, then βG contains no nontrivial finite groups. We also extend in some sense this result to the case where G is an arbitrary countable group. To this end, we develop a special technique based on the concepts of a local left group and a local homomorphism.

8.1   Local Left Groups and Local Homomorphisms

Definition 8.1. A *local left group* is a T_1-space X with a partial binary operation $\cdot$ and a distinguished point $1 \in X$ such that

(i) $x \cdot 1 = x$ for all $x \in X$,

(ii) $(x \cdot y) \cdot z = x \cdot (y \cdot z)$ whenever all the products in the equality are defined, and

(iii) for every $x \in X \setminus \{1\}$, there is a neighborhood U of 1 such that $x \cdot y$ is defined for all $y \in U$, $x \cdot U$ is a neighborhood of x and $\lambda_x : U \ni y \mapsto x \cdot y \in x \cdot U$ is a homeomorphism.

A basic example of a local left group is an open neighborhood of the identity of a T_1 left topological group.

Let X be a local left group and let X_d be the partial semigroup X reendowed with the discrete topology. As in the case of βS, the partial operation of X_d can be naturally extended to βX_d by

$$ pq = \lim_{x \to p} \lim_{y \to q} xy, $$

where $x, y \in X$, making βX_d a right topological partial semigroup. The product pq is defined if and only if

$$ \{x \in X : \{y \in X : xy \text{ is defined}\} \in q\} \in p, $$

in particular, if the ultrafilter q converges to $1 \in X$.

Definition 8.2. Given a local left group X,

$$ \mathrm{Ult}(X) = \{p \in X_d^* : p \text{ converges to } 1 \in X\}. $$

It is straightforward to check that $\mathrm{Ult}(X)$ is a closed subsemigroup in X_d^*. If X is an open neighborhood of the identity of a left topological group $(G, \mathcal{T})$, we identify $\mathrm{Ult}(X)$ with $\mathrm{Ult}(\mathcal{T})$.

Definition 8.3. Let X and Y be local left groups. A mapping $f : X \to Y$ is a *local homomorphism* if $f(1_X) = 1_Y$ and for every $x \in X \setminus \{1\}$, there is a neighborhood U of $1 \in X$ such that $f(xy) = f(x)f(y)$ for all $y \in U$. We say that $f : X \to Y$ is a *local isomorphism* if f is a local homomorphism and homeomorphism.

Note that if $f : X \to Y$ is a local isomorphism, so is $f^{-1} : Y \to X$.

Lemma 8.4. *Let $f : X \to Y$ be a continuous local homomorphism. Define $f^* : \mathrm{Ult}(X) \to \mathrm{Ult}(Y)$ by $f^* = \overline{f}|_{\mathrm{Ult}(X)}$, where $\overline{f} : \beta X_d \to \beta Y_d$ is the continuous extension of f. Then f^* is a continuous homomorphism. Furthermore, if f is injective, f^* is a continuous injective homomorphism, and if f is a local isomorphism, f^* is a topological and algebraic isomorphism.*

Proof. Apply Lemma 7.22. $\square$

Definition 8.5. Let X be a local left group and let S be a semigroup. A mapping $f : X \to S$ is a *local homomorphism* if for every $x \in X \setminus \{1\}$, there is a neighborhood U of 1 such that $f(xy) = f(x)f(y)$ for all $y \in U \setminus \{1\}$.

Lemma 8.6. *Let T be a compact right topological semigroup and let $f : X \to T$ be a local homomorphism such that $f(X) \subseteq \Lambda(T)$. Define $f^* : \mathrm{Ult}(X) \to T$ by $f^* = \overline{f}|_{\mathrm{Ult}(X)}$, where $\overline{f} : \beta X_d \to T$ is the continuous extension of f. Then f^* is a continuous homomorphism. Furthermore, if for every neighborhood U of $1 \in X$, $f(U \setminus \{1\})$ is dense in T, then f^* is surjective.*

Proof. That f^* is a homomorphism follows from Lemma 7.22. To check the second statement, let $t \in T$. We have that for every neighborhood U of $1 \in X$ and for every neighborhood V of $t \in T$, there exists $x \in U \setminus \{1\}$ such that $f(x) \in V$. It follows from this that there exists $p \in \mathrm{Ult}(\mathcal{T})$ such that $\overline{f}(p) = t$. $\square$

Definition 8.7. We say that an element $a \in \bigoplus_\omega \mathbb{Z}_2$ is *basic* if $\mathrm{supp}(a)$ is a nonempty interval in ω. Each nonzero $a \in \bigoplus_\omega \mathbb{Z}_2$ can be uniquely written in the form

$$a = a_0 + \cdots + a_k$$

where

  (a) for each $i \le k$, a_i is basic, and

  (b) for each $i \le k - 1$, $\max \mathrm{supp}(a_i) + 2 \le \min \mathrm{supp}(a_{i+1})$.

We call such a decomposition *canonical* .

   Endow $\bigoplus_\omega \mathbb{Z}_2$ with the group topology by taking as a neighborhood base at 0 the subgroups

$$H_n = \left\{ x \in \bigoplus_\omega \mathbb{Z}_2 : x(i) = 0 \text{ for all } i < n \right\}$$

where $n < \omega$.

Lemma 8.8. *Let S be a semigroup and let $f : \bigoplus_\omega \mathbb{Z}_2 \to S$. The first two of the following statements are equivalent and imply the third:*

(1) *$f(a) = f(a_0) \cdots f(a_k)$ whenever $a = a_0 + \cdots + a_k$ is a canonical decomposition,*

(2) *$f(a + b) = f(a) f(b)$ whenever $\max \operatorname{supp}(a) + 2 \leq \min \operatorname{supp}(b)$,*

(3) *f is a local homomorphism.*

Proof. (1) $\Rightarrow$ (2) Let $\max \operatorname{supp}(a) + 2 \leq \min \operatorname{supp}(b)$ and let $a = a_0 + \cdots + a_k$ and $b = b_0 + \cdots + b_l$ be the canonical decompositions. Since $\max \operatorname{supp}(a) = \max \operatorname{supp}(a_k)$ and $\min \operatorname{supp}(b) = \min \operatorname{supp}(b_0)$, one has $\max \operatorname{supp}(a_k) + 2 \leq \min \operatorname{supp}(b_0)$, so $a + b = a_0 + \cdots + a_k + b_0 + \cdots + b_l$ is the canonical decomposition. Hence

$$f(a + b) = f(a_0 + \cdots + a_k + b_0 + \cdots + b_l)$$
$$= f(a_0) \cdots f(a_k) f(b_0) \cdots f(b_l) = f(a) f(b).$$

(2) $\Rightarrow$ (1) If $a = a_0 + \cdots + a_k$ is a canonical decomposition, then

$$f(a) = f(a_0 + \cdots + a_{k-1}) f(a_k) = \cdots = f(a_0) \cdots f(a_k).$$

(2) $\Rightarrow$ (3) Let $0 \neq a \in \bigoplus_\omega \mathbb{Z}_2$ and let $n = \max \operatorname{supp}(a) + 2$. Then for every $b \in H_n \setminus \{0\}$, $\max \operatorname{supp}(a) + 2 \leq \min \operatorname{supp}(b)$, and so $f(a + b) = f(a) f(b)$. $\square$

We now come to the main result of this section.

Theorem 8.9. *Let X be a countable regular local left group and let $\{U_x : x \in X \setminus \{1\}\}$ be a family of neighborhoods of $1 \in X$. Then there is a continuous bijection $h : X \to \bigoplus_\omega \mathbb{Z}_2$ with $h(1) = 0$ such that*

(1) *$h^{-1}(H_n) \subseteq U_x$ whenever $\max \operatorname{supp}(h(x)) + 2 \leq n$, and*

(2) *$h(xy) = h(x) + h(y)$ whenever $\max \operatorname{supp}(h(x)) + 2 \leq \min \operatorname{supp}(h(y))$.*

Notice that condition (2) in Theorem 8.9 may be rewritten as

(2′) *$h^{-1}(ab) = h^{-1}(a) + h^{-1}(b)$ whenever $\max \operatorname{supp}(a) + 2 \leq \min \operatorname{supp}(b)$,*

so $h^{-1} : \bigoplus_\omega \mathbb{Z}_2 \to X$ is a local homomorphism, by Lemma 8.8. Since h is continuous, it follows that h also is a local homomorphism. Finally, if $\{U_x : x \in X \setminus \{1\}\}$ is a neighborhood base at $1 \in X$, then h is open, and so h is a local isomorphism.

To see that h is open, let $x \in X$. Put $n = \max \operatorname{supp}(h(x)) + 2$ (if $x = 1$, put $n = 0$). Since h is continuous, there is a neighborhood V of 1 such that $h(V) \subseteq H_n$. Now let U be any neighborhood of 1 contained in V. Then $h(xU) = h(x) + h(U)$ by (2). Pick $y \in X \setminus \{1\}$ such that $U_y \subseteq U$ and put $m = \max \operatorname{supp}(h(y)) + 2$. Then $h(U_y) \supseteq H_m$ by (1). Hence, $h(xU) \supseteq h(x) + H_m$.

Definition 8.10. Given $m \geq 2$, let $W = W(\mathbb{Z}_m)$ denote the set of all words on the alphabet $\mathbb{Z}_m$ including the empty word $\emptyset$. A nonempty word $w \in W$ is *basic* if all nonzero letters in w form a final subword. In particular, every nonempty *zero* word (= all the letters are zeros) is basic. For every $v, w \in W$ such that $|v| + 2 \leq |w|$ and the first $|v| + 1$ letters in w are zeros, define $v + w \in W$ to be the result of substituting v for the initial subword of length $|v|$ in w. Each nonempty $w \in W$ can be uniquely written in the form $w = w_0 + \cdots + w_k$ where

(a) for each $i \leq k$, w_i is basic, and

(b) for each $i < k$, w_i is nonzero.

We call such a decomposition *canonical*.

Proof of Theorem 8.9. Enumerate X without repetitions as $\{1, x_1, x_2, \ldots\}$ and let $W = W(\mathbb{Z}_2)$. We shall assign to each $w \in W$ a point $x(w) \in X$ and a clopen neighborhood $X(w)$ of $x(w)$ such that

(i) $x(0^n) = 1$, $X(\emptyset) = X$, and $X(0^n) \subseteq U_{x(v)}$ for all $v \in W$ with $|v| \leq n - 2$,

(ii) $X(w^\frown 0) \cap X(w^\frown 1) = \emptyset$ and $X(w^\frown 0) \cup X(w^\frown 1) = X(w)$,

(iii) $x(w) = x(w_0)\cdots x(w_k)$ and $X(w) = x(w_0)\cdots x(w_{k-1})X(w_k)$ where $w = w_0 + \cdots + w_k$ is the canonical decomposition, and

(iv) $x_n \in \{x(v) : v \in W \text{ and } |v| = n\}$.

Choose as $X(0)$ a clopen neighborhood of 1 such that $x_1 \notin X(0)$. Put $X(1) = X \setminus X(0)$, $x(0) = 1$ and $x(1) = x_1$.

Fix $n \geq 2$ and suppose that $X(w)$ and $x(w)$ have been constructed for all w with $|w| < n$ such that conditions (i)–(iv) hold.

Notice that the subsets $X(w)$, $|w| = n - 1$, form a partition of X. So, one of them, say $X(u)$, contains x_n. Let $u = u_0 + \cdots + u_r$ be the canonical decomposition. Then $X(u) = x(u_0)\cdots x(u_{r-1})(X(u_r))$ and $x_n = x(u_0)\cdots x(u_{r-1})y_n$ for some $y_n \in X(u_r)$. If $y_n = x(u_r)$, choose as $X(0^n)$ a clopen neighborhood of 1 such that

(a) $X(0^n) \subseteq U_{x(w)}$ for all $w \in W$ with $|w| \leq n - 2$, and

(b) for each basic w with $|w| = n - 1$, $X(w) \setminus x(w)X(0^n) \neq \emptyset$.

For each basic w with $|w| = n - 1$, put

$$X(w^\frown 1) = X(w) \setminus x(w)X(0^n)$$

and pick as $x(w^\frown 1)$ any element of $X(w^\frown 1)$. Also put $x(0^n) = 1$. If $y_n \neq x(u_r)$, choose $X(0^n)$ in addition so that

(c) $y_n \notin x(u_r)X(0^n)$

and put $x(u_r^\frown 1) = y_n$.

For all nonbasic w with $|w| = n$, define $X(w)$ and $x(w)$ by condition (iii). Then

$$x(w) = x(w_0)\cdots x(w_k) \in x(w_0)\cdots x(w_{k-1})X(w_k) = X(w)$$

and if $x_n \notin \{x(w) : |w| = n - 1\}$,

$$x_n = x(u_0) \cdots x(u_{r-1}) x(u_r \widehat{\ } 1) = x(u \widehat{\ } 1) \in \{x(w) : |w| = n\}.$$

To check (ii), let $|w| = n - 1$. If w is basic, $X(w \widehat{\ } 0) = x(w) X(0^n) \subset X(w)$ and $X(w \widehat{\ } 1) = X(w) \setminus x(w) X(0^n)$. Suppose that w is nonbasic and let $w = w_0 + \cdots + w_k$ be the canonical decomposition. If w_k is zero, then $w \widehat{\ } 0 = w_0 + \cdots + w_{k-1} + 0^n$ is the canonical decomposition, and consequently,

$$X(w \widehat{\ } 0) = x(w_0) \cdots x(w_{k-1}) X(0^n) = x(w_0) \cdots x(w_{k-1}) X(w_k \widehat{\ } 0).$$

Otherwise $w \widehat{\ } 0 = w_0 + \cdots + w_{k-1} + w_k + 0^n$ is the canonical decomposition, and then

$$X(w \widehat{\ } 0) = x(w_0) \cdots x(w_{k-1}) x(w_k) X(0^n) = x(w_0) \cdots x(w_{k-1}) X(w_k \widehat{\ } 0).$$

In any case,

$$X(w \widehat{\ } 0) = x(w_0) \cdots x(w_{k-1}) X(w_k \widehat{\ } 0).$$

Next, since $w \widehat{\ } 1 = w_0 + \cdots + w_{k-1} + w_k \widehat{\ } 1$ is the canonical decomposition,

$$X(w \widehat{\ } 1) = x(w_0) \cdots x(w_{k-1}) X(w_k \widehat{\ } 1).$$

It follows that

$$\begin{aligned}
X(w \widehat{\ } 0) \cup X(w \widehat{\ } 1) &= x(w_0) \cdots x(w_{k-1})[X(w_k \widehat{\ } 0) \cup X(w_k \widehat{\ } 1)] \\
&= x(w_0) \cdots x(w_{k-1}) X(w_k) \\
&= X(w)
\end{aligned}$$

and $X(w \widehat{\ } 0) \cap X(w \widehat{\ } 1) = \emptyset$.

Now for every $x \in X$, there is $w \in W$ with nonzero last letter such that $x = x(w)$, so $\{v \in W : x = x(v)\} = \{w \widehat{\ } 0^n : n < \omega\}$. Consequently, we can define $h : X \to \bigoplus_\omega \mathbb{Z}_2$ by putting for every $w = \xi_0 \cdots \xi_n \in W$,

$$h(x(w)) = \overline{w} = (\xi_0, \ldots, \xi_n, 0, 0, \ldots).$$

Obviously, $h(1) = 0$. It is clear also that h is bijective. Since for every $z = (\xi_i)_{i < \omega} \in \bigoplus_\omega \mathbb{Z}_2$,

$$h^{-1}(z + H_n) = X(\xi_0 \cdots \xi_{n-1}),$$

h is continuous. And since

$$h^{-1}(H_n) = X(0^n) \subseteq U_{x(w)}$$

for all $w \in W$ with $|w| \leq n - 2$, (1) is satisfied.

To check (2), let max supp($h(x)$) + 2 $\leq$ min supp($h(y)$). Pick $w, v \in W$ with nonzero last letters such that $x = x(w)$ and $y = x(v)$. Let $w = w_0 + \cdots + w_k$ and $v = v_0 + \cdots + v_t$ be the canonical decompositions. Then $y \in X(0^{|w|+1})$, consequently $w + v = w_0 + \cdots + w_k + v_0 + \cdots + v_t$ is the canonical decomposition, and so
$$xy = x(w_0) \cdots x(w_k) x(v_0) \cdots x(v_t) = x(w + v).$$
Hence,
$$h(xy) = h(x(w + v))$$
$$= \overline{w + v}$$
$$= \overline{w} + \overline{v}$$
$$= h(x(w)) + h(x(v))$$
$$= h(x) + h(y). \qquad \square$$

Corollary 8.11. *Let X be a countable nondiscrete regular local left group. Then there is a continuous bijective local homomorphism $f : X \to \bigoplus_\omega \mathbb{Z}_2$, and consequently, $\mathrm{Ult}(X)$ is topologically and algebraically isomorphic to a subsemigroup of $\mathbb{H}$. If X is first countable, f can be chosen to be a local isomorphism, and consequently, $\mathrm{Ult}(X)$ is topologically and algebraically isomorphic to $\mathbb{H}$.*

Proof. It is immediate from Theorem 8.9 and Lemma 8.4. $\qquad \square$

Corollary 8.12. *Let X and Y be countable nondiscrete regular local left groups. Then there is a bijection $f : X \to Y$ such that both f and f^{-1} are local homomorphisms.*

Proof. Let $h : X \to \bigoplus_\omega \mathbb{Z}_2$ and $g : Y \to \bigoplus_\omega \mathbb{Z}_2$ be bijections guaranteed by Theorem 8.9. Put $f = g^{-1} \circ h$. $\qquad \square$

We say that a homomorphisms of an ultrafilter semigroup is *proper* if it can be induced by a local homomorphism as in Lemmas 8.4 and 8.6.

Corollary 8.13. *Let X be a countable nondiscrete regular local left group and let Q and T be finite semigroups. Then for every local homomorphism $f : X \to Q$ and for every surjective homomorphism $g : T \to Q$, there is a local homomorphism $h : X \to T$ such that $f = g \circ h$. Consequently, for every proper homomorphism $\alpha : \mathrm{Ult}(X) \to Q$ and for every surjective homomorphism $\beta : T \to Q$, there is a proper homomorphism $\gamma : \mathrm{Ult}(X) \to T$ such that $\alpha = \beta \circ \gamma$.*

Proof. For every $x \in X \setminus \{1\}$, pick a neighborhood U_x of $1 \in X$ such that $f(xy) = f(x)f(y)$ for all $y \in U_x \setminus \{1\}$, and let $\varphi : X \to \bigoplus_\omega \mathbb{Z}_2$ be a bijection guaranteed by by Theorem 8.9. It then follows that $f\varphi^{-1}(a + b) = f\varphi^{-1}(a)f\varphi^{-1}(b)$

whenever $\max \operatorname{supp}(a) + 2 \leq \min \operatorname{supp}(b)$. For every basic $a \in \bigoplus_\omega \mathbb{Z}_2$, pick $\psi(a) \in g^{-1} f \varphi^{-1}(a) \subseteq T$, so $g \psi(a) = f \varphi^{-1}(a)$. Define $\psi : \bigoplus_\omega \mathbb{Z}_2 \to T$ by

$$\psi(a) = \psi(a_0) \cdots \psi(a_k)$$

where $a = a_0 + \cdots + a_k$ is a canonical decomposition. Let $h = \psi \circ \varphi$.

It is clear that h is a local homomorphism. To see that $g \circ h = f$, let $x \in X \setminus \{1\}$ and let $\varphi(x) = a_0 + \cdots + a_k$ be the canonical decomposition. Then

$$\begin{aligned}
gh(x) &= g \psi \varphi(x) \\
&= g \psi(a_0 + \cdots + a_k) \\
&= g(\psi(a_0) \cdots \psi(a_k)) \\
&= g \psi(a_0) \cdots g \psi(a_k) \\
&= f \varphi^{-1}(a_0) \cdots f \varphi^{-1}(a_k) \\
&= f \varphi^{-1}(a_0 + \cdots + a_k) \\
&= f(x).
\end{aligned}$$

Finally, it follows from $f = g \circ h$ that $f^* = g \circ h^*$.   □

Let X be a local left group and let S be a finite semigroup. We say that a local homomorphism $f : X \to S$ is *surjective* if for every neighborhood U of $1 \in X$, $f(U \setminus \{1\}) = S$.

Corollary 8.14. *Let X be a countable nondiscrete regular local left group and let Q be a finite semigroup. Then for every local homomorphism $f : X \to Q$ and for every surjective local homomorphism $g : \bigoplus_\omega \mathbb{Z}_2 \to Q$, there is a continuous local homomorphism $h : X \to \bigoplus_\omega \mathbb{Z}_2$ such that $f = g \circ h$. Consequently, for every proper homomorphism $\alpha : \operatorname{Ult}(X) \to Q$ and for every surjective proper homomorphism $\beta : \mathbb{H} \to Q$, there is a proper homomorphism $\gamma : \operatorname{Ult}(X) \to \mathbb{H}$ such that $\alpha = \beta \circ \gamma$.*

Proof. For every $x \in X \setminus \{1\}$, pick a neighborhood U_x of $1 \in X$ such that $f(xy) = f(x)f(y)$ for all $y \in U_x \setminus \{1\}$, and let $\varphi : X \to \bigoplus_\omega \mathbb{Z}_2$ be a bijection guaranteed by Theorem 8.9. For every $n < \omega$, pick $n_g > n$ such that $g(a + b) = g(a)g(b)$ whenever $\max \operatorname{supp}(a) \leq n$ and $\min \operatorname{supp}(b) \geq n_g$. Then for every basic $a \in \bigoplus_\omega \mathbb{Z}_2$, pick a nonzero $\psi(a) \in g^{-1} f \varphi^{-1}(a) \subseteq \bigoplus_\omega \mathbb{Z}_2$ such that the following condition is satisfied:

If $m < \max \operatorname{supp}(a)$ and

$$n = \max\{\max \operatorname{supp}(\psi(b)) : b \text{ is basic and } \max \operatorname{supp}(b) \leq m\},$$

then $\min \operatorname{supp}(\psi(a)) \geq n_g$.

Define $\psi : \bigoplus_\omega \mathbb{Z}_2 \to \bigoplus_\omega \mathbb{Z}_2$ by

$$\psi(a) = \psi(a_0) + \cdots + \psi(a_k),$$

where $a = a_0 + \cdots + a_k$ is a canonical decomposition, and let $h = \psi \circ \varphi$. It follows from the condition that

$$g(\psi(a_0) + \cdots + \psi(a_k)) = g\psi(a_0) \cdots g\psi(a_k),$$

and so $f = g \circ h$. The condition also implies that

$$\min \operatorname{supp}(\psi(a)) \geq \max \operatorname{supp}(a)$$

for every basic a, which gives us that ψ is continuous.

To see this, suppose that $\max \operatorname{supp}(a) = m + 1$ and the statement holds for all basic b with $\max \operatorname{supp}(b) = m$. Pick any such b. Then by the inductive assumption, $\min \operatorname{supp}(\psi(b)) \geq m$. It follows that $n \geq m$. Now, applying the condition, we obtain that

$$\min \operatorname{supp}(\psi(a)) \geq n_g \geq n + 1 \geq m + 1 = \max \operatorname{supp}(a). \qquad \square$$

8.2   Triviality of Finite Groups in $\beta\mathbb{Z}$

Lemma 8.15. *Let G be a group and let Q be a group in βG. Then Q is contained either in G or in G^*.*

Proof. It is immediate from the fact that G^* is an ideal of βG (Lemma 6.8). $\qquad \square$

Definition 8.16. Given a group G and a group Q in G^*,

$$G(Q) = \{x \in G : xQ = Q\}.$$

If $x, y \in G(Q)$, then $xy^{-1}Q = xy^{-1}yQ = xQ = Q$, and so $xy^{-1} \in G(Q)$. Hence, $G(Q)$ is a subgroup of G. Also note that

$$G(Q) = \{x \in G : xu \in Q\}$$

where u is the identity of Q. Indeed, if $xu \in Q$, then $xQ = x(uQ) = (xu)Q = Q$.

Lemma 8.17. *$G(Q) \ni x \mapsto xu \in Q$ is an injective homomorphism.*

Proof. That this mapping is injective follows from Lemma 6.9. To see that this is a homomorphism, let $x, y \in G(Q)$. Then $(xy)u = x(yu) = x(u(yu)) = (xu)(yu)$. $\qquad \square$

Theorem 8.18. *Let G be a countable group and let Q be a finite group in G^*. If $G(Q)$ is trivial, so is Q.*

Proof. Assume on the contrary that Q is nontrivial while $G(Q) = \{1\}$. Without loss of generality one may suppose that Q is a cyclic group of order $n > 1$. Let u be the identity of Q. Define $C \subseteq G^*$ by

$$C = \{x \in G^* : xu \in Q\}$$

Equivalently,

$$C = \{x \in G^* : xQ = Q\}.$$

It is clear that C is a closed subsemigroup of G^* and Q is a minimal left ideal of C. Furthermore, C is left saturated.

Indeed, let $xy = z$ for some $x \in \beta G$ and $y, z \in C$. Then $xyQ = zQ$, and so $xQ = Q$. Consequently, $x \in C \cup G(Q)$. Since $G(Q) = \{1\}$, $x \in C \cup \{1\}$.

By Theorem 7.13, there is a regular left invariant topology $\mathcal{T}$ on G such that $\mathrm{Ult}(\mathcal{T}) = C$. Since Q is a minimal left ideal of C, it follows that C has only one minimal right ideal. Consequently, $\mathcal{T}$ is extremally disconnected, by Proposition 7.7. Being regular extremally disconnected, $\mathcal{T}$ is zero-dimensional. (Note that we showed zero-dimensionality of $\mathcal{T}$ not using the fact that G is countable.)

Next for every $p \in Q$, let

$$C_p = \{x \in C : xu = p\}.$$

It is easy to see that $\{C_p : p \in Q\}$ is a partition of C into closed subsets and $p \in C_p$ for each $p \in Q$. Let $\mathcal{F}_p$ be the filter on G such that $\overline{\mathcal{F}_p} = C_p$. For every $p \in Q$, choose $V_p \in \mathcal{F}_p$ such that $V_p \cap V_q = \emptyset$ if $p \neq q$. We now show that for each $p \in Q$, there is $W_p \in \mathcal{F}_p$ such that

$$W_p C_q \subseteq \overline{V_{pq}}$$

for all $q \in Q$.

Indeed, let $\tau = \rho_u|_{\beta G \setminus \{1\}}$. Then $C_{pq} = \tau^{-1}(pq)$. It follows that there exists $A_{pq} \in pq$ such that

$$\tau^{-1}(\overline{A_{pq}}) \subseteq \overline{V_{pq}},$$

or equivalently,

$$\{x \in \beta G \setminus \{1\} : xu \in \overline{A_{pq}}\} \subseteq \overline{V_{pq}}.$$

Since $C_p q = C_p uq = pq$ and Q is finite, there is $W_p \in \mathcal{F}_p$ such that $W_p q \subseteq \overline{A_{pq}}$ for all $q \in Q$. Then

$$W_p C_q u = W_p q \subseteq \overline{A_{pq}},$$

and consequently, $W_p C_q \subseteq \overline{V_{pq}}$.

Choose the subsets W_p in addition so that

$$W_p \subseteq V_p \quad \text{and}$$

$$X = \bigcup_{p \in Q} W_p \cup \{1\}$$

is open in $\mathcal{T}$. Then define $f : X \to Q$ by

$$f(x) = p \quad \text{if } x \in W_p.$$

The value $f(1)$ does not matter. We claim that f is a local homomorphism.

To see this, let $x \in X \setminus \{1\}$. Then $x \in W_p$ for some $p \in Q$. For each $q \in Q$, choose $U_{x,q} \in \mathcal{F}_q$ such that

$$U_{x,q} \subseteq W_q \quad \text{and}$$

$$xU_{x,q} \subseteq V_{pq}.$$

This can be done because $W_p C_q \subseteq \overline{V_{pq}}$. Then choose a neighborhood U_x of $1 \in X$ such that

$$U_x \subseteq \bigcup_{q \in Q} U_{x,q} \cup \{1\} \quad \text{and} \quad xU_x \subseteq X.$$

Now let $y \in U_x \setminus \{1\}$. Then $y \in U_{x,q}$ for some $q \in Q$. Since $xU_{x,q} \subseteq V_{pq}$, one has $xy \in V_{pq}$. But then, since $xU_x \subseteq X$, $xy \in W_{pq}$. Hence

$$f(xy) = pq = f(x)f(y).$$

Having checked that f is a local homomorphism, let $\alpha = f^*$. Then $\alpha : \mathrm{Ult}(X) \to Q$ is a proper homomorphism with the property that $\alpha|_Q = \mathrm{id}_Q$. Now let T be a cyclic group of order n^2 and let $\beta : T \to Q$ be a surjective homomorphism. By Corollary 8.13, there is a proper homomorphism $\gamma : \mathrm{Ult}(X) \to T$ such that $\alpha = \beta \circ \gamma$. It follows that $\gamma(Q) \subseteq T$ is a subgroup of order n. But T has only one subgroup of order n and this is the kernel of β, so $\beta(\gamma(Q)) = \{0\}$, a contradiction. $\square$

Corollary 8.19. *Let G be a countable torsion free group. Then βG contains no nontrivial finite groups.*

Proof. By Lemma 8.15, every group in βG is contained either in G or in G^*. Let Q be a finite group in G^*. By Lemma 8.17, Q contains an isomorphic copy of $G(Q)$. Consequently, $G(Q)$ is finite. Since G is torsion free, it follows that $G(Q)$ is trivial. Then by Theorem 8.18, Q is trivial as well. $\square$

As an immediate consequence of Corollary 8.19 we obtain that

Corollary 8.20. *$\beta\mathbb{Z}$ contains no nontrivial finite groups.*

Corollary 8.20 and Corollary 8.11 give us the following.

Corollary 8.21. *Let X be a countable regular local left group. Then* $\mathrm{Ult}(X)$ *contains no nontrivial finite groups.*

Proof. Pick a nondiscrete first countable group topology $\mathcal{T}$ on $\mathbb{Z}$. By Corollary 8.11, $\mathrm{Ult}(\mathcal{T})$ is isomorphic to $\mathbb{H}$ and $\mathrm{Ult}(X)$ is isomorphic to a subgroup of $\mathbb{H}$, and by Corollary 8.20, $\mathrm{Ult}(\mathcal{T})$ contains no nontrivial finite groups. $\square$

8.3   Local Automorphisms of Finite Order

Let X be a set and let $f : X \to X$. A subset $Y \subseteq X$ is *invariant* (with respect to f) if $f(Y) \subseteq Y$. We say that a family $\mathcal{F}$ of subsets of X is *invariant* if for every $Y \in \mathcal{F}$, $f(Y) \in \mathcal{F}$. For every $x \in X$, let $O(x) = \{f^n(x) : n < \omega\}$.

Lemma 8.22. *Let X be a space, let $f : X \to X$ be a homeomorphism, and let $x \in X$ with $|O(x)| = s \in \mathbb{N}$. Let U be a neighborhood of x such that the family $\{f^j(U) : j < s\}$ is disjoint and suppose that there is $n \in \mathbb{N}$ such that $f^n|_U = \mathrm{id}_U$. Then there is an open neighborhood V of x contained in U such that the family $\{f^j(V) : j < s\}$ is invariant. If X is zero-dimensional, then V can be chosen to be clopen.*

Proof. Clearly, $n = sl$ for some $l \in \mathbb{N}$. Choose an open neighbourhood W of x such that $f^{j+is}(W) \subseteq f^j(U)$ for all $j < s$ and $i < l$, in particular, $f^{is}(W) \subseteq U$ for all $i < l$. This can be done because $f^s(x) = x$. Now let

$$V = \bigcup_{i < l} f^{is}(W).$$

Then V is an open neighborhood of x contained in U and

$$f^s(V) = \bigcup_{i < l} f^{(i+1)s}(W).$$

Since $f^{ls} = f^0$,

$$f^s(V) = \bigcup_{i < l} f^{is}(W) = V.$$

It follows that $\{f^j(V) : j < s\}$ is invariant. $\square$

A bijection $f : X \to X$ has *finite order* if there is $n \in \mathbb{N}$ such that $f^n = \mathrm{id}_X$, and the smallest such n is the *order* of f.

Corollary 8.23. *Let X be a space with a distinguished point $1 \in X$ and let $f :$
$X \to X$ be a homeomorphism with $f(1) = 1$. Then exactly one of the following two
possibilities holds:*

(1) *for every neighborhood U of 1 and $n \in \mathbb{N}$, there is $x \in U$ with $|O(x)| > n$,*

(2) *there is an open invariant neighborhood U of 1 such that $f|_U$ has finite order.*

Definition 8.24. Let X be a space with a distinguished point $1 \in X$ and let $f : X \to$
X be a homeomorphism of finite order with $f(1) = 1$. Define the *spectrum* of f by

$$\mathrm{spec}(f) = \{|O(x)| : x \in X \setminus \{1\}\},$$

and more generally, for any subset $Y \subseteq X$,

$$\mathrm{spec}(f, Y) = \{|O(x)| : x \in Y \setminus \{1\}\}.$$

We say that f is *spectrally irreducible* if for every neighborhood U of 1,

$$\mathrm{spec}(f, U) = \mathrm{spec}(f).$$

Also, a neighborhood U of a point $x \in X$ is *spectrally minimal* if for every neighborhood V of x contained in U,

$$\mathrm{spec}(f, V) = \mathrm{spec}(f, U).$$

Corollary 8.25. *Let X be a space with a distinguished point $1 \in X$ and let $f : X \to$
X be a homeomorphism of finite order with $f(1) = 1$. Then there is an open invariant
neighborhood U of 1 such that $f|_U$ is spectrally irreducible.*

Corollary 8.25 tells us that we can restrict ourselves in the study of homeomorphisms of finite order in a neighborhood of a fixed point to considering spectrally
irreducible ones.

Definition 8.26. A *local automorphism* of a local left group X is a local isomorphism
of X onto itself.

In other words, a mapping $f : X \to X$ is a local automorphism if f is both
a homeomorphism with $f(1) = 1$ and a local homomorphism, that is, for every
$x \in X \setminus \{1\}$, there is a neighborhood U of 1 such that $f(xy) = f(x)f(y)$ for all
$y \in U$.

The next lemma says that the spectrum of a spectrally irreducible local automorphism of finite order is a finite subset of $\mathbb{N}$ closed under taking the least common
multiple lcm, that is, $\mathrm{lcm}(s, t) \in \mathrm{spec}(f)$ for all $s, t \in \mathrm{spec}(f)$.

Lemma 8.27. *Let X be a Hausdorff local left group and let $f : X \to X$ be a spectrally irreducible local automorphism of finite order. Let $x_0 \in X \setminus \{1\}$ with $|O(x_0)| = s$ and let U be a spectrally minimal neighbourhood of x_0. Then*

$$\mathrm{spec}(f, U) = \{\mathrm{lcm}(s, t) : t \in \{1\} \cup \mathrm{spec}(f)\}.$$

Proof. For each $x \in O(x_0)$, let V_x be a neighborhood of 1 such that $V_x \ni y \mapsto xy \in xV_x$ is a homeomorphism and $f(xy) = f(x)f(y)$ for all $y \in V_x$. Choose a neighborhood V of 1 such that

$$V \subseteq \bigcap_{x \in O(x_0)} V_x,$$

$x_0 V \subseteq U$, and the subsets xV, where $x \in O(x_0)$, are pairwise disjoint. Let n be the order of f. Choose a neighborhood W of 1 such that $f^i(W) \subseteq V$ for all $i < n$. Then clearly this inclusion holds for all $i < \omega$, and furthermore, for every $y \in W$, $f^i(x_0 y) = f^i(x_0)f^i(y)$. Indeed, it is trivial for $i = 0$, and further, by induction, we obtain that

$$f^i(x_0 y) = ff^{i-1}(x_0 y) = f(f^{i-1}(x_0)f^{i-1}(y)) = f^i(x_0)f^i(y).$$

Now let $y \in W$, $|O(y)| = t$ and $k = \mathrm{lcm}(s, t)$. We claim that $|O(x_0 y)| = k$. Indeed,

$$f^k(x_0 y) = f^k(x_0)f^k(y) = x_0 y.$$

On the other hand, suppose that $f^i(x_0 y) = x_0 y$ for some i. Then $f^i(x_0)f^i(y) = x_0 y$. Since the subsets xV, $x \in O(x_0)$, are pairwise disjoint, it follows from this that $f^i(x_0) = x_0$, so $s | i$. But then also $f^i(y) = y$, and so $t | i$. Hence $k | i$.   □

Now, given any finite subset of $\mathbb{N}$ closed under lcm, we produce a spectrally irreducible local automorphism of the corresponding spectrum.

Example 8.28. Let S be a finite subset of $\mathbb{N}$ closed under lcm and let $m = 1 + \sum_{s \in S} s$. Consider the direct sum $\bigoplus_\omega \mathbb{Z}_m$ of ω copies of the group $\mathbb{Z}_m$. Endow $\bigoplus_\omega \mathbb{Z}_m$ with the group topology by taking as a neighbourhood base at 0 the subgroups

$$H_n = \left\{ x \in \bigoplus_\omega \mathbb{Z}_m : x(i) = 0 \text{ for all } i < n \right\}$$

where $n < \omega$. Write the elements of S as $s_1 < \cdots < s_t$. Define the permutation π_0 on $\mathbb{Z}_m$ by the product of disjoint cycles

$$\pi_0 = (1, \ldots, s_1)(s_1 + 1, \ldots, s_1 + s_2) \cdots (s_1 + \cdots + s_{t-1} + 1, \ldots, s_1 + \cdots + s_t).$$

Let π be the coordinatewise permutation on $\bigoplus_\omega \mathbb{Z}_m$ induced by π_0, that is, $\pi(x)(n) = \pi_0(x(n))$. Then π is a homeomorphism with $\pi(0) = 0$, $\mathrm{spec}(\pi, H_n) = S$ for each

$n < \omega$, and $\pi(x + y) = \pi(x) + \pi(y)$ whenever $\max \operatorname{supp}(x) < \min \operatorname{supp}(y)$. Hence, π is a spectrally irreducible local automorphism of spectrum S. We call π the *standard* permutation of spectrum S.

We now come to the main result of this section.

Theorem 8.29. *Let X be a countable nondiscrete regular local left group, let $f : X \to X$ be a spectrally irreducible local automorphism of finite order, let $S = \operatorname{spec}(f)$, and let $m = 1 + \sum_{s \in S} s$. Let π be the standard permutation on $\bigoplus_\omega \mathbb{Z}_m$ of spectrum S. Then there is a continuous bijection $h : X \to \bigoplus_\omega \mathbb{Z}_m$ with $h(1) = 0$ such that*

(1) $f = h^{-1}\pi h$, *and*

(2) $h(xy) = h(x) + h(y)$ *whenever* $\max \operatorname{supp}(h(x)) + 2 \le \min \operatorname{supp}(h(y))$.

If X is first countable, then h can be chosen to be a homeomorphism.

Proof. The proof is similar to that of Theorem 8.9.

Enumerate X as $\{x_n : n < \omega\}$ with $x_0 = 1$ and let $W = W(\mathbb{Z}_m)$. The permutation π_0 on $\mathbb{Z}_m$, which induces the standard permutation π on $\bigoplus_\omega \mathbb{Z}_m$, also induces the permutation π_1 on W. If $w = \xi_0 \cdots \xi_n$, then $\pi_1(w) = \pi_0(\xi_0) \cdots \pi_0(\xi_n)$. We will write π instead of π_0 and π_1.

We shall assign to each $w \in W$ a point $x(w) \in X$ and a clopen spectrally minimal neighborhood $X(w)$ of $x(w)$ such that

(i) $x(0^n) = 1$ and $X(\emptyset) = X$,

(ii) $\{X(w^\frown\xi) : \xi \in \mathbb{Z}_m\}$ is a partition of $X(w)$,

(iii) $x(w) = x(w_0) \cdots x(w_{k-1})x(w_k)$ and $X(w) = x(w_0) \cdots x(w_{k-1})X(w_k)$ where $w = w_0 + \cdots + w_k$ is the canonical decomposition,

(iv) $f(x(w)) = x(\pi(w))$ and $f(X(w)) = X(\pi(w))$, and

(v) $x_n \in \{x(v) : v \in W \text{ and } |v| = n\}$.

For this, we need the following

Lemma 8.30. *Let X be a countable nondiscrete regular space with a distinguished point $1 \in X$ and let $f : X \to X$ be a homeomorphism of finite order with $f(1) = 1$. Let U be a clopen invariant subset of X, let K be a finite invariant subset of U, and let $\mathcal{P}$ be a clopen invariant partition of U such that for each $C \in \mathcal{P}$, $\operatorname{spec}(f, K \cap C) = \operatorname{spec}(f, C)$. Then there is a clopen invariant partition $\{U(x) : x \in K\}$ of U inscribed into $\mathcal{P}$ such that for each $x \in K$, $U(x)$ is a spectrally minimal neighborhood of x.*

Proof. Enumerate U as $\{x_n : n < \omega\}$ with $x_0 \in K$. For each $x \in K$, we shall construct an increasing sequence $(U_n(x))_{n<\omega}$ of clopen spectrally minimal neighborhoods of x such that for every $n < \omega$, the family $\{U_n(x) : x \in K\}$ is disjoint, inscribed into $\mathcal{P}$ and invariant, and $x_n \in U_n = \bigcup_{x \in K} U_n(x)$. Then the subsets $U(x) = \bigcup_{n < \omega} U_n(x)$, $x \in K$, will be as required. We proceed by induction on n.

Let y_i, $i < l$, be representatives of all orbits in K and let $|O(y_i)| = s_i$. For each $i < l$, choose a neighborhood W_i of y_i such that $f^j(W_i)$ is a spectrally minimal neighborhood of $f^j(y_i)$ for all $j < s_i$, and the family $\{f^j(W_i) : i < l, j < s_i\}$ is disjoint and inscribed into $\mathcal{P}$. By Lemma 8.22, for each $i < l$, there is a clopen neighborhood V_i of y_i contained in W_i such that the family $\{f^j(V_i) : j < s_i\}$ is invariant. Put $U_0(f^j(y_i)) = f^j(V_i)$.

Fix $n > 0$ and suppose that we have constructed required $U_{n-1}(x)$, $x \in K$. Without loss of generality one may suppose also that $x_n \notin U_{n-1}$. Let $|O(x_n)| = s$ and let $x_n \in C_n \in \mathcal{P}$. Using Lemma 8.22, choose a clopen neighborhood V_n of x_n such that for each $j < s$, $f^j(V_n)$ is a spectrally minimal neighborhood of $f^j(x_n)$, and the family

$$\{f^j(V_n) : j < s\} \cup \{U_{n-1}(x) : x \in K\}$$

is disjoint, inscribed into $\mathcal{P}$ and invariant. Pick $z_n \in K \cap C_n$ with $|O(z_n)| = s$. For each $j < s$, put

$$U_n(f^j(z_n)) = U_{n-1}(f^j(z_n)) \cup f^j(V_n).$$

For each $x \in K \setminus O(z_n)$, put $U_n(x) = U_{n-1}(x)$.   □

Now write the elements of S as $s_1 < \cdots < s_t$. For each $i = 1, \ldots, t$, pick a representative ζ_i of the orbit in $\mathbb{Z}_m \setminus \{0\}$ of lengths s_i. Choose a clopen invariant neighborhood U_1 of $1 \in X$ such that $x_1 \notin U_1$ and $\mathrm{spec}(f, X \setminus U_1) = \mathrm{spec}(f)$. Put $x(0) = 1$ and $X(0) = U_1$. Then pick points $a_i \in X \setminus U_1$, $i = 1, \ldots, t$, with pairwise disjoint orbits of lengths s_i such that $x_1 \in \bigcup_{i=1}^t O(a_i)$. For each $i = 1, \ldots, t$ and $j < s_i$, put

$$x(\pi^j(\zeta_i)) = f^j(a_i).$$

By Lemma 8.30, there is an invariant partition $\{X(\xi) : \xi \in \mathbb{Z}_m \setminus \{0\}\}$ of $X \setminus U_1$ such that $X(\xi)$ is a clopen spectrally minimal neighborhood of $x(\xi)$.

Fix $n > 1$ and suppose that $X(w)$ and $x(w)$ have been constructed for all $w \in W$ with $|w| < n$ so that conditions (i)–(v) are satisfied.

Notice that the subsets $X(w)$, $|w| = n - 1$, form a partition of X. So one of them, say $X(u)$, contains x_n. Let $u = u_0 + \cdots + u_q$ be the canonical decomposition. Then $X(u) = x(u_0) \cdots x(u_{q-1}) X(u_q)$ and $x_n = x(u_0) \cdots x(u_{q-1}) y_n$ for some $y_n \in X(u_q)$. Choose a clopen invariant neighborhood U_n of 1 such that for all basic w with $|w| = n - 1$,

(a) $x(w) U_n \subset X(w)$,

(b) $f(xy) = f(x) f(y)$ for all $y \in U_n$, and

(c) $\mathrm{spec}(f, X(w) \setminus x(w) U_n) = \mathrm{spec}(X(w))$.

If $y_n \neq x(u_q)$, choose U_n in addition so that

(d) $y_n \notin x(u_q)U_n$.

Put $x(0^n) = 1$ and $X(0^n) = U_n$.

Let $w \in W$ be an arbitrary nonzero basic word with $|w| = n - 1$ and let $O(w) = \{w_j : j < s\}$, where $w_{j+1} = \pi(w_j)$ for $j < s - 1$ and $\pi(w_{s-1}) = w_0$. Put $Y_j = X(w_j) \setminus x(w_j)U_n$. Using Lemma 8.27, choose points $b_i \in Y_0$, $i = 1, \ldots, t$, with pairwise disjoint orbits of lengths $\mathrm{lcm}(s_i, s)$. If $u_q \in O(w)$, choose b_i in addition so that

$$y_n \in \bigcup_{i=1}^{t} O(b_i).$$

For each $i = 1, \ldots, t$ and $j < s_i$, put

$$x(\pi^j(w^\frown \xi_i)) = f^j(b_i).$$

Then, using Lemma 8.30, inscribe an invariant partition

$$\{X(v^\frown \xi) : v \in O(w), \xi \in \mathbb{Z}_m \setminus \{0\}\}$$

into the partition $\{Y_j : j < s\}$ such that $X(v^\frown \xi)$ is a clopen spectrally minimal neighborhood of $x(v^\frown \xi)$.

For nonbasic $w \in W$ with $|w| = n$, define $x(w)$ and $X(w)$ by condition (iii).

To check (ii) and (iv), let $|w| = n - 1$ and let $w = w_0 + \cdots + w_k$ be the canonical decomposition. Then

$$X(w^\frown 0) = x(w_0) \cdots x(w_k)X(0^n) = x(w_0) \cdots x(w_{k-1})x(w_k)X(0^n)$$

and

$$X(w^\frown \xi) = x(w_0) \cdots x(w_{k-1})X(w_k^\frown \xi),$$

so (ii) is satisfied. Next,

$$\begin{aligned}
f(x(w)) &= f(x(w_0) \cdots x(w_{k-1})x(w_k)) \\
&= f(x(w_0))f(x(w_1) \cdots x(w_{k-1})x(w_k)) \\
&\quad\vdots \\
&= f(x(w_0)) \cdots f(x(w_{k-1}))f(x(w_k)) \\
&= x(\pi(w_0)) \cdots x(\pi(w_{k-1}))x(\pi(w_k)) \\
&= x(\pi(w_0) \cdots \pi(w_{k-1})\pi(w_k)) \\
&= x(\pi(w)),
\end{aligned}$$

so (iv) is satisfied as well.

To check (v), suppose that $x_n \notin \{x(w) : |w| = n - 1\}$. Then

$$x_n = x(u_0) \cdots x(u_{q-1}) y_n$$
$$= x(u_0) \cdots x(u_{q-1}) x(u_q^\frown \xi)$$
$$= x(u^\frown \xi).$$

Now, for every $x \in X$, there is $w \in W$ with nonzero last letter such that $x = x(w)$, so $\{v \in W : x = x(v)\} = \{w^\frown 0^n : n < \omega\}$. Hence, we can define $h : X \to \bigoplus_\omega \mathbb{Z}_m$ by putting for every $w = \xi_0 \cdots \xi_n \in W$,

$$h(x(w)) = \overline{w} = (\xi_0, \dots, \xi_n, 0, 0, \dots).$$

Obviously, $h(1) = 0$. It is clear also that h is bijective. Since for every $z = (\xi_i)_{i<\omega} \in \bigoplus_\omega \mathbb{Z}_m$,

$$h^{-1}(z + H_n) = X(\xi_0 \cdots \xi_{n-1}),$$

h is continuous.

To see (1), let $x = x(w)$. Then

$$h(f(x(w))) = h(x(\pi(w)))$$
$$= \overline{\pi(w)}$$
$$= \pi(\overline{w})$$
$$= \pi(h(x(w))).$$

To see (2), let $x = x(w)$, $w = w_0 + \cdots + w_k$ and $n = \max \operatorname{supp}(h(x)) + 2$. Let $y \in h^{-1}(H_n)$, $y = x(v)$ and $v = v_0 + \cdots + v_l$. Then

$$h(xy) = h(x(w_0) \cdots x(w_k) x(v_0) \cdots x(v_l))$$
$$= h(x(w + v))$$
$$= \overline{w + v}$$
$$= \overline{w} + \overline{v}$$
$$= h(x(w)) + h(x(v))$$
$$= h(x) + h(y).$$

Finally, if X is first countable, $\{X(0^n) : n < \omega\}$ can be chosen to be a neighborhood base at 1 and then h will be a homeomorphism. $\square$

We conclude the section by the following application of Theorem 8.29.

Theorem 8.31. *Let X be a countable regular local left group, let $f : X \to X$ be a local automorphism of finite order, and let $p \in \operatorname{Ult}(X)$ be an idempotent. If $f(p) \neq p$, then the subset $\{p, f(p)\} \subset \operatorname{Ult}(X)$ generates algebraically a free product of one-element semigroups $\{p\}$ and $\{f(p)\}$.*

Proof. Consider an arbitrary relation

$$p_1 \cdots p_k = q_1 \cdots q_s$$

in $\mathrm{Ult}(X)$, where $p_i, q_j \in \{p, f(p)\}$, $p_i \neq p_{i+1}$ and $q_j \neq q_{j+1}$. We prove that $p_1 = q_1$ and $k = s$.

Without loss of generality one may suppose that f is spectrally irreducible. Let

$$h : X \to \bigoplus_\omega \mathbb{Z}_m$$

be a bijection guaranteed by Theorem 8.29. Denote C the set of all nonfixed points in $\mathbb{Z}_m$ (with respect to π_0) and let

$$Y = \{x \in X : \text{ there is a coordinate of } h(x) \text{ belonging to } C\}.$$

(Equivalently, Y consists of all nonfixed points in X.) Note that $\overline{Y} \cap \mathrm{Ult}(X)$ is a subsemigroup containing p and $f(p)$. For every $x \in Y$, consider the sequence of coordinates of $h(x)$ belonging to C and denote $\alpha(x)$ and $\gamma(x)$ the first and the last elements in this sequence. Then for every $u, v \in \overline{Y} \cap \mathrm{Ult}(X)$, $\alpha(uv) = \alpha(u)$.

Indeed, let $\alpha(u) = c \in C$ and let $A = \{x \in Y : \alpha(x) = c\}$. Then $A \in u$. For every $x \in A$, put $n(x) = \max \mathrm{supp}(h(x)) + 2$ and $U_x = h^{-1}(H_{n(x)})$. We have that $\bigcup_{x \in A} x U_x \in uv$ and for every $y \in U_x$, $\alpha(xy) = \alpha(x) = c$, so $\alpha(uv) = c$.

Similarly, $\gamma(uv) = \gamma(v)$, and if $f(u) \neq u$, then $\alpha(u) \neq \alpha(f(u))$ and $\gamma(u) \neq \gamma(f(u))$. Applying α and γ to the relation gives us that $\alpha(p_1) = \alpha(q_1)$ and $\gamma(p_k) = \gamma(q_s)$, so $p_1 = q_1$ and $p_k = q_s$.

We now show that $k = s$. Define the subset $F \subseteq C^2$ by

$$F = \{(\gamma(q), \alpha(q)) : q \in \{p, f(p)\}\}$$

and let $n \geq \max\{k, s\}$. For every $x \in X$, consider the sequence of coordinates of $h(x)$ belonging to C and define $v(x) \in \mathbb{Z}(n)$ to be the number modulo n of pairs of neighbouring elements in this sequence other than pairs from F. Then for every $u, v \in \mathrm{Ult}(X)$,

$$v(uv) = \begin{cases} v(u) + v(v) & \text{if } (\gamma(u), \alpha(v)) \in F \\ v(u) + v(v) + 1 & \text{otherwise.} \end{cases}$$

It follows from this that

$$v(p_1 \cdots p_k) = v(p_1) + \cdots + v(p_k) + k - 1$$

and

$$v(q_1 \cdots q_s) = v(q_1) + \cdots + v(q_s) + s - 1.$$

Also we have that for every $q \in \{p, f(p)\}$, $v(qq) = 2v(q)$. Consequently, since q is an idempotent, $v(q) = v(qq) = 2v(q)$. Hence, $v(q) = 0$. Finally, we obtain that

$$v(p_1 \cdots p_k) = k - 1 \quad \text{and} \quad v(q_1 \cdots q_s) = s - 1,$$

so $k = s$. $\qquad\qquad\qquad\qquad\qquad\qquad\qquad\qquad\qquad\qquad\qquad\qquad\qquad\qquad\square$

8.4   Finite Groups in G^*

Finite groups in G^* can be constructed in the following trivial way.

Example 8.32. Let G be a group, let F be a finite subgroup of G, and let u be an idempotent in G^* which commutes with each element of F. Then Fu is a finite subgroup of G^* isomorphic to F. The isomorphism is given by

$$F \ni x \mapsto xu \in Fu.$$

Indeed, that this mapping is injective follows from Lemma 6.9. To see that this is a homomorphism, let $x, y \in F$. Then $xyu = xyuu = xuyu$.

In this section we show that if G is a countable group, then all finite groups in G^* have such a trivial structure.

Theorem 8.33. *Let G be a countable group, let Q be a finite group in G^* with identity u, and let $F = G(Q)$. Then u commutes with each element of F and $Q = Fu$.*

Proof. We first show that u commutes with each element of F. Let $a \in F$ and assume on the contrary that $a^{-1}ua \neq u$. Note that both idempotents u and $a^{-1}ua$ belong to the semigroup $C(u) = \{x \in G^* : xu = u\}$. Indeed, since $a \in G(Q)$, one has $au \in Q$, so $uau = au$ and then $a^{-1}uau = a^{-1}au = u$. By Theorem 7.17, there is a regular left invariant topology $\mathcal{T}$ on G with $\mathrm{Ult}(\mathcal{T}) = C(u)$. Consider the conjugation

$$f : x \mapsto a^{-1}xa$$

on $(G, \mathcal{T})$. Clearly, $f(u) = a^{-1}ua$. We claim that f is a homeomorphism. To see this, let $p \in C(u)$. Then

$$a^{-1}pau = a^{-1}puau = a^{-1}uau = a^{-1}au = u,$$

so $a^{-1}pa \in C(u)$. Consequently, f is continuous. Similarly, f^{-1} is continuous. It follows that f is a local automorphism of finite order. Since $f(u) \neq u$, idempotents u and $f(u) = a^{-1}ua$ generate a free product by Theorem 8.31. But this contradicts the equality $a^{-1}uau = u$.

We now show that $Q = Fu$. Without loss of generality one may suppose that Q is a finite cyclic group with a generator q. Then for every $x \in F$, $xq = qx$. Indeed, since $xu = ux$ and Q is Abelian,

$$xq = xuq = qxu = qux = qx.$$

Choose $A \in q$ such that $xy = yx$ for all $x \in F$ and $y \in A$. Let H be the subgroup of G generated by the subset $A \cup F$. We have that $Q \subset H^*$ and F is central in H. Let $L = H/F$ and let $g : \beta H \to \beta L$ be the continuous extension of the canonical homomorphism $H \to L$. Then g is a homomorphism and the elements of βL are the subsets of the form Fp where $p \in \beta H$. Consider the group

$$R = g(Q) = Q/(Fu)$$

in L^*. We claim that the subgroup $L(R) = \{x \in L : xR = R\}$ in L is trivial.

To see this, let $x \in L(R)$. Pick $y \in H$ with $g(y) = x$. Then $g(yQ) = g(Q)$. Since yQ and Q are complete preimages with respect to g, it follows that $yQ = Q$. Consequently $y \in F$, and so $x = 1 \in L$.

Since $L(R)$ is trivial, R is trivial as well by Theorem 8.18, and then $Q = Fu$. $\square$

References

The results of Sections 8.1 and 8.2 are from [85] (announced in [86]). Theorem 8.33 is due to I. Protasov [58]. Its proof is based on Theorem 8.18 and Theorem 8.31. The latter and Theorem 8.29 were proved in [87]. The exposition of this chapter is based on the treatment in [92].

Chapter 9
Ideal Structure of βG

In this chapter we show that for every infinite group G of cardinality κ, the ideal $U(G)$ of βG consisting of uniform ultrafilters can be decomposed ($=$ partitioned) into $2^{2^{\kappa}}$ closed left ideals of βG. We also prove that if G is Abelian, then βG contains $2^{2^{\kappa}}$ minimal right ideals and the structure group of $K(\beta G)$ contains a free group on $2^{2^{\kappa}}$ generators. We conclude by showing that if κ is not Ulam-measurable, then $K(\beta G)$ is not closed.

9.1 Left Ideals

Let G be an arbitrary infinite group of cardinality κ. For every $A \subseteq G$, let $U(A)$ denote the set of uniform ultrafilters from $\overline{A}$. It is easy to see that the set $U(G)$ of all uniform ultrafilters on G is a closed two-sided ideal of βG.

Definition 9.1. Let $\mathcal{J}(G)$ denote the finest decomposition of $U(G)$ into closed left ideals of βG with the property that the corresponding quotient space of $U(G)$ is Hausdorff.

Definition 9.1 can be justified by noting that the family of all such decompositions is nonempty (it contains the trivial decomposition $\{U(G)\}$) and considering the diagonal of the corresponding quotient mappings.

Definition 9.2. For every $p \in U(G)$, define $I_p \subseteq \beta G$ by

$$I_p = \bigcap_{A \in p} \mathrm{cl}(GU(A)).$$

The next theorem is the main result of this section.

Theorem 9.3. *If κ is a regular cardinal, then $\mathcal{J}(G) = \{I_p : p \in U(G)\}$.*

Before proving Theorem 9.3 we establish several auxiliary statements.

Lemma 9.4. *For every $p \in U(G)$, I_p is a closed left ideal of βG contained in $U(G)$.*

Proof. Clearly, I_p is a closed subset of $U(G)$. In order to show that I_p is a left ideal of βG, it suffices to show that for every $x \in G$, $xI_p \subseteq I_p$. Since

$$xI_p = \bigcap_{A \in p} x\mathrm{cl}(GU(A)) \quad \text{and} \quad x\mathrm{cl}(GU(A)) = \mathrm{cl}(xGU(A)) = \mathrm{cl}(GU(A)),$$

it follows that $xI_p = I_p$. $\qquad\square$

A decomposition $\mathcal{D}$ of a space X into closed subsets is called *upper semicontinuous* if for every open $U \subseteq X$, $\bigcup\{A \in \mathcal{D} : A \subseteq U\}$ is open in X. Equivalently, $\mathcal{D}$ is upper semicontinuous if for every $A \in \mathcal{D}$ and for every neighborhood U of $A \subseteq X$, there is a neighborhood V of $A \subseteq X$ such that if $B \in \mathcal{D}$ and $B \cap V \neq \emptyset$, then $B \subseteq U$.

Lemma 9.5. *Let $\mathcal{D}$ be a decomposition of a compact Hausdorff space X into closed subsets and let Y be the corresponding quotient space of X. Then Y is Hausdorff if and only if $\mathcal{D}$ is upper semicontinuous.*

Proof. Let $f : X \to Y$ denote the natural quotient mapping.

Suppose that Y is Hausdorff and let $U \subseteq X$ be open. It then follows that $f(X \setminus U) \subseteq Y$ is closed, so $f^{-1}(Y \setminus f(X \setminus U)) \subseteq X$ is open. It remains to notice that

$$f^{-1}(Y \setminus f(X \setminus U)) = \bigcup\{A \in \mathcal{D} : A \subseteq U\}.$$

Conversely, suppose that $\mathcal{D}$ is upper semicontinuous and let A_1, A_2 be distinct members of $\mathcal{D}$. Pick disjoint neighborhoods U_1, U_2 of A_1, A_2 in X. For each $i = 1, 2$, let $V_i = \bigcup\{B \in \mathcal{D} : B \subseteq U_i\}$. Then V_i is a neighborhood of $A_i \subseteq X$ and $V_i = f^{-1}(f(V_i))$. It follows that $f(V_1), f(V_2)$ are disjoint neighborhoods of $f(A_1), f(A_2) \in Y$. □

Lemma 9.6. *Let $\mathcal{J}$ be a decomposition of $U(G)$ into closed left ideals such that the corresponding quotient space of $U(G)$ is Hausdorff. Then for every $J \in \mathcal{J}$ and $p \in J$, $I_p \subseteq J$.*

Proof. It suffices to show that for every neighborhood V of $J \subseteq U(G)$, $I_p \subseteq \mathrm{cl}(V)$. By Lemma 9.5, $\mathcal{J}$ is upper semicontinuous. Therefore, one may suppose that for every $I \in \mathcal{J}$, if $I \cap V \neq \emptyset$, then $I \subseteq V$. It follows from this that $GV \subseteq V$. Since V is a neighborhood of $p \in U(G)$, there is $A \in p$ such that $U(A) \subseteq V$. Consequently, $GU(A) \subseteq V$, so $\mathrm{cl}(GU(A)) \subseteq \mathrm{cl}(V)$. Hence, $I_p \subseteq \mathrm{cl}(V)$. □

Recall that given a set X and a cardinal λ,

$$[X]^{\lambda} = \{A \subseteq X : |A| = \lambda\} \quad \text{and} \quad [X]^{<\lambda} = \{A \subseteq X : |A| < \lambda\}.$$

Definition 9.7. For every $p \in U(G)$, let $\mathcal{F}_p$ denote the filter on G with a base consisting of subsets of the form

$$\bigcup_{x \in G} x(A \setminus F_x)$$

where $A \in p$ and $F_x \in [G]^{<\kappa}$ for each $x \in G$.

Lemma 9.8. *For every $p \in U(G)$,*

$$I_p = \bigcap_{C \in \mathcal{F}_p} \overline{C}.$$

Proof. To see that $I_p \subseteq \bigcap_{C \in \mathcal{F}_p} \overline{C}$, let $A \in [G]^\kappa$ and $F_x \in [G]^{<\kappa}$ for each $x \in G$ and let $C = \bigcup_{x \in G} x(A \setminus F_x)$. For every $x \in G$, $xU(A) \subseteq x\overline{(A \setminus F_x)} = \overline{x(A \setminus F_x)} \subseteq \overline{C}$. Consequently, $\mathrm{cl}(GU(A)) \subseteq \overline{C}$.

To see the converse inclusion, let $B \subseteq G$ and $I_p \subseteq \overline{B}$. It then follows that there is $A \in p$ such that $\mathrm{cl}(GU(A)) \subseteq \overline{B}$. (Indeed, $G \setminus \overline{B} = \overline{G \setminus B}$ is compact and for every $y \in \overline{G \setminus B}$, there is $A_y \in p$ such that $y \notin \mathrm{cl}(GU(A_y))$.) For every $x \in G$, one has $xU(A) \subseteq \overline{B}$, consequently, there is $F_x \in [G]^{<\kappa}$ such that $x(A \setminus F_x) \subseteq B$. Let $C = \bigcup_{x \in G} x(A \setminus F_x)$. Then $C \in \mathcal{F}_p$ and $C \subseteq B$.   □

Lemma 9.9. *Suppose that κ is a regular cardinal. Let $A \in [G]^\kappa$ and $F_x \in [G]^{<\kappa}$ for every $x \in G$ and let $B = G \setminus \bigcup_{x \in G} x(A \setminus F_x)$. Then there are $H_x, K_x \in [G]^{<\kappa}$ for every $x \in G$ such that*

$$\left(\bigcup_{x \in G} x(A \setminus H_x) \right) \cap \left(\bigcup_{x \in G} x(B \setminus K_x) \right) = \emptyset.$$

Proof. Enumerate G as $\{x_\alpha : \alpha < \kappa\}$. For every $\alpha < \kappa$, define $H_{x_\alpha}, K_{x_\alpha} \in [G]^{<\kappa}$ by

$$H_{x_\alpha} = \bigcup_{\beta \le \alpha} F_{x_\beta^{-1} x_\alpha} \quad \text{and} \quad K_{x_\alpha} = \bigcup_{\beta \le \alpha} x_\alpha^{-1} x_\beta F_{x_\alpha^{-1} x_\beta}.$$

Then for every $\alpha < \kappa$ and $\beta \le \alpha$,

$$x_\beta^{-1} x_\alpha (A \setminus H_{x_\alpha}) \cap B = \emptyset \quad \text{and} \quad x_\alpha^{-1} x_\beta A \cap (B \setminus K_{x_\alpha}) = \emptyset,$$

and so

$$x_\alpha (A \setminus H_{x_\alpha}) \cap x_\beta B = \emptyset \quad \text{and} \quad x_\beta A \cap x_\alpha (B \setminus K_{x_\alpha}) = \emptyset.$$

Consequently, for every $\alpha, \beta < \kappa$,

$$x_\alpha (A \setminus H_{x_\alpha}) \cap x_\beta (B \setminus K_{x_\beta}) = \emptyset.   □$$

Now we are in a position to prove Theorem 9.3.

Proof of Theorem 9.3. Let $\mathcal{J} = \{I_p : p \in U(G)\}$. By Lemma 9.4, all members of $\mathcal{J}$ are closed left ideals of βG contained in $U(G)$. To show that $\mathcal{J}$ is an upper semicontinuous decomposition of $U(G)$, let $p \in U(G)$, $A \in p$, and $F_x \in [G]^{<\kappa}$ for every $x \in G$, and let $B = \bigcup_{x \in G} x(A \setminus F_x)$. By Lemma 9.9, there are $H_x, K_x \in [G]^{<\kappa}$ for every $x \in G$ such that $Q \cap R = \emptyset$ where

$$Q = \bigcup_{x \in G} x(A \setminus H_x) \quad \text{and} \quad R = \bigcup_{x \in G} x(B \setminus K_x).$$

By Lemma 9.8, $I_p \subseteq \overline{Q}$ and for every $r \in U(B)$, $I_r \subseteq \overline{R}$, consequently, $I_r \subseteq \overline{G \setminus Q}$. This shows that $\mathcal{J}$ is a decomposition. It follows from this also that for every

$q \in U(Q)$, $I_q \subseteq \overline{G \setminus B}$, which shows that $\mathcal{J}$ is upper semicontinuous. Thus, $\mathcal{J}$ is a decomposition of $U(G)$ into closed left ideals such that the corresponding quotient space of $U(G)$ is Hausdorff. That $\mathcal{J}$ is the finest decomposition of this kind follows from Lemma 9.6. ☐

Now we consider decompositions of $U(G)$ with an additional property that for every member I of the decomposition, $IG \subseteq I$.

Definition 9.10. Let $\mathcal{J}'(G)$ denote the finest decomposition of $U(G)$ into closed left ideals of βG with the property that the corresponding quotient space of $U(G)$ is Hausdorff and for every member I of the decomposition, $IG \subseteq I$.

Definition 9.11. For every $p \in U(G)$, define $I'_p \subseteq \beta G$ by

$$I'_p = \bigcap_{A \in p} \mathrm{cl}(GU(A)G).$$

As in the proof of Lemma 9.8, one shows that

$$I'_p = \bigcap_{C \in \mathcal{F}'_p} \overline{C}$$

where $\mathcal{F}'_p$ denotes the filter on G with a base consisting of subsets of the form

$$\bigcup_{x,y \in G} x(A \setminus F_{x,y})y,$$

where $A \in p$ and $F_{x,y} \in [G]^{<\kappa}$ for every $x, y \in G$. The next lemma is the corresponding version of Lemma 9.9.

Lemma 9.12. *Suppose that κ is a regular cardinal. Let $A \in [G]^{\kappa}$ and $F_{x,y} \in [G]^{<\kappa}$ for every $x, y \in G$ and let $B = G \setminus \bigcup_{x,y \in G} x(A \setminus F_{x,y})y$. Then there are $H_{x,y}, K_{x,y} \in [G]^{<\kappa}$ for every $x, y \in G$ such that*

$$\left(\bigcup_{x,y \in G} x(A \setminus H_{x,y})y \right) \cap \left(\bigcup_{x,y \in G} x(B \setminus K_{x,y})y \right) = \emptyset.$$

Proof. Enumerate $G \times G$ as $\{(x_\alpha, y_\alpha) : \alpha < \kappa\}$ and for every $\alpha < \kappa$, define $H_{x_\alpha, y_\alpha}, K_{x_\alpha, y_\alpha} \in [G]^{<\kappa}$ by

$$H_{x_\alpha, y_\alpha} = \bigcup_{\beta \leq \alpha} F_{x_\beta^{-1} x_\alpha, y_\alpha y_\beta^{-1}} \quad \text{and} \quad K_{x_\alpha, y_\alpha} = \bigcup_{\beta \leq \alpha} x_\alpha^{-1} x_\beta F_{x_\alpha^{-1} x_\beta, y_\beta y_\alpha^{-1}} y_\beta y_\alpha^{-1}.$$

Then for every $\alpha < \kappa$ and $\beta \leq \alpha$,

$$x_\beta^{-1} x_\alpha (A \setminus H_{x_\alpha, y_\alpha}) y_\alpha y_\beta^{-1} \cap B = \emptyset \quad \text{and} \quad x_\alpha^{-1} x_\beta A y_\beta y_\alpha^{-1} \cap (B \setminus K_{x_\alpha, y_\alpha}) = \emptyset,$$

so

$$x_\alpha(A \setminus H_{x_\alpha,y_\alpha})y_\alpha \cap x_\beta By_\beta = \emptyset \quad \text{and} \quad x_\beta Ay_\beta \cap x_\alpha(B \setminus K_{x_\alpha,y_\alpha})y_\alpha = \emptyset,$$

and consequently, for every $\alpha, \beta < \kappa$,

$$x_\alpha(A \setminus H_{x_\alpha,y_\alpha})y_\alpha \cap x_\beta(B \setminus K_{x_\beta,y_\beta})y_\beta = \emptyset. \qquad \square$$

It is easy to see that the corresponding versions of Lemma 9.4 and Lemma 9.6 also hold. Hence, we obtain the following analogue of Theorem 9.3.

Theorem 9.13. *If κ is a regular cardinal, then $\mathcal{J}'(G) = \{I'_p : p \in U(G)\}$.*

The next lemma will allow us to compute the cardinality of $\mathcal{J}'(G)$.

Lemma 9.14. *Let $A \in [G]^\kappa$. Then there are $B \in [A]^\kappa$ and $F_{x,y} \in [G]^{<\kappa}$ for every $x, y \in G$ such that whenever $B_0, B_1 \in [B]^\kappa$ and $B_0 \cap B_1 = \emptyset$, one has*

$$\left(\bigcup_{x,y \in G} x(B_0 \setminus F_{x,y})y \right) \cap \left(\bigcup_{x,y \in G} x(B_1 \setminus F_{x,y})y \right) = \emptyset.$$

Proof. Enumerate $G \times G$ as $\{(x_\alpha, y_\alpha) : \alpha < \kappa\}$. Construct inductively a κ-sequence $(a_\gamma)_{\gamma<\kappa}$ in A such that for every $\gamma < \kappa$ and $\alpha \le \gamma$,

$$x_\alpha a_\gamma y_\alpha \notin \{x_\beta a_\delta y_\beta : \beta \le \delta < \gamma\}.$$

Define $B \in [A]^\kappa$ and $F_{x_\alpha,y_\alpha} \in [G]^{<\kappa}$ for every $\alpha < \kappa$ by

$$B = \{a_\gamma : \gamma < \kappa\} \quad \text{and} \quad F_{x_\alpha,y_\alpha} = \{a_\beta : \beta < \alpha\}.$$

We claim that these are as required. Indeed, assume the contrary. Then

$$x_\alpha a_\gamma y_\alpha = x_\beta a_\delta y_\beta$$

for some $\alpha, \beta < \kappa$ and some distinct $\gamma, \delta < \kappa$ such that $\alpha \le \gamma$ and $\beta \le \delta$. But this is a contradiction. $\qquad \square$

Corollary 9.15. *If κ is a regular cardinal, then $|\mathcal{J}'(G)| = 2^{2^\kappa}$, and for every $I \in \mathcal{J}'(G)$, I is nowhere dense in $U(G)$.*

Proof. Let $A \in [G]^\kappa$ and let B be a subset of A guaranteed by Lemma 9.14. Then $|U(B)| = 2^{2^\kappa}$ and for any distinct $p, q \in U(B)$, $I_p \cap I_q = \emptyset$.

To see that I is nowhere dense in $U(G)$, suppose that $U(A) \cap I \ne \emptyset$. If $U(B) \cap I = \emptyset$, we are done. Otherwise $I = I_p$ for some $p \in U(B)$. Pick $C \in [B]^\kappa$ such that $C \notin p$. Then $U(C) \cap I = \emptyset$. $\qquad \square$

The next theorem covers in some sense the case where κ is a singular cardinal.

Theorem 9.16. *If* $\kappa > \omega$, *then there is a decomposition* $\mathcal{J}$ *of* $U(G)$ *into closed left ideals of* βG *such that*

(1) *the corresponding quotient space of* $U(G)$ *is homeomorphic to* $U(\kappa)$,

(2) *for every* $J \in \mathcal{J}$, $JG \subseteq J$, *and*

(3) *for every* $J \in \mathcal{J}$, J *is nowhere dense in* $U(G)$.

The proof of Theorem 9.16 is based on the following lemma.

Lemma 9.17. *Let* $\kappa > \omega$. *Then there is a surjective function* $f : G \to \kappa$ *such that*

(a) *for every* $\alpha < \kappa$, $|f^{-1}(\alpha)| < \kappa$, *and*

(b) *whenever* $x, y \in G$ *and* $f(x) < f(y)$, *one has* $f(xy) = f(yx) = f(y)$.

Proof. Construct inductively a κ-sequence $(G_\alpha)_{\alpha < \kappa}$ of subgroups of G such that

(i) for every $\alpha < \kappa$, $|G_\alpha| < \kappa$,

(ii) for every $\alpha < \kappa$, $G_\alpha \subset G_{\alpha+1}$,

(iii) for every limit ordinal $\alpha < \kappa$, $G_\alpha = \bigcup_{\beta < \alpha} G_\beta$, and

(iv) $\bigcup_{\alpha < \kappa} G_\alpha = G$.

Note that G is a disjoint union of nonempty sets $G_{\alpha+1} \setminus G_\alpha$, where $\alpha < \kappa$, and G_0. Define $f : G \to \kappa$ by

$$f(x) = \begin{cases} \alpha & \text{if } x \in G_{\alpha+1} \setminus G_\alpha \\ 0 & \text{if } x \in G_0. \end{cases}$$

Clearly, f is surjective and satisfies (a). To check (b), let $x, y \in G$ and $f(x) < f(y)$. Then $x \in G_\beta$ and $y \in G_{\alpha+1} \setminus G_\alpha$ for some $\beta \le \alpha < \kappa$. It follows that both xy and yx also belong to $G_{\alpha+1} \setminus G_\alpha$. Hence, $f(xy) = f(yx) = f(y)$. □

Proof of Theorem 9.16. Let $f : G \to \kappa$ be a function guaranteed by Lemma 9.17 and let $\overline{f} : \beta G \to \beta \kappa$ be the continuous extension of f. Then

(i) $\overline{f}(U(G)) = U(\kappa)$ and $\overline{f}^{-1}(U(\kappa)) = U(G)$,

(ii) $\overline{f}(qp) = \overline{f}(p)$ for all $p \in U(G)$ and $q \in \beta G$,

(iii) $\overline{f}(px) = \overline{f}(p)$ for all $p \in U(G)$ and $x \in G$, and

(iv) for every $u \in U(\kappa)$, $\overline{f}^{-1}(u)$ is nowhere dense in $U(G)$.

Indeed, (i) follows from surjectivity of f and condition (a). To see (ii), let $A \in p$. For every $x \in G$, let $A_x = A \setminus \{y \in G : f(y) \leq f(x)\}$. Then $A_x \in p$ and by condition (b), $f(xy) = f(y) \in f(A)$ for all $y \in A_x$. Consequently, $B = \bigcup_{x \in G} x A_x \in qp$ and $f(B) \subseteq f(A)$. Hence, $\overline{f}(qp) = \overline{f}(p)$. The check of (iii) is similar. Finally, to see (iv), let $A \in [G]^\kappa$ and suppose that $U(A) \cap \overline{f}^{-1}(u) \neq \emptyset$. Then $E = f(A) \in u$. Pick $D \in [E]^\kappa$ such that $D \notin u$ and let $B = f^{-1}(D) \cap A$. Then $B \subseteq A$, $U(B) \neq \emptyset$, but $f(B) \notin u$, and so $U(B) \cap \overline{f}^{-1}(u) = \emptyset$. Hence, $\overline{f}^{-1}(u)$ is nowhere dense in $U(G)$.

Now let $\mathcal{J} = \{\overline{f}^{-1}(u) : u \in U(\kappa)\}$. It then follows from (i)-(iv) that $\mathcal{J}$ is as required.   □

Applying Theorem 9.16 in the case $\kappa > \omega$ and Corollary 9.15 in the case $\kappa = \omega$, we obtain the following result.

Theorem 9.18. $|\mathcal{J}'(G)| = 2^{2^\kappa}$, and for every $I \in \mathcal{J}'(G)$, I is nowhere dense in $U(G)$.

We conclude this section by the following consequence of Theorem 9.18.

Corollary 9.19. $\Lambda(\beta G) = G$.

Proof. We have to show that for every $p \in G^*$, $\lambda_p : \beta G \ni x \mapsto px \in \beta G$ is not continuous. Without loss of generality one may suppose that $p \in U(G)$. By Theorem 9.18, there is a nontrivial decomposition $\mathcal{J}$ of $U(G)$ into closed left ideals of βG such that for every $I \in \mathcal{J}$, $IG \subseteq I$. Let J be the member of $\mathcal{J}$ containing p. Pick any $K \in \mathcal{J}$ different from J and any $q \in K$. Then $pq \in K$ and $\mathrm{cl}(pG) \subseteq J$. Consequently, λ_p is discontinuous at q.   □

9.2   Right Ideals

In this section we prove the following result.

Theorem 9.20. *For every infinite Abelian group G of cardinality κ, βG contains 2^{2^κ} minimal right ideals.*

The proof of Theorem 9.20 involves some additional concepts.

The *Bohr compactification* of a topological group G is a compact group bG together with a continuous homomorphism $e : G \to bG$ such that $e(G)$ is dense in bG and the following universal property holds: For every continuous homomorphism $h : G \to K$ from G into a compact group K there is a continuous homomorphism $h^b : bG \to K$ such that $h = h^b \circ e$. In the case where G is a discrete Abelian group, the Bohr compactification can be naturally defined in terms of the Pontryagin duality as follows. Let $\hat{G}$ be the dual group of G and let $\hat{G}_d$ be the group $\hat{G}$ reendowed with the discrete topology. Then bG is the dual group of $\hat{G}_d$. The mapping $e : G \rightarrow bG$ is given by $e(x)(\chi) = \chi(x)$, where $x \in G$ and $\chi \in \hat{G}_d$. It is injective. (See [34, 26.11 and 26.12].)

Recall that filters $\mathcal{F}$ and $\mathcal{G}$ on a set X are *incompatible* if $A \cap B = \emptyset$ for some $A \in \mathcal{F}$ and $B \in \mathcal{G}$. A filter $\mathcal{F}$ on a space X is *open* if $\mathcal{F}$ has a base of open subsets of X.

In order to prove Theorem 9.20, we show the following.

Theorem 9.21. *For every infinite Abelian group G of cardinality κ, there are $2^{2^{\kappa}}$ pairwise incompatible open filters on bG converging to zero.*

Before proving Theorem 9.21, let us show how it implies Theorem 9.20.

Proof of Theorem 9.20. Let $\mathcal{T}$ denote the Bohr topology on G, that is, the topology induced by the mapping $e : G \rightarrow bG$, and let $S = \text{Ult}(\mathcal{T})$. By Theorem 9.21, there are $2^{2^{\kappa}}$ pairwise incompatible open filters on bG converging to zero. Considering the restriction of the filters to $e(G)$, we conclude that there are pairwise incompatible open filters $\mathcal{F}_\alpha$, $\alpha < 2^{2^{\kappa}}$, on $(G, \mathcal{T})$ converging to zero. For each $\alpha < 2^{2^{\kappa}}$, let $J_\alpha = \overline{\mathcal{F}_\alpha} \setminus \{0\}$. Then by Lemma 7.3, each J_α is a closed right ideal of S, and since the filters $\mathcal{F}_\alpha$ are pairwise incompatible, the ideals J_α are pairwise disjoint. Furthermore, since $(G, \mathcal{T})$ is a subgroup of a compact group, by Lemma 7.10, S contains all the idempotents of G^*, in particular, the idempotents of $K(\beta G)$. Consequently, $S \cap K(\beta G) \neq \emptyset$. But then, by Proposition 6.24, $K(S) = K(\beta G) \cap S$. It follows from this that every minimal right ideal R of S is contained in a minimal right ideal R' of βG, and the correspondence $R \mapsto R'$ is injective. Consequently, the number of minimal right ideals of βG is greater than or equal to that of S, and so it is $2^{2^{\kappa}}$. □

To prove Theorem 9.21, we need three lemmas. The first of them is an elementary fact on infinite Abelian groups.

Lemma 9.22. *Let G be an infinite Abelian group of cardinality κ. Then G admits a homomorphism onto one of the following groups:*

(1) $\mathbb{Z}$, $\bigoplus_\omega \mathbb{Z}_p$, $\mathbb{Z}_{p^\infty}$ and $\bigoplus_{p \in Q} \mathbb{Z}_p$ if $\kappa = \omega$,

(2) $\bigoplus_\kappa \mathbb{Z}_p$ and $\bigoplus_\kappa \mathbb{Z}_{p^\infty}$ if $\kappa > \omega$ and $\text{cf}(\kappa) > \omega$,

(3) $\bigoplus_\kappa \mathbb{Z}_p$, $\bigoplus_\kappa \mathbb{Z}_{p^\infty}$, $\bigoplus_{p \in Q} \bigoplus_{\kappa_p} \mathbb{Z}_p$ and $\bigoplus_{p \in Q} \bigoplus_{\kappa_p} \mathbb{Z}_{p^\infty}$ if $\kappa > \omega$ and $\text{cf}(\kappa) = \omega$.

Here, p is a prime number and Q is an infinite subset of the primes. $\mathbb{Z}_{p^\infty}$ denotes the quasicyclic p-group. If $\kappa > \omega$ and $\text{cf}(\kappa) = \omega$, $(\kappa_p)_{p \in Q}$ is an infinite increasing sequence of uncountable cardinals cofinal in κ, that is, $\sup_{p \in Q} \kappa_p = \kappa$.

Proof. If G is finitely generated, then $\kappa = \omega$ and G admits a homomorphism onto $\mathbb{Z}$. Therefore, one may assume that G is not finitely generated. We first prove that G admits a homomorphism onto a periodic group of cardinality κ.

Let $\{a_i : i \in I\}$ be a maximal independent subset of G and let $A = \langle a_i : i \in I \rangle$ be the subgroup generated by $\{a_i : i \in I\}$. Then $A = \bigoplus_{i \in I} \langle a_i \rangle$, and for every nonzero $g \in G$, one has $\langle g \rangle \cap A \neq \{0\}$, so G/A is periodic. If $|G/A| = \kappa$, we are done. Suppose that $|G/A| < \kappa$. Then $|A| = \kappa$ and $|I| = \kappa$, because G is not finitely generated. We show that there is a subgroup H of G and a subset $I_1 \subseteq I$ with $|I_1| = \kappa$ such that $G = H \oplus \bigoplus_{i \in I_1} \langle a_i \rangle$.

To this end, choose a complete set S for representatives of the cosets of A in G, and let $H_0 = \langle S \rangle \cap A$. Define $I_0 \subset I$ by

$$I_0 = \{i \in I : x(i) \neq 0 \text{ for some } x \in H_0\}$$

and put $I_1 = I \setminus I_0$. If G/A is finite, I_0 is finite as well. If G/A is infinite, $|I_0| \leq |G/A|$, because $|\langle S \rangle| = |G/A|$ and then $|H_0| \leq |G/A|$. In any case, $|I_0| < \kappa$, and consequently $|I_1| = \kappa$. Let

$$A_0 = \langle a_i : i \in I_0 \rangle, \quad A_1 = \langle a_i : i \in I_1 \rangle \quad \text{and} \quad H = \langle S \cup A_0 \rangle.$$

We claim that $G = H \oplus A_1$. Indeed, since $G = \langle S \cup A_0 \cup A_1 \rangle$, one has $H + A_1 = G$. To see that $H \cap A_1 = \{0\}$, let $g \in H \cap A_1$. Then $g = d + c_0 = c_1$ for some $d \in \langle S \rangle$, $c_0 \in A_0$ and $c_1 \in A_1$. Consequently, $d = -c_0 + c_1 \in A$. But then $d \in H_0 \subseteq A_0$. Hence, $c_1 = 0$, and $g = 0$.

Having established that $G = H \oplus \bigoplus_{i \in I_1} \langle a_i \rangle$, we obtain that G admits a homomorphism onto $\bigoplus_{i \in I_1} \langle a_i \rangle$, and so onto a periodic group of cardinality κ.

Now let G be a p-group. Then there is a so-called basic subgroup B of G (see [28, Theorem 32.3]). We have that B is a direct sum of cyclic groups, say $B = \bigoplus_{j \in J} \langle b_j \rangle$, and G/B is divisible, that is, isomorphic to $\bigoplus_\lambda \mathbb{Z}_{p^\infty}$, where $0 \leq \lambda \leq \kappa$. Suppose that $|G/B| = \kappa$. Then $\lambda > 0$, and $\lambda = \kappa$ if $\kappa > \omega$. It follows that G admits a homomorphism onto $\mathbb{Z}_{p^\infty}$ if $\kappa = \omega$, and onto $\bigoplus_\kappa \mathbb{Z}_{p^\infty}$ if $\kappa > \omega$. Now suppose that $|G/B| < \kappa$. Then $|B| = \kappa$, and consequently $|J| = \kappa$. It follows that $G = C \oplus \bigoplus_{j \in J_1} \langle b_j \rangle$ for some subgroup C of G and a subset $J_1 \subset J$ with $|J_1| = \kappa$ (see the first part of the proof). Hence, G admits a homomorphism onto $\bigoplus_{j \in J_1} \langle b_j \rangle$, and so onto $\bigoplus_\kappa \mathbb{Z}_p$.

Finally, let G be periodic. Then $G = \bigoplus_{p \in M} G_p$, where M is the set of all primes p such that the p-primary component G_p of G is nontrivial. If $|G_p| = \kappa$ for some $p \in M$, we are done, because then G admits a homomorphism onto G_p, a p-group of cardinality κ. Suppose that $|G_p| < \kappa$ for each $p \in M$. Then M is infinite and $\mathrm{cf}(\kappa) = \omega$. If $\kappa = \omega$, all G_p are finite, and so G admits a homomorphism onto $\bigoplus_{p \in M} \mathbb{Z}_p$. Suppose that $\kappa > \omega$. For each $p \in M$, put $\kappa_p = |G_p|$. Clearly $\sup_{p \in M} \kappa_p = \kappa$. Choose an infinite subset $N \subseteq M$ such that $(\kappa_p)_{p \in N}$ is an increasing sequence of uncountable cardinals cofinal in κ. By the previous paragraph, for each $p \in N$, G_p admits a homomorphism onto a group K_p of cardinality κ_p which is isomorphic to $\bigoplus_{\kappa_p} \mathbb{Z}_p$ or $\bigoplus_{\kappa_p} \mathbb{Z}_{p^\infty}$. It follows that there is an infinite subset $Q \subseteq N$ such that either K_p is isomorphic to $\bigoplus_{\kappa_p} \mathbb{Z}_p$ for all $p \in Q$ or K_p is isomorphic to $\bigoplus_{\kappa_p} \mathbb{Z}_{p^\infty}$ for all $p \in Q$. Then the group $K = \bigoplus_{p \in Q} K_p$ is isomorphic to $\bigoplus_{p \in Q} \bigoplus_{\kappa_p} \mathbb{Z}_p$ or $\bigoplus_{p \in Q} \bigoplus_{\kappa_p} \mathbb{Z}_{p^\infty}$, $|K| = \kappa$, and G admits a homomorphism onto K. $\qquad\square$

Now, using Lemma 9.22 and the Pontrjagin duality, we prove the following statement on bG.

Lemma 9.23. *For every infinite discrete Abelian group G of cardinality κ, bG admits a continuous homomorphism onto $\prod_{2^\kappa} \mathbb{T}$ or $\prod_{2^\kappa} \mathbb{Z}_p$.*

Here, both products $\prod_{2^\kappa} \mathbb{T}$ and $\prod_{2^\kappa} \mathbb{Z}_p$ are endowed with the product topology.

Proof. The dual groups of continuous homomorphic images of bG are the subgroups of $\hat{G}_d$ and the dual groups of homomorphic images of G are the closed subgroups of $\hat{G}$ (see [34, Theorems 23.25 and 24.8]). The dual groups of $\prod_{2^\kappa} \mathbb{T}$ and $\prod_{2^\kappa} \mathbb{Z}_p$ are $\bigoplus_{2^\kappa} \mathbb{Z}$ and $\bigoplus_{2^\kappa} \mathbb{Z}_p$, respectively. Consequently, in order to prove the lemma, it suffices to show that G admits a homomorphism onto a group whose dual group contains an isomorphic copy of $\bigoplus_{2^\kappa} \mathbb{Z}$ or $\bigoplus_{2^\kappa} \mathbb{Z}_p$. Consider two cases.

Case 1: $\kappa = \omega$. Then G admits a homomorphism onto one of the following groups: $\mathbb{Z}$, $\bigoplus_\omega \mathbb{Z}_p$, $\mathbb{Z}_{p^\infty}$ and $\bigoplus_{p \in Q} \mathbb{Z}_p$. Their dual groups are $\mathbb{T}$, $\prod_\omega \mathbb{Z}_p$, $\mathbb{Z}(p^\infty)$ and $\prod_{p \in Q} \mathbb{Z}_p$, respectively. Here, $\mathbb{Z}(p^\infty)$ denotes the group of p-adic integers. The second group is algebraically isomorphic to $\bigoplus_{2^\omega} \mathbb{Z}_p$. The others contain torsion-free subgroups of cardinality 2^ω, and so contain an isomorphic copy of $\bigoplus_{2^\omega} \mathbb{Z}$.

Case 2: $\kappa > \omega$. Then G admits a homomorphism onto one of the following groups: $\bigoplus_\kappa \mathbb{Z}_p$, $\bigoplus_\kappa \mathbb{Z}_{p^\infty}$, $\bigoplus_{p \in Q} \bigoplus_{\kappa_p} \mathbb{Z}_p$ and $\bigoplus_{p \in Q} \bigoplus_{\kappa_p} \mathbb{Z}_{p^\infty}$ (the two latter groups appear if $\mathrm{cf}(\kappa) = \omega$). Their dual groups are $\prod_\kappa \mathbb{Z}_p$, $\prod_\kappa \mathbb{Z}(p^\infty)$, $\prod_{p \in Q} \prod_{\kappa_p} \mathbb{Z}_p$ and $\prod_{p \in Q} \prod_{\kappa_p} \mathbb{Z}(p^\infty)$, respectively. The first group is algebraically isomorphic to $\bigoplus_{2^\kappa} \mathbb{Z}_p$. The others contain torsion-free subgroups of cardinality 2^κ, and so contain an isomorphic copy of $\bigoplus_{2^\kappa} \mathbb{Z}$. $\qquad\square$

The third lemma deals with products of topological spaces.

Lemma 9.24. *Let κ be an infinite cardinal. For each $\alpha < \kappa$, let X_α be a space having at least two disjoint nonempty open sets, and let $X = \prod_{\alpha < \kappa} X_\alpha$. Then there are at least 2^κ many pairwise incompatible open filters on X converging to the same point.*

Before proving Lemma 9.24 note that if each factor in an infinite product $X = \prod_{n < \omega} X_n$ has at least two disjoint nonempty open sets, then X is not extremally disconnected. Indeed, for each $n < \omega$, let U_n and V_n be disjoint nonempty open subsets of X_n, let $x_n \in V_n$, and let $x = (x_n)_{n<\omega} \in X$. For every $m < \omega$, define an open subset $W_m = \prod_{n<\omega} W_{m,n} \subset X$ by

$$W_{m,n} = \begin{cases} U_n & \text{if } n = m \\ V_n & \text{if } n < m \\ X_n & \text{if } n > m. \end{cases}$$

It follows that $U = \bigcup_{m<\omega} W_{2m}$ and $V = \bigcup_{m<\omega} W_{2m+1}$ are disjoint open subsets of X with $x \in (\text{cl } U) \cap (\text{cl } V)$.

Proof of Lemma 9.24. Let L be the set of limit ordinals $< \kappa$ including 0. Then

$$X = \prod_{\alpha \in L} \prod_{n<\omega} X_{\alpha+n},$$

$|L| = \kappa$, and for each $\alpha \in L$, $\prod_{n<\omega} X_{\alpha+n}$ is not extremally disconnected. Therefore, one may suppose that each X_α in the product $X = \prod_{\alpha<\kappa} X_\alpha$ is not extremally disconnected, so there are disjoint open subsets $U_\alpha, V_\alpha \subset X_\alpha$ and $x_\alpha \in X_\alpha$ such that $x_\alpha \in (\text{cl } U_\alpha) \cap (\text{cl } V_\alpha)$. For every $Y = (Y_\alpha)_{\alpha<\kappa} \in \prod_{\alpha<\kappa}\{U_\alpha, V_\alpha\}$, define the filter $\mathcal{F}(Y)$ on X by declaring as a base the subsets of the form $\prod_{\alpha<\kappa} Z_\alpha$, where

(i) for each $\alpha < \kappa$, Z_α is a nonempty open subset of X_α,

(ii) for all but finitely many $\alpha < \kappa$, $Z_\alpha = X_\alpha$, and

(iii) if $Z_\alpha \neq X_\alpha$, then $Z_\alpha = Y_\alpha \cap W$ for some neighborhood W of $x_\alpha \in X_\alpha$.

Then $\mathcal{F}(Y)$, where $Y \in \prod_{\alpha<\kappa}\{U_\alpha, V_\alpha\}$, are pairwise incompatible open filters on X converging to $x = (x_\alpha)_{\alpha<\kappa}$.   □

Now we are in a position to prove Theorem 9.21.

Proof of Theorem 9.21. By Lemma 9.23, there is a continuous surjective homomorphism $f : bG \to K$, where K is $\prod_{2^\kappa} \mathbb{T}$ or $\prod_{2^\kappa} \mathbb{Z}_p$. By Lemma 9.24, there are pairwise incompatible open filters $\mathcal{F}_\alpha$, $\alpha < 2^{2^\kappa}$, on K converging to zero. For each $\alpha < 2^{2^\kappa}$, let $\mathcal{H}_\alpha$ be the filter on bG with a base consisting of subsets of the form $f^{-1}(A) \cap U$, where $A \in \mathcal{F}_\alpha$ and U runs over neighborhoods of zero. Then $\mathcal{H}_\alpha$, $\alpha < 2^{2^\kappa}$, are pairwise incompatible open filters on bG converging to zero.   □

9.3   The Structure Group of $K(\beta G)$

In this section we prove the following result.

Theorem 9.25. *Let G be an infinite group of cardinality κ embeddable into a direct sum of countable groups. Then the structure group of $K(\beta G)$ contains a free group on 2^{2^κ} generators.*

Since every Abelian group can be isomorphically embedded into a direct sum of groups isomorphic to $\mathbb{Q}$ or $\mathbb{Z}_{p^\infty}$, we obtain from Theorem 9.25 as a consequence that

Corollary 9.26. *For every infinite Abelian group G of cardinality κ, the structure group of $K(\beta G)$ contains a free group on 2^{2^κ} generators.*

Recall that if a semigroup S has a smallest ideal which is a completely simple semigroup, then for every $p \in E(K(S))$, $pSp \subseteq K(S)$ is a maximal group in S with identity p.

The first step in the proof of Theorem 9.25 is the following result.

Theorem 9.27. *Let $A = \{x \in \bigoplus_\omega \mathbb{Z}_2 : |\mathrm{supp}(x)| = 1\}$ and let $p \in E(K(\mathbb{H}))$. Then the elements $p + q + p \in K(\mathbb{H})$, where $q \in A^*$, generate a free group.*

Proof. Let $q_1, \ldots, q_n$ be distinct elements of A^*, let G be the subgroup of $p + \mathbb{H} + p$ generated by $p + q_i + p$, $i = 1, \ldots, n$, and let F be a free group on n generators, say $x_1, \ldots, x_n$. It suffices to show that there exists a homomorphism of G into F sending $p + q_i + p$ to x_i for each $i = 1, \ldots, n$. By Corollary 1.24, F can be algebraically embedded into a compact group K. Without loss of generality one may suppose that $F \subseteq K$. Partition A into subsets A_i, $i = 1, \ldots, n$, such that $A_i \in q_i$. Define $f_0 : A \to K$ by

$$f_0(a) = x_i \quad \text{if } a \in A_i.$$

For every $n < \omega$, let a_n denote the element of A with $\mathrm{supp}(a_n) = \{n\}$. Extend f_0 to a local homomorphism $f : \bigoplus_\omega \mathbb{Z}_2 \to K$ by

$$f(a_{n_1} + \cdots + a_{n_k}) = f_0(a_{n_1}) \cdots f_0(a_{n_k})$$

where $1 \leq k < \omega$ and $n_1 < \cdots < n_k < \omega$. Then $f^* : \mathbb{H} \to K$ is a homomorphism such that $f^*(q_i) = x_i$. Since $f^*(p)$ is the identity, it follows that

$$f^*(p + q_i + p) = f^*(p) f^*(q_i) f^*(p) = f^*(q_i) = x_i. \qquad \square$$

Lemma 9.28. *Let G be a countable group, let $\mathcal{T}_0$ be a regular left invariant topology on G, and let $(U_n)_{1 \leq n < \omega}$ be any sequence of neighborhoods of 1 in $\mathcal{T}_0$. Then $\mathcal{T}_0$ can be weakened to a regular left invariant topology $\mathcal{T}$ on G with a countable base in which each U_n remains a neighborhood of 1.*

Proof. Enumerate $G \setminus \{1\}$ as $\{x_n : 1 \leq n < \omega\}$. Construct inductively a sequence $(V_n)_{n<\omega}$ of clopen neighborhoods of 1 in $\mathcal{T}_0$ with $V_0 = G$ such that for every $n \geq 1$ the following conditions are satisfied:

(1) $V_n \subseteq U_n$,

(2) $x_n \notin V_n$,

(3) $V_n \subseteq V_{n-1}$, and

(4) $y_{k,m} V_n \subseteq V_k \setminus V_{k+1}$ for all $k, m < \omega$ with $k + m = n - 2$, where $\{y_{k,m} : m < \omega\}$ is an enumeration of $V_k \setminus V_{k+1}$ fixed immediately after V_{k+1} has been chosen.

Then $\bigcap_{n<\omega} V_n = \{1\}$ and for every $n < \omega$ and $x \in V_n \setminus V_{n+1}$, there is $l < \omega$ such that $xV_l \subseteq V_n \setminus V_{n+1}$. It follows that there is a zero-dimensional left invariant topology $\mathcal{T}$ on G for which $\{V_n : n < \omega\}$ is a neighborhood base at 1. Obviously, we have also that $\mathcal{T}_0 \subseteq \mathcal{T}$ and each U_n is a neighborhood of 1 in $\mathcal{T}$. $\square$

The next result implies Theorem 9.25 in case $\kappa = \omega$.

Theorem 9.29. *Let G be a countably infinite group and let $p \in E(K(\beta G))$. Then there is an infinite $A \subseteq G$ such that the elements $pqp \in K(\beta G)$, where $q \in A^*$, generate a free group.*

Proof. By Theorem 7.17 and Lemma 9.28, there is a first countable regular left invariant topology $\mathcal{T}$ on G in which p converges to 1. By Corollary 8.11, there is a local isomorphism $f : (G, \mathcal{T}) \to \bigoplus_\omega \mathbb{Z}_2$, so $f^* : \mathrm{Ult}(\mathcal{T}) \to \mathbb{H}$ is an isomorphism, and consequently, $f^*(p) \in K(\mathbb{H})$. Let $B = \{x \in \bigoplus_\omega \mathbb{Z}_2 : |\mathrm{supp}(x)| = 1\}$ and let $A = f^{-1}(B)$. Then for any distinct $q_1, \ldots, q_n \in A^*$, $f^*(q_1), \ldots, f^*(q_n)$ are distinct elements from B^*, and by Theorem 9.27, the elements

$$f^*(p) + f^*(q_i) + f^*(p) = f^*(pq_i p), \quad i = 1, \ldots, n,$$

generate a free group. Hence, the elements $pq_i p$, $i = 1, \ldots, n$, also generate a free group. $\square$

Given a set D and $A \subseteq D$, let $Q(A)$ denote the set of countably incomplete ultrafilters from $\overline{A} \subseteq \beta D$.

Lemma 9.30. *If $|A| = \kappa \geq \omega$, then $|Q(A)| = 2^{2^\kappa}$.*

Proof. We show that there are 2^{2^κ} uniform countably incomplete ultrafilters on κ.

Let $\{A_n : n < \omega\}$ be a partition of κ such that $|A_n| = \kappa$ for every $n < \omega$ and let U be the set of uniform ultrafilters u on κ such that for every $m < \omega$, $\bigcup_{m \leq n < \omega} A_n \in u$. It is clear that each $u \in U$ is countably incomplete. We claim that $|U| = 2^{2^\kappa}$.

Indeed, for every $n < \omega$, there are 2^{2^κ} uniform ultrafilters on κ containing A_n. Enumerate them without repetitions as $u_{n,\alpha}$, $\alpha < 2^{2^\kappa}$. For every $\alpha < 2^{2^\kappa}$, pick any ultrafilter u_α on κ such that $\bigcup_{m \leq n < \omega} B_{n,\alpha} \in u_\alpha$, whenever $m < \omega$ and $B_{n,\alpha} \in u_{n,\alpha}$. Then $u_\alpha \in U$ and $u_\alpha \neq u_\beta$ if $\alpha \neq \beta$. $\square$

The uncountable case of Theorem 9.25 is a consequence of the following result.

Theorem 9.31. *Let G be a group of cardinality $\kappa > \omega$ embeddable into a direct sum of countable groups. Then there is $A \subseteq G$ with $|A| = \kappa$ such that for every $p \in E(K(\beta G))$, the elements $pqp \in K(\beta G)$, where $q \in Q(A)$, generate a free group.*

Proof. Let $H = \bigoplus_{\alpha < \kappa} H_\alpha$ be a direct sum of countable groups and let G be a subgroup of H. Note that for every $\alpha < \kappa$, there is $x \in G \setminus \{1\}$ with min supp$(x) \geq \alpha$.

Indeed, since $|\bigoplus_{\beta < \alpha} H_\beta| < \kappa$, there are distinct $y, z \in G$ such that $y(\beta) = z(\beta)$ for all $\beta < \alpha$. Put $x = yz^{-1}$. Then $x \neq 1$ and for each $\beta < \alpha$, $x(\beta) = 1_\beta$.

Choose inductively a κ-sequence $(x_\gamma)_{\gamma < \kappa}$ in $G \setminus \{1\}$ such that

$$\max \operatorname{supp}(x_\beta) < \min \operatorname{supp}(x_\gamma)$$

whenever $\beta < \gamma < \kappa$. For every $\gamma < \kappa$, $\bigoplus_{\beta \in \operatorname{supp}(x_\gamma)} H_\beta$ is countable. Therefore, without loss of generality one may suppose that $|\operatorname{supp}(x_\gamma)| = 1$. Let

$$A = \{x_\gamma : \gamma < \kappa\}.$$

We have to show that for any $p \in E(K(\beta G))$ and for any distinct $q_1, \ldots, q_n \in Q(A)$, the elements $pq_1 p, \ldots, pq_n p$ generate a free group.

For every $\alpha < \kappa$, define the idempotent $p_\alpha \in \beta H_\alpha$ by $p_\alpha = \overline{\pi_\alpha}(p)$, where $\pi_\alpha : H \to H_\alpha$ is the projection and $\overline{\pi_\alpha} : \beta H \to \beta H_\alpha$ is the continuous extension of π_α. Endow H_α with a regular left invariant topology in which p_α converges to 1_α (Theorem 7.17). Now endow H with the topology induced by the product topology on $\prod_{\alpha < \kappa} H_\alpha$ and let $\mathcal{T}$ denote the topology on G induced from H. Note that $p \in \operatorname{Ult}(\mathcal{T})$ and $A^* \subseteq \operatorname{Ult}(\mathcal{T})$.

Partition A into infinite subsets A_i, $i = 1, \ldots, n$, such that $A_i \in q_i$, and for each $i = 1, \ldots, n$, partition A_i into infinite subsets A_{ij}, $j < \omega$, such that $A_{ij} \notin q_i$. Also partition the set

$$B = \left\{x \in \bigoplus_\omega \mathbb{Z}_2 : |\operatorname{supp}(x)| = 1\right\}$$

into infinite subsets B_i, $i = 1, \ldots, n$, and enumerate B_i as $\{b_{ij} : j < \omega\}$ without repetitions.

For every $\alpha < \kappa$, define a local homomorphism $h_\alpha : H_\alpha \to \bigoplus_\omega \mathbb{Z}_2$ as follows. If there is $a \in A$ with supp$(a) = \{\alpha\}$, then pick a clopen neighborhood U_α of 1_α such that $\pi_\alpha(a) \notin U_\alpha$, pick i, j such that $a \in A_{ij}$, and define h_α by

$$h_\alpha(x) = \begin{cases} 0 & \text{if } x \in U_\alpha \\ b_{ij} & \text{otherwise.} \end{cases}$$

If there is no $a \in A$ with supp$(a) = \{\alpha\}$, define h_α by $h_\alpha(x) = 0$.

Now define a local homomorphism $h : H \to \bigoplus_\omega \mathbb{Z}_2$ by

$$h(x_1 + \cdots + x_t) = h_{\alpha_1}(x_1) + \cdots + h_{\alpha_t}(x_t),$$

where $t \in \mathbb{N}$, $\alpha_1 < \cdots < \alpha_t < \kappa$, and $x_s \in H_{\alpha_s}$ for each $s = 1,\ldots,t$, and let $f = h|_G$. Then $f : (G, \mathcal{T}) \to \bigoplus_\omega \mathbb{Z}_2$ is a local homomorphism such that $f(A_{ij}) = \{b_{ij}\}$ for every $i = 1,\ldots,n$ and $j < \omega$. It follows that

$$f^* : \mathrm{Ult}(\mathcal{T}) \to \beta \left(\bigoplus_\omega \mathbb{Z}_2 \right)$$

is a surjective homomorphism and $f^*(q_1),\ldots,f^*(q_n)$ are distinct elements from B^*. Since f^* is surjective, $f^*(p) \in E(K(\beta(\bigoplus_\omega \mathbb{Z}_2)))$ (Lemma 6.25). Note that

$$E\left(K\left(\beta\left(\bigoplus_\omega \mathbb{Z}_2\right)\right)\right) = E(K(\mathbb{H})).$$

Consequently, by Theorem 9.27, the elements

$$f^*(p) + f^*(q_i) + f^*(p) = f^*(pq_i\,p), \quad i = 1,\ldots,n,$$

generate a free group. Hence, the elements $pq_i\,p$, $i = 1,\ldots,n$, also generate a free group. $\square$

9.4 $K(\beta G)$ is not Closed

In this section we prove the following result.

Theorem 9.32. *Let G be an infinite group of cardinality κ and assume that κ is not Ulam-measurable. Then both $K(\beta G)$ and $E(\mathrm{cl}\,K(\beta G))$ are not closed.*

Pick $p \in E(K(\beta G))$ and let $T = C(p) = \{x \in G^* : xp = p\}$. By Theorem 7.17, there is a regular extremally disconnected left invariant topology $\mathcal{T}$ on G such that $\mathrm{Ult}(\mathcal{T}) = T$. Note that $K(T)$ is a right zero semigroup consisting of the idempotents from the minimal right ideal $p(\beta G)$ of βG, that is, $K(T) = E(p(\beta G))$. It follows that in order to prove Theorem 9.32, it suffices to show that

Theorem 9.33. *There are elements in* $\mathrm{cl}\,K(T)$ *which are not in* $(\mathrm{cl}\,K(T))T$.

Indeed, let $q \in (\mathrm{cl}\,K(T)) \setminus ((\mathrm{cl}\,K(T))T)$. Then clearly, $q \in \mathrm{cl}\,K(\beta G)$.

To see that $q \notin K(\beta G)$, assume the contrary. Then $q \in K(\beta G) \cap T = K(T)$, and so q is an idempotent. Consequently, $q = qq \in (\mathrm{cl}\,K(T))T$, a contradiction.

To see that $q \notin E(\mathrm{cl}\,K(\beta G))$, assume the contrary. Then $q = qq \in (\mathrm{cl}\,K(T))T$, a contradiction.

To prove Theorem 9.33, we need the following lemma.

Lemma 9.34. *Let X be a nondiscrete regular extremally disconnected space and assume that $|X|$ is not Ulam-measurable. Then there is a sequence $(U_n)_{n<\omega}$ of clopen sets in X such that $\bigcap_{n<\omega} U_n$ is a nonempty nowhere dense set.*

Proof. Choose a family $\mathcal{B}$ of clopen sets in X such that $Y = \bigcap \mathcal{B}$ is not open and $\lambda = |\mathcal{B}|$ is as small as possible. Since X is extremally disconnected and Y is closed, it follows that cl int $Y = \text{int } Y$, that is, int Y is clopen. Let $\mathcal{C} = \{U \setminus \text{int } Y : U \in \mathcal{B}\}$. Then $|\mathcal{C}| = \lambda$, all members of $\mathcal{C}$ are clopen, and $\bigcap \mathcal{C} = Z$ is a nonempty closed nowhere dense set.

Enumerate $\mathcal{C}$ as $\{U_\alpha : \alpha < \lambda\}$. Define a decreasing λ-sequence $(W_\alpha)_{\alpha < \lambda}$ of clopen subsets of X by putting $W_0 = X$ and $W_\alpha = \bigcap_{\beta < \alpha} U_\beta$ for $\alpha > 0$. Then $W_\alpha = \bigcap_{\beta < \alpha} W_\beta$ if α is a limit ordinal and $\bigcap_{\alpha < \lambda} W_\alpha = Z$. Define $f : X \setminus Z \to \lambda$ by

$$f(x) = \alpha \text{ if } x \in W_\alpha \setminus W_{\alpha+1}.$$

Pick $x \in Z$ and let $\mathcal{F}$ denote the filter on λ with a base consisting of subsets of the form $f(U \setminus Z)$ where U is a neighborhood of $x \in X$. Clearly, $\bigcap \mathcal{F} = \emptyset$. We claim that $\mathcal{F}$ is a λ-complete ultrafilter, and so $\lambda = \omega$.

To see that $\mathcal{F}$ is an ultrafilter, let $A \subseteq \lambda$. Then $X \setminus Z$ is a disjoint union of open sets $V_A = f^{-1}(A) = \bigcup_{\alpha \in A} W_\alpha \setminus W_{\alpha+1}$ and $V_{\lambda \setminus A} = f^{-1}(\lambda \setminus A)$. Since X is extremally disconnected, there is a neighborhood U of x such that either $U \cap V_{\lambda \setminus A} = \emptyset$ or $U \cap V_A = \emptyset$, so either $U \setminus Z \subseteq V_A$ or $U \setminus Z \subseteq V_{\lambda \setminus A}$. It follows that either $A \in \mathcal{F}$ or $\lambda \setminus A \in \mathcal{F}$.

To see that $\mathcal{F}$ is λ-complete, let $\mu < \lambda$, and for every $\alpha < \mu$, let $A_\alpha \in \mathcal{F}$. Then for every $\alpha < \mu$, $V_\alpha = f^{-1}(A_\alpha) \cup Z$ is a neighborhood of x. Consequently, by the minimality of λ, $V = \bigcap_{\alpha < \mu} V_\alpha$ is also a neighborhood of x. Define $A \in \mathcal{F}$ by $A = f(V \setminus Z)$. Then $A \subseteq \bigcap_{\alpha < \mu} A_\alpha$.   □

Proof of Theorem 9.33. By Theorem 9.18, there is an infinite decomposition $\mathcal{I}$ of $U(G)$ into closed left ideals of βG. Pick a sequence $(I_n)_{n < \omega}$ of distinct members of $\mathcal{I}$, and for every $n < \omega$, pick an idempotent $q_n \in I_n \cap K(T)$. Let $q \in \text{cl } \{q_n : n < \omega\}$ and let I denote the member of $\mathcal{I}$ such that $q \in I$. Without loss of generality one may suppose that I is distinct from all I_n. We show that $q \notin (\text{cl } K(T))T$.

By Lemma 9.34, there is a decreasing sequence $(W_n)_{n < \omega}$ of clopen neighborhoods of the identity of $(G, \mathcal{T})$ such that $Z = \bigcap_{n < \omega} W_n$ is a nowhere dense set. For every $n < \omega$, let $V_n = W_n \setminus W_{n+1}$. Assume on the contrary that $q = rs$ for some $r \in \text{cl } K(T)$ and $s \in T$. Clearly, $s \in I$. Note also that r is dense (Proposition 7.7), and consequently, $\bigcup_{n < \omega} V_n \in r$. Thus, we have that

$$(\text{cl } \{q_m : m < \omega\}) \cap \left(\text{cl } \bigcup_{n < \omega} V_n s \right) \neq \emptyset.$$

Then by Corollary 2.24, either $q_m \in \text{cl } \bigcup_{n < \omega} V_n s$ for some $m < \omega$ or $q' \in \overline{V_n s}$ for some $q' \in \text{cl } \{q_m : m < \omega\}$ and $n < \omega$. In the first case we obtain that $q_m \in I$, since $\text{cl } \bigcup_{n < \omega} V_n s \subseteq I$, a contradiction. The second case gives us that $V_n \in q'$, since $\overline{V_n s} \subseteq \overline{V_n}$, again a contradiction.   □

Remark 9.35. In the case $\kappa = \omega$ Theorem 9.33 can be strengthened as follows: There are elements in cl $K(T)$ which are not in T^2. Indeed, in this case the sequence $(W_n)_{n<\omega}$ of clopen neighborhoods of 1 may be chosen so that $\bigcap_{n<\omega} W_n = \{1\}$. Then for every $r \in T$, one has $\bigcup_{n<\omega} V_n \in r$.

References

It has long been known that $U(G)$ can be decomposed into 2^{2^κ} left ideals [14]. That it can be decomposed into 2^{2^κ} closed left ideals was established by I. Protasov [61] in the case where κ is a regular cardinal and by M. Filali and P. Salmi [24] for all κ. Theorems 9.3, 9.13, and 9.16 are from [109]. Corollary 9.19 is a partial case of the result of A. Lau and J. Pym [44]. Our proof of Corollary 9.19 is from [24].

Theorem 9.20 is from [104]. An introduction to the Pontryagin duality can be found in [55] and [34]. See also [50] and [17].

Theorem 9.25 is from [111]. Corollary 9.26 was proved also independently by S. Ferri, N. Hindman, and D. Strauss [23]. Theorem 9.27 is due to N. Hindman and J. Pym [36].

Theorem 9.32 complements the result from [106] which says that, for every infinite semigroup S embeddable algebraically into a compact group, both $K(\beta S)$ and $E(\text{cl } K(\beta S))$ are not closed.

Chapter 10
Almost Maximal Topological Groups

In this chapter almost maximal topological groups and their ultrafilter semigroups are studied. A topological (or left topological) group is said to be *almost maximal* if the underlying space is almost maximal. We show that the ultrafilter semigroup of any countable regular almost maximal left topological group is a projective in the category $\mathfrak{F}$ of finite semigroups. Assuming MA, for every projective S in $\mathfrak{F}$, we construct a group topology $\mathcal{T}$ on $\bigoplus_\omega \mathbb{Z}_2$ such that $\mathrm{Ult}(\mathcal{T})$ is isomorphic to S. We show that every countable almost maximal topological group contains an open Boolean subgroup and its existence cannot be established in ZFC. We then describe projectives in $\mathfrak{F}$. These are certain chains of rectangular bands. We conclude by showing that the ultrafilter semigroup of a countable regular almost maximal left topological group is its topological invariant.

10.1 Construction

Definition 10.1. An object S in some category is an *absolute coretract* if for every surjective morphism $f : T \rightarrow S$ there exists a morphism $g : S \rightarrow T$ such that $f \circ g = \mathrm{id}_S$.

Let $\mathfrak{C}$ denote the category of compact Hausdorff right topological semigroups. The next lemma gives us some simple examples of absolute coretracts in $\mathfrak{C}$.

Lemma 10.2. *Finite left zero semigroups, right zero semigroups and chains of idempotents are absolute coretracts in $\mathfrak{C}$.*

Proof. Let S be a finite left zero semigroup, let T be a compact Hausdorff right topological semigroup, and let $f : T \rightarrow S$ be a continuous surjective homomorphism. Pick a minimal left ideal L of T. For each $e \in S$, $f^{-1}(e)$ is a right ideal of T, so pick a minimal right ideal $R_e \subseteq f^{-1}(e)$ of T, and let $g(e)$ be the identity of the group $R_e \cap L$. Then $\{g(e) : e \in S\}$ is a left zero semigroup and $f \circ g = \mathrm{id}_S$.

The proof for finite right zero semigroups is similar.

Finally, suppose that S is a finite chain of idempotents, say $e_1 > \cdots > e_n$. Construct inductively a chain of idempotents $g(e_1) > \cdots > g(e_n)$ in T such that $f(g(e_i)) = e_i$ for each $i = 1, \ldots, n$. As $g(e_1)$ pick any idempotent in $T_1 = f^{-1}(e_1)$. Now fix $k < n$ and assume that we have constructed idempotents $g(e_1) > \cdots > g(e_k)$ in T such that $f(g(e_i)) = e_i$. Let $T_{k+1} = f^{-1}(\{e_1, \ldots, e_{k+1}\})$. Pick a minimal right ideal $R_{k+1} \subseteq g(e_k)T_{k+1}$ of T_{k+1}. Note that $R_{k+1} \subseteq f^{-1}(e_{k+1})$ (Lemma 6.25), so $g(e_k) \notin R_{k+1}$. And pick a minimal left ideal $L_{k+1} \subseteq T_{k+1}g(e_k)$ of T_{k+1}. Define $g(e_{k+1})$ to be the identity of the group $R_{k+1} \cap L_{k+1}$.   □

Definition 10.3. Let S be a finite semigroup. We say that S is an *absolute* $\mathbb{H}$-*coretract* if for every surjective proper homomorphism $\alpha : \mathbb{H} \to S$ there exists a homomorphism $\beta : S \to \mathbb{H}$ such that $\alpha \circ \beta = \mathrm{id}_S$.

Clearly, every finite absolute coretract in $\mathfrak{C}$ is an absolute $\mathbb{H}$-coretract.

Theorem 10.4. *Assume* $\mathfrak{p} = \mathfrak{c}$. *Let S be a finite absolute* $\mathbb{H}$-*coretract. Then there exists a group topology* $\mathcal{T}$ *on* $\bigoplus_\omega \mathbb{Z}_2$ *such that* $\mathrm{Ult}(\mathcal{T})$ *is isomorphic to* S.

The proof of Theorem 10.4 is based on the following lemma.

Lemma 10.5. *Let G be an infinite group, let $S = \{p_i : i < m\}$ be a finite semigroup in G^*, and let $\mathcal{F}$ be the filter on G such that $\overline{\mathcal{F}} = S$. For every $i < m$, let $A_i \in p_i$, and for every $x \in G$, let $B_x \in \mathcal{F}$. Then there is a sequence $(x_n)_{n<\omega}$ in G such that*

(1) $x_n \in A_{i_n}$ *where* $i_n \equiv n$ (mod m),

(2) *whenever* $k < \omega$ *and* $n_0 < \cdots < n_k < \omega$, *one has* $x_{n_0} \cdots x_{n_k} \in A_i$ *where* i *is defined by* $p_i = p_{i_{n_0}} \cdots p_{i_{n_k}}$,

(3) *whenever* $0 < k < \omega$, *one has* $\mathrm{FP}((x_n)_{k \leq n < \omega}) \subseteq \bigcap\{B_x : x \in \mathrm{FP}((x_n)_{n<k})\}$.

Proof. We shall construct a sequence $(x_n)_{n<\omega}$ satisfying (1)–(3) and, in addition, the following two conditions:

(4) *whenever* $k < \omega$, $n_0 < \cdots < n_k < \omega$ *and* $p \in S$, *one has* $x_{n_0} \cdots x_{n_k} p \subseteq \overline{A_i}$ *where* i *is defined by* $p_i = p_{i_{n_0}} \cdots p_{i_{n_k}} p$, *and*

(5) *whenever* $0 < k < \omega$ *and* $p \in S$,

$$\mathrm{FP}((x_n)_{k \leq n < \omega}) \cdot p \subseteq \overline{\bigcap\{B_x : x \in \mathrm{FP}((x_n)_{n<k})\}}.$$

Choose $C_0 \in p_0$ such that $C_0 \subseteq A_0$ and for each $p \in S$, one has $\overline{C_0}p \subseteq \overline{A_i}$ where i is defined by $p_0 p = p_i$. Pick $x_0 \in C_0$. Now fix $l > 0$ and assume that we have constructed a sequence $(x_n)_{n<l}$ satisfying (1)–(5).

Let $j \equiv l$ (mod m) and $j < m$. Choose $C_l \in p_j$ such that

(i) $C_l \subseteq A_j$,

(ii) whenever $k < l$ and $n_0 < \cdots < n_k < l$, $x_{n_0} \cdots x_{n_k}\overline{C_l} \subseteq \overline{A_i}$ where i is defined by $p_i = p_{i_{n_0}} \cdots p_{i_{n_k}} p_j$,

(iii) whenever $0 < k < l$, $\mathrm{FP}((x_n)_{k \leq n < l})\overline{C_l} \subseteq \overline{\bigcap\{B_x : x \in \mathrm{FP}((x_n)_{n<k})\}}$,

(iv) whenever $p \in S$, $\overline{C_l}p \subseteq \overline{A_i}$ where i is defined by $p_i = p_j p$,

(v) whenever $k < l, n_0 < \cdots < n_k < l$ and $p \in S$, one has $x_{n_0} \cdots x_{n_k} \overline{C_l} p \subseteq \overline{A_i}$ where i is defined by $p_i = p_{in_0} \cdots p_{in_k} p_j p$,

(vi) whenever $p \in S, \overline{C_l} p \subseteq \bigcap \{B_x : x \in FP((x_n)_{n<l})\}$, and

(vii) whenever $k < l$ and $p \in S$,

$$FP((x_n)_{k \leq n < l}) \overline{C_l} p \subseteq \bigcap \{B_x : x \in FP((x_n)_{n<k})\}.$$

Then pick $x_l \in C_l$. $\square$

Proof of Theorem 10.4. Let $G = \bigoplus_\omega \mathbb{Z}_2$ and let $Y = \{y_n : n < \omega\}$ be the standard basis of G, that is,

$$y_n(m) = \begin{cases} 1 & \text{if } m = n \\ 0 & \text{otherwise.} \end{cases}$$

Let $\mathcal{T}_0$ denote the group topology on G with a neighborhood base at 0 consisting of subgroups

$$FS((y_n)_{m \leq n < \omega}) \cup \{0\},$$

where $m < \omega$, so $\mathrm{Ult}(\mathcal{T}_0) = \mathbb{H}$. Pick any mapping $f_0 : Y \to S$ such that $|f_0^{-1}(p)| = \omega$ for each $p \in S$ and extend f_0 to a local homomorphism $f : (G, \mathcal{T}_0) \to S$ by

$$f(y_{n_0} + \cdots + y_{n_k}) = f_0(y_{n_0}) \cdots f_0(y_{n_k})$$

where $k < \omega$ and $n_0 < \cdots < n_k < \omega$. Let $\pi = f^*$. Then $\pi : \mathrm{Ult}(\mathcal{T}_0) \to S$ is a surjective proper homomorphism. For every $p \in S$, let $B_p = f^{-1}(p) \setminus \{0\}$. Equivalently,

$$B_p = \{y_{n_0} + \cdots + y_{n_k} : k < \omega, n_0 < \cdots < n_k < \omega, f(y_{n_0}) \cdots f(y_{n_k}) = p\}.$$

Enumerate the subsets of G as $\{Z_\alpha : \alpha < \mathfrak{c}\}$ with $Z_0 = G$. For every $\alpha < \mathfrak{c}$, we shall construct a sequence $(x_{\alpha,n})_{n<\omega}$ in $G \setminus \{0\}$ such that

(i) $\max \mathrm{supp}(x_{\alpha,n}) < \min \mathrm{supp}(x_{\alpha,n+1})$,

(ii) for every $\beta < \alpha, |\{x_{\alpha,n} : n < \omega\} \setminus FS((x_{\beta,n})_{n<\omega})| < \omega$,

(iii) for every $p \in S, |\{x_{\alpha,n} : n < \omega\} \cap B_p| = \omega$, and

(iv) for every $p \in S$, either $FS((x_{\alpha,n})_{n<\omega}) \cap B_p \subseteq Z_\alpha$ or $FS((x_{\alpha,n})_{n<\omega}) \cap B_p \subseteq G \setminus Z_\alpha$.

Note that these conditions imply that

(v) for every $k < \omega, \alpha_0 < \cdots < \alpha_k < \mathfrak{c}$ and $n_0, \ldots, n_{k-1} \in \mathbb{N}$, there is $n_k \in \mathbb{N}$ such that $FS((x_{\alpha_k,n})_{n_k \leq n < \omega}) \subseteq FS((x_{\alpha_i,n})_{n_i \leq n < \omega})$ for each $i < k$, and

(vi) for every $\alpha < \mathfrak{c}$ and $m < \omega$, there is $\beta < \mathfrak{c}$ such that $FS((x_{\beta,n})_{n<\omega}) \subseteq FS((x_{\alpha,n})_{m \leq n < \omega})$.

As $(x_{0,n})_{n<\omega}$ pick any sequence in $G \setminus \{0\}$ satisfying (i) and (iii) ((ii) and (iv) are satisfied trivially), for example, the sequence $(y_n)_{n<\omega}$ itself.

Fix $\alpha < \mathfrak{c}$ and assume that we have constructed sequences $(x_{\beta,n})_{n<\omega}$, $\beta < \alpha$, satisfying (i)–(iv). Since $\mathfrak{p} = \mathfrak{c}$ and $\alpha < \mathfrak{c}$, there is $Y_\alpha \subseteq G \setminus \{0\}$ such that for every $\beta < \alpha$ and $m < \omega$, $|Y_\alpha \setminus \mathrm{FS}((x_{\beta,n})_{m\leq n<\omega})| < \omega$, and for every $p \in S$, $|Y_\alpha \cap B_p| = \omega$. Pick a sequence $(y_{\alpha,n})_{n<\omega}$ in Y_α such that

$$\max \mathrm{supp}(y_{\alpha,n}) < \min \mathrm{supp}(y_{\alpha,n+1})$$

for all $n < \omega$ and $|\{y_{\alpha,n} : n < \omega\} \cap B_p| = \omega$ for every $p \in S$. Define the group topology $\mathcal{T}_\alpha$ on G by taking as a neighborhood base at 0 the subgroups

$$\mathrm{FS}((y_{\alpha,n})_{m\leq n<\omega}) \cup \{0\}$$

where $m < \omega$. Clearly, $\mathrm{Ult}(\mathcal{T}_\alpha) \subseteq \mathrm{Ult}(\mathcal{T}_0)$ and the restriction of π to $\mathrm{Ult}(\mathcal{T}_\alpha)$ is a proper homomorphism onto S. By Corollary 8.11, there is a proper isomorphism $\varphi_\alpha : \mathrm{Ult}(\mathcal{T}_0) \to \mathrm{Ult}(\mathcal{T}_\alpha)$, so $\pi \circ \varphi_\alpha : \mathrm{Ult}(\mathcal{T}_0) \to S$ is a surjective proper homomorphism. Since S is an absolute $\mathbb{H}$-coretract, there is a homomorphism $\delta_\alpha : S \to \mathrm{Ult}(\mathcal{T}_0)$ such that $\pi \circ \varphi_\alpha \circ \delta_\alpha = \mathrm{id}_S$. Let $\varepsilon = \varphi_\alpha \circ \delta_\alpha$. Then $\varepsilon_\alpha : S \to \mathrm{Ult}(\mathcal{T}_\alpha)$ is a homomorphism such that $\pi \circ \varepsilon_\alpha = \mathrm{id}_S$. For every $p \in S$, choose $A_{\alpha,p} \in \varepsilon_\alpha(p)$ such that either $A_{\alpha,p} \subseteq Z_\alpha$ or $A_{\alpha,p} \subseteq G \setminus Z_\alpha$, and put $A_\alpha = \bigcup_{p\in S} A_{\alpha,p}$. Then, using Lemma 10.2, choose a sequence $(x_{\alpha,n})_{n<\omega}$ satisfying (i) and (iii) and such that $\mathrm{FS}((x_{\alpha,n})_{n<\omega}) \subseteq A_\alpha$.

The subgroups $\mathrm{FS}((x_{\alpha,n})_{n<\omega}) \cup \{0\}$, $\alpha < \mathfrak{c}$, form a neighborhood base at 0 in a group topology $\mathcal{T}$ on G. For every $p \in S$, the subsets $\mathrm{FS}((x_{\alpha,n})_{n<\omega}) \cap B_p$, $\alpha < \mathfrak{c}$, form a base for an ultrafilter p' on G. Let $S' = \{p' : p \in S\}$. Then $\mathrm{Ult}(\mathcal{T}) = S'$ and $\pi|_{S'} : S' \ni p' \mapsto p \in S$ is an isomorphism. □

From Theorem 10.4 and Lemma 10.2 we obtain the following.

Corollary 10.6. *Assume* $\mathfrak{p} = \mathfrak{c}$. *Then for every* $n \in \mathbb{N}$, *there exist group topologies* $\mathcal{L}_n$, $\mathcal{R}_n$, *and* $\mathcal{C}_n$ *on* $\bigoplus_\omega \mathbb{Z}_2$ *whose ultrafilter semigroups are n-element left zero semigroup, right zero semigroup, and chain of idempotents, respectively.*

10.2   Properties

Definition 10.7. An object S in some category is a *projective* if for every morphism $f : S \to Q$ and for every surjective morphism $g : T \to Q$ there exists a morphism $h : S \to T$ such that $g \circ h = f$.

Note that every projective is an absolute coretract. In many categories these notions coincide but not in all.

Definition 10.8. Let S be a finite semigroup. We say that S is an $\mathbb{H}$-*projective* if for every homomorphism $\alpha : S \to Q$ of S into a finite semigroup Q and for every surjective proper homomorphism $\beta : \mathbb{H} \to Q$ there exists a homomorphism $\gamma : S \to \mathbb{H}$ such that $\alpha = \beta \circ \gamma$.

Clearly, every $\mathbb{H}$-projective is an absolute $\mathbb{H}$-coretract.
Let $\mathfrak{F}$ denote the category of finite semigroups.

Lemma 10.9. *Every $\mathbb{H}$-projective (absolute $\mathbb{H}$-coretract) is a projective (absolute coretract) in $\mathfrak{F}$.*

Proof. Let S be an $\mathbb{H}$-projective, let Q and T be finite semigroups, let $\alpha : S \to Q$ be a homomorphism, and let $\beta : T \to Q$ be a surjective homomorphism. Pick a surjective proper homomorphism $\delta : \mathbb{H} \to T$. Then $\beta \circ \delta : \mathbb{H} \to Q$ is a surjective proper homomorphism. Consequently, since S is an $\mathbb{H}$-projective, there is a homomorphism $\varepsilon : S \to \mathbb{H}$ such that $\alpha = \beta \circ \delta \circ \varepsilon$. Define the homomorphism $\gamma : S \to T$ by $\gamma = \delta \circ \varepsilon$. Then $\alpha = \beta \circ \gamma$.

The proof for an absolute $\mathbb{H}$-coretract is similar.   □

Theorem 10.10. *The ultrafilter semigroup of a countable regular almost maximal left topological group is an $\mathbb{H}$-projective.*

Before proving Theorem 10.10 we establish the following fact.

Lemma 10.11. *Let $(G, \mathcal{T})$ be an almost maximal T_1 left topological group and let $S = \mathrm{Ult}(\mathcal{T})$. Then for every homomorphism $\alpha : S \to Q$, there are an open neighborhood X of the identity of $(G, \mathcal{T})$ and a local homomorphism $f : X \to Q$ such that $f^* = \alpha$.*

Proof. For each $p \in S$, choose $A_p \in p$ such that $A_p \cap A_q = \emptyset$ if $p \neq q$. Then, for each $p \in S$, choose $B_p \in p$ such that $\overline{B_p q} \subseteq \overline{A_{pq}}$ for all $q \in S$. This can be done because the mapping $\beta G \ni x \mapsto xq \in \beta G$ is continuous and S is finite. Choose the subsets B_p in addition so that $B_p \subseteq A_p$ and

$$X = \bigcup_{p \in S} B_p \cup \{1\}$$

is open in $(G, \mathcal{T})$. Define $f : X \to Q$ by putting for every $p \in S$ and $x \in B_p$, $f(x) = \alpha(p)$. The value $f(1)$ does not matter. We claim that f is a local homomorphism and $f^* = \alpha$.

It suffices to check the first statement. Let $x \in X \setminus \{1\}$. Then $x \in B_p$ for some $p \in S$. For each $q \in S$, choose $C_q \in q$ such that $C_q \subseteq A_q$ and $x C_q \subseteq A_{pq}$. Choose a neighborhood U of $1 \in X$ such that $U \subseteq \bigcup_{q \in S} C_q \cup \{1\}$ and $xU \subseteq X$. Now let

$y \in U \setminus \{1\}$. Then $y \in C_q$ for some $q \in S$, so $y \in A_q$ and then $y \in B_q$. Hence $f(x)f(y) = \alpha(p)\alpha(q)$. On the other hand, $xy \in A_{pq}$, then $xy \in B_{pq}$, and so

$$f(xy) = \alpha(pq) = \alpha(p)\alpha(q).$$

Hence $f(xy) = f(x)f(y)$.   □

Proof of Theorem 10.10. Let $(G, \mathcal{T})$ be a countable regular almost maximal left topo-logical group and let $S = \mathrm{Ult}(\mathcal{T})$. Let Q be a finite semigroup, let $\alpha : S \to Q$ be a homomorphism, and let $\beta : \mathbb{H} \to Q$ be a surjective proper homomorphism. By Lemma 10.11, there are an open neighborhood X of the identity of $(G, \mathcal{T})$ and a local homomorphism $f : X \to Q$ such that $f^* = \alpha$, so α is proper. Now by Corol-lary 8.14, there is a proper homomorphism $\gamma : \mathrm{Ult}(\mathcal{T}) \to \mathbb{H}$ such that $\alpha = \beta \circ \gamma$. Hence, S is an $\mathbb{H}$-projective.   □

Recall that a *band* is a semigroup of idempotents.

Theorem 10.12. *The ultrafilter semigroup of a countable regular almost maximal left topological group is a band.*

Proof. Let $(G, \mathcal{T})$ be a countable regular almost maximal left topological group and let $S = \mathrm{Ult}(\mathcal{T})$. Assume on the contrary that S is not a band. By Corollary 8.21, S contains no nontrivial finite groups. Consequently, S contains a 2-element null subsemigroup, that is, there are distinct $p, q \in S$ such that

$$q^2 = qp = pq = p^2 = p.$$

It follows from $q^2 = p^2$ that for any $A \in p$ and $B \in q$, $\overline{B}q \cap \overline{A}p \neq \emptyset$, and hence by Corollary 2.23, either $Bq \cap \overline{A}p \neq \emptyset$ or $\overline{B}q \cap Ap \neq \emptyset$. Consider two cases.

Case 1: $Bq \cap \overline{A}p \neq \emptyset$ for some $A \in p$ and $B \in q$. Then $yq = p'p$ for some $y \in B$ and $p' \in \overline{A}$, so $q = y^{-1}p'p = y^{-1}p'pp = qp = p$, a contradiction.

Case 2: $\overline{B}q \cap Ap \neq \emptyset$ for all $A \in p$ and $B \in q$. Then $q_{A,B}q = x_{A,B}p$ for some $q_{A,B} \in \overline{B}$ and $x_{A,B} \in A$, so $p = x_{A,B}^{-1}q_{A,B}q = r_{A,B}q$ where $r_{A,B} = x_{A,B}^{-1}q_{A,B}$. Since $\mathcal{T}$ is regular, it follows from $p = r_{A,B}q$ that $r_{A,B} \in S$. And since S is finite, it then follows from $q_{A,B} = x_{A,B}r_{A,B}$ that $q_B = pr_B$ for some $r_B \in S$ and $q_B \in \overline{B}$. Consequently, $q = pr$ for some $r \in S$. But then $q = pr = ppr = pq = p$, again a contradiction.   □

Theorem 10.12, Theorem 10.10 and Theorem 10.4 raise the question of character-izing finite bands which are $\mathbb{H}$-projectives. We address this question in Section 10.4.

Now we consider the structure of a countable almost maximal topological group.

Let X be a space with a distinguished point $1 \in X$ and let $f : X \to X$ be a homeomorphism with $f(1) = 1$. We say that f is *trivial* if the set of fixed points of f is a neighborhood of f.

Lemma 10.13. *Let X be a countable regular local left group and suppose that $K(\mathrm{Ult}(X))$ is finite. Then every local automorphism on X is trivial.*

Proof. Assume on the contrary that there is a nontrivial local automorphism $f : X \to X$. For every $n \in \mathbb{N}$, let $X_n = \{x \in X : |O(x)| > n\}$. Note that X_n is open, $f^n(x) \neq x$ for all $x \in X_n$, and $X_{n+1} \subseteq X_n$. By Corollary 8.23, it suffices to consider the following two cases.

Case 1: for every $n \in \mathbb{N}$, $1 \in \mathrm{cl}\, X_n$. Let $\mathcal{F}$ be the filter on X with a base consisting of subsets $U \cap X_n$, where U runs over neighborhoods of $1 \in X$ and $n \in \mathbb{N}$, and let $R = \overline{\mathcal{F}}$. Then for every $p \in R$, all elements $p, f^*(p), (f^*)^2(p), \ldots$ are distinct. Indeed, otherwise $(f^*)^n(p) = p$ for some $n \in \mathbb{N}$, and consequently by Corollary 2.18, $\{x \in X : f^n(x) = x\} \in p$, which contradicts $X_n \in p$. Next, by Corollary 7.3, R is a right ideal of $\mathrm{Ult}(X)$, so there is $p \in R \cap K(\mathrm{Ult}(X))$. Since f^* is an automorphism on $\mathrm{Ult}(X)$, it follows that $(f^*)^n(p) \in K(\mathrm{Ult}(X))$ for every n. Hence, $K(\mathrm{Ult}(X))$ is infinite, a contradiction.

Case 2: f has finite order. Since f is nontrivial, $1 \in \mathrm{cl}\, X_1$. Let $R_1 = \overline{X_1} \cap \mathrm{Ult}(X)$. Then R_1 is a right ideal of $\mathrm{Ult}(X)$, so there is an idempotent $p \in R_1 \cap K(\mathrm{Ult}(X))$. We have also that $f^*(p) \neq p$ and clearly $f^*(p) \in K(\mathrm{Ult}(X))$. But then, applying Theorem 8.31, we obtain that the structure group of $K(\mathrm{Ult}(X))$ is infinite, again a contradiction. $\qquad\square$

Lemma 10.14. *Let G be a countable group endowed with an invariant topology. Suppose that for every $a \in G$, the conjugation $G \ni x \mapsto axa^{-1} \in G$ is a trivial local automorphism. Then the inversion $\iota : G \ni x \mapsto x^{-1} \in G$ is a local automorphism.*

Proof. To see that ι is a local homomorphism, let $a \in G \setminus \{1\}$. Since $G \ni x \mapsto axa^{-1} \in G$ is a trivial, $U = \{x \in G : axa^{-1} = x\}$ is a neighborhood of 1. Pick a neighborhood V of 1 such that $V = V^{-1} \subseteq U$. Then for every $x \in V$,

$$\iota(ax) = (ax)^{-1} = x^{-1}a^{-1} = a^{-1}ax^{-1}a^{-1} = a^{-1}x^{-1} = \iota(a)\iota(x). \qquad\square$$

Theorem 10.15. *Every countable almost maximal topological group contains an open Boolean subgroup.*

Proof. Let G be a countable almost maximal topological group. By Lemma 10.13, the conjugations of G are trivial. Then by Lemma 10.14, the inversion $\iota : G \ni x \mapsto x^{-1} \in G$ is a local automorphism. Consequently by Lemma 10.13, ι is trivial, so $\{x \in G : \iota(x) = x\}$ is a neighborhood of 1. Since $\{x \in G : \iota(x) = x\} = B(G)$, G contains an open Boolean subgroup by Lemma 5.3. $\qquad\square$

We conclude this section by showing that the existence of a countable almost maximal topological group cannot be established in ZFC.

Theorem 10.16. *The existence of a countable almost maximal topological group implies the existence of a P-point in ω^*.*

Proof. Let $(G, \mathcal{T})$ be a countable almost maximal topological group. By Theorem 10.15, one may suppose that $G = \bigoplus_\omega \mathbb{Z}_2$. Let $\mathcal{F}$ be the neighborhood filter of 0 in $\mathcal{T}$ and let $\mathcal{T}_0$ denote the topology on G induced by the product topology on $\prod_\omega \mathbb{Z}_2$. By Theorem 10.12 and Lemma 7.10, $\mathcal{T}_0 \subseteq \mathcal{T}$. Then by Theorem 5.19, each point of $\overline{\phi(\mathcal{F})}$ is a P-point.   □

Combining Theorem 10.16 and Theorem 2.38 gives us that

Corollary 10.17. *It is consistent with ZFC that there is no countable almost maximal topological group.*

Remark 10.18. It is easy to see that Theorem 10.15 and Theorem 10.16 remain to be true if to replace 'countable almost maximal topological group' by 'countable topological group $(G, \mathcal{T})$ with finite $K(\mathrm{Ult}(\mathcal{T}))$'.

10.3   Semilattice Decompositions and Burnside Semigroups

This is a preliminary section for the next one.

A partially ordered set is a *semilattice* if every 2-element subset $\{a, b\}$ has a greatest lower bound $a \wedge b$. Obviously, every semilattice is a commutative band with respect to the operation $\wedge$. Conversely, every commutative band is a semilattice with respect to the standard ordering $\leq$ on idempotents. That is, $a \leq b$ if and only if $ab = ba = a$, and then $a \wedge b = ab$. We shall identify semilattices with commutative bands.

A *semilattice decomposition* of a semigroup S is a homomorphism $f : S \to \Gamma$ of S onto a semilattice Γ. Equivalently, a semilattice decomposition of S is a partition $\{S_\alpha : \alpha \in \Gamma\}$ of S into subsemigroups with the property that for every $\alpha, \beta \in \Gamma$ there is $\gamma \in \Gamma$ such that $S_\alpha S_\beta \subseteq S_\gamma$ and $S_\beta S_\alpha \subseteq S_\gamma$.

A semigroup which is a union of groups is called *completely regular*.

Theorem 10.19. *Every completely regular semigroup decomposes into a semilattice of completely simple semigroups.*

Proof. Let S be a completely regular semigroup. For every $a \in S$, let $J(a) = SaS$. We first note that if H is a subgroup of S containing a, then $H \subseteq J(a)$ and for every $h \in H$, $J(h) = J(a)$. For $h = haa^{-1} \in J(a)$ and $a = ahh^{-1} \in J(h)$. It follows from $J(a^2) = J(a)$ that $J(ab) = J(ba)$ for all $a, b \in S$. For $ab \cdot ab = a \cdot ba \cdot b \in J(ba)$, so $ab \in J(ba)$, and similarly $ba \in J(ab)$.

Now we show that $J(ab) = J(a) \cap J(b)$. That $J(ab) \subseteq J(a) \cap J(b)$ is obvious. Conversely, let $c \in J(a) \cap J(b)$. Write $c = uav = xby$ for some $u, v, x, y \in S$.

Then $c^2 = xbyuav \in J(byua) = J(abyu)$. Consequently, $c \in J(abyu) \subseteq J(ab)$, and so $J(a) \cap J(b) \subseteq J(ab)$. Hence $J(ab) = J(a) \cap J(b)$.

It follows that $\Gamma = \{J(a) : a \in S\}$ is the semilattice of principal ideals of S under intersection and $S \ni a \mapsto J(a) \in \Gamma$ is a surjective homomorphism. For every $a \in S$, the preimage of the element $J(a) \in \Gamma$ is the set $J_a = \{b \in S : J(b) = J(a)\}$. Being the preimage of an idempotent, J_a is a subsemigroup of S.

To see that J_a is simple, let $b \in J_a$. Write $a = ubv$ for some $u, v \in S$. Then $a = eae = eubve$ where e is the identity of a group containing a. It follows that $J(a) \subseteq J(eu) \subseteq J(e) = J(a)$, so $eu \in J_a$. Similarly $ve \in J_a$. Hence $a \in J_a b J_a$.

Finally, to show that J_a is completely simple, it suffices to show that $J_a e$ is a minimal left ideal of J_a. Let $b \in J_a e$. Write $b = ue$ for some $u \in J_a$ and, since J_a is simple, $e = vbw$ for some $v, w \in J_a$. Then $e = vuew = xew$ where $x = vu \in J_a$. Let f be the identity of a group containing x and x^{-1} the inverse of x with respect to f. Then $fe = fxew = xew = e$ and $e = fe = x^{-1}xe = x^{-1}vue = x^{-1}vb \in J_a b$. $\qquad \square$

A band is a completely simple semigroup if and only if it is *rectangular*, that is, isomorphic to the direct product of a left zero semigroup and a right zero semigroup.

Corollary 10.20. *Every band decomposes into a semilattice of rectangular bands.*

The next lemma tells us that a semilattice decomposition of a completely regular semigroup is in fact unique.

Lemma 10.21. *Let S be a completely regular semigroup and let $\{S_\alpha : \alpha \in Y\}$ be a decomposition of S into a semilattice of completely simple semigroups. Then the semigroups S_α are precisely the maximal completely simple subsemigroups of S.*

Proof. Let $f : S \to Y$ be a homomorphism of S onto a commutative band Y and let T be a simple subsemigroup of S. Then $f(T)$ is a simple subsemigroup of Y. But simple commutative bands are trivial. Hence, $T \subseteq f^{-1}(\alpha)$ for some $\alpha \in Y$. $\qquad \square$

Maximal completely simple subsemigroups of a completely regular semigroup S are called *completely simple components* of S. If S is a band, we say *rectangular components* instead of completely simple components.

Recall that Green's relations $\mathcal{R}, \mathcal{L}, \mathcal{J}$ on any semigroup S are defined by

$$a\mathcal{R}b \Leftrightarrow aS^1 = bS^1, \quad a\mathcal{L}b \Leftrightarrow S^1a = S^1b, \quad a\mathcal{J}b \Leftrightarrow S^1aS^1 = S^1bS^1.$$

The proof of Theorem 10.19 shows that $\mathcal{J}$-classes of a completely regular semigroup are its completely simple components, $\mathcal{R}$-classes and $\mathcal{L}$-classes are minimal right and left ideals of the components.

Definition 10.22. Let $1 \leq k < \omega$ and $0 \leq m < n < \omega$. The *Burnside semigroup* $B(k, m, n)$ is the largest semigroup on k generators satisfying the identity $x^m = x^n$.

Note that $B(k, 1, 2)$ is the free band on k generators and $B(k, 0, n)$ is the Burnside group $B(k, n)$ (see Example 1.25).

Theorem 10.23. *For every fixed $n \geq 2$, the following statements are equivalent:*

(1) *the semigroups $B(k, 1, n)$ are finite for all $k \geq 1$, and*

(2) *the groups $B(k, 0, n - 1)$ are finite for all $k \geq 1$.*

Proof. The implication (1) $\Rightarrow$ (2) is obvious. We need to prove (2) $\Rightarrow$ (1).

Let $B = B(k, 1, n)$, let F be the free semigroup on a k-element alphabet A, and let $h : F \rightarrow B$ be the canonical homomorphism. Note that for any $u, v \in F$, $h(u) = h(v)$ if and only if v can be obtained from u by a succession of elementary operations in each of which a subword w of a word is replaced by w^n, or vice versa.

For every $w \in F$, let $\mathrm{ct}(w)$ denote the set of letters of A appearing in w.

Lemma 10.24. *Let $u, v \in F$. Then $h(u) \mathcal{J} h(v)$ if and only if $\mathrm{ct}(u) = \mathrm{ct}(v)$.*

Proof. If $h(u) \mathcal{J} h(v)$, then $h(u) = h(v_1 v v_2)$ and $h(v) = h(u_1 u u_2)$. Since the function ct is invariant under elementary operations, $\mathrm{ct}(v_1 v v_2) = \mathrm{ct}(u)$, so $\mathrm{ct}(v) \subseteq \mathrm{ct}(u)$. Similarly, $\mathrm{ct}(u) \subseteq \mathrm{ct}(v)$. Hence, $\mathrm{ct}(u) = \mathrm{ct}(v)$.

Conversely, suppose that $\mathrm{ct}(u) = \mathrm{ct}(v) = \{x_1, \ldots, x_n\}$. Since B is a semilattice of its $\mathcal{J}$-classes, it then follows that $h(u) \mathcal{J} h(x_1) \cdots h(x_n)$ and $h(v) \mathcal{J} h(x_1) \cdots h(x_n)$. Hence, $h(u) \mathcal{J} h(v)$. $\square$

By Lemma 10.24, the mapping $B \ni h(w) \mapsto \mathrm{ct}(w) \subseteq A$ gives us the semilattice decomposition of B into completely simple components ($= \mathcal{J}$-classes). The semilattice involved is the set of nonempty subsets of A under union. We need to show that $\mathcal{J}$-classes of B are finite. For every nonempty $C \subseteq A$, let J_C denote the $\mathcal{J}$-class of B corresponding to C. Clearly, for every $a \in A$, $J_{\{a\}}$ is just the group $\{a, a^2, \ldots, a^{n-1}\}$ (with identity a^{n-1}).

For every $w \in F$, let $\rho^*(w)$ denote the letter of A which has the latest first appearance in w, $\rho'(w)$ the subword of w that precedes the first appearance of $\rho^*(w)$, and $\rho(w)$ the subword $\rho'(w) \rho^*(w)$. Let $\lambda^*(w)$ denote the letter of A which has the earliest last appearance in w, $\lambda'(w)$ the subword of w that follows the last appearance of $\lambda^*(w)$, and $\lambda(w)$ the subword $\lambda^*(w) \lambda'(w)$.

Lemma 10.25. *Let $C \subseteq A$ with $|C| > 1$ and let $u, v \in F$ with $\mathrm{ct}(u) = \mathrm{ct}(v) = C$. Then*

(1) *$h(u) \mathcal{R} h(v)$ if and only if $\rho^*(u) = \rho^*(v)$ and $h(\rho'(u)) = h(\rho'(v))$,*

(2) *$h(u) \mathcal{L} h(v)$ if and only if $\lambda^*(u) = \lambda^*(v)$ and $h(\lambda'(u)) = h(\lambda'(v))$.*

Proof. (1) If $h(u)\mathcal{R}h(v)$, then $h(u) = h(vw)$ for some $w \in F$ with $\mathrm{ct}(w) = C$. Clearly, $\rho(vw) = \rho(v)$. It is easy to see that the functions ρ^* and $h\rho'$ are invariant under elementary operations. Consequently, $\rho^*(u) = \rho^*(v)$ and $h(\rho'(u)) = h(\rho'(v))$.

Conversely, suppose that $\rho^*(u) = \rho^*(v)$ and $h(\rho'(u)) = h(\rho'(v))$. Then $h(\rho(u)) = h(\rho(v))$. Consequently, in order to finish the proof it suffices to show that for every $w \in F$, $h(w)\mathcal{R}h(\rho(w))$.

We proceed by the induction on the length $|w|$ of w. The statement is obviously true if $|w| = 1$. Let $|w| > 1$ and suppose that the statement holds for all words of length $< |w|$. If $w = \rho(w)$, there is nothing to prove. Let $w \neq \rho(w)$. Then $w = w_1 a w_2 a$ for some words w_1, w_2, possibly empty. Consequently, $\rho(w) = \rho(w_1 a w_2)$ and by the inductive assumption, $h(\rho(w_1 a w_2))\mathcal{R}h(w_1 a w_2)$, so $h(\rho(w))\mathcal{R}h(w_1 a w_2)$. But

$$h(w_1 a w_2) = h(w_1 (a w_2)^n) = h(w_1 a w_2 a)h(w_2 (a w_2)^{n-2}) = h(w)h(w_2 (a w_2)^{n-2})$$

and $h(w) = h(w_1 a w_2)h(a)$, so $h(w_1 a w_2)\mathcal{R}h(w)$. Hence $h(\rho(w))\mathcal{R}h(w)$.

The proof of (2) is similar. $\square$

Lemma 10.25 gives us a one-to-one correspondence between $\mathcal{R}$-classes ($\mathcal{L}$-classes) of the $\mathcal{J}$-class J_C and the pairs (x, a) where $a \in C$ and $x \in J_{C\setminus\{a\}}$.

Lemma 10.26. *Let S be a finitely generated completely regular semigroup and let C be a be a completely simple component of S. Suppose that C contains a finite number of minimal left (right) ideals. Then the structure group of C is finitely generated.*

Proof. Let T be the subsemigroup of S consisting of all completely simple components of S over C, let X be a finite generating subset of S, and let $Y = X \cap T$. Then Y is a generating subset of T and $C = K(T)$. Suppose that C contains a finite number of minimal left ideals. Let H be a maximal subgroup of C, let R be the minimal right ideal of C containing H, and let $E = E(R)$. Then Te, where $e \in E$, are the minimal left ideals of C and $R \cap (Te)$, where $e \in E$, the maximal subgroups of R. Let $e_1 \in E$ be the identity of H. We claim that $\{exe_1 : x \in Y \cup Y^{-1}, e \in E\}$ is a generating subset of H. To see this, let $h \in H$. Write $h = x_1 \cdots x_n$ for some $x_1, \ldots, x_n \in Y \cup Y^{-1}$. Then $h = e_1 h e_1 = e_1 x_1 \cdots x_n e_1$. Define inductively $e_2, \ldots, e_n \in E$ by $e_i x_i \in Te_{i+1}$. Then $e_i x_i = e_i x_i e_{i+1} = e_i x_i e_1 e_{i+1}$, and so $h = (e_1 x_1 e_1)(e_2 x_2 e_1) \cdots (e_n x_n e_1)$. $\square$

Now for each $i = 1, \ldots, k$, pick $C_i \subseteq A$ with $|C_i| = i$ and let $c_i = |J_{C_i}|$. Clearly, $c_1 = n - 1$. Fix $i > 1$ and assume that c_{i-1} is finite. Then by Lemma 10.25, the number of minimal right (left) ideals of J_{C_i} is $i c_{i-1}$, and by Lemma 10.26, the cardinality g_i of the structure group of J_{C_i} is finite, so $c_i = (i c_{i-1})^2 g_i$ is also finite. $\square$

From Theorem 10.23 we obtain the following.

Corollary 10.27. *The semigroups $B(k, 1, 2)$ and $B(k, 1, 3)$ are finite for all $k \geq 1$.*

Proof. Since $B(k, 0, 2)$ is the Boolean group on k generators, $|B(k, 0, 2)| = 2^k$. Hence, the result follows from Theorem 10.23. $\square$

10.4   Projectives

In this section we describe finite bands which are $\mathbb{H}$-projectives.

Let V denote the set of words of the form

$$i_1 i_2 \cdots i_p \lambda_p \lambda_{p-1} \cdots \lambda_1$$

where $p \in \mathbb{N}$ and $i_q, \lambda_q \in \omega$ for each $q = 1, \ldots, p$. Define the operation on V by

$$i_1 \cdots i_p \lambda_p \cdots \lambda_1 \cdot j_1 \cdots j_q \rho_q \cdots \rho_1 = \begin{cases} i_1 \cdots i_p \rho_p \cdots \rho_1 & \text{if } p = q \\ i_1 \cdots i_p \lambda_p \cdots \lambda_{q+1} \rho_q \cdots \rho_1 & \text{if } p > q \\ i_1 \cdots i_p j_{p+1} \cdots j_q \rho_q \cdots \rho_1 & \text{if } p < q. \end{cases}$$

It is easy to see that V is a band being decomposed into a decreasing chain of its rectangular components V_p whose elements are words of length $2p$.

Now let **P** denote the family of finite subsemigroups of V satisfying the following conditions for every $p \in \mathbb{N}$:

(1) if $i_1 \cdots i_p \lambda_p \cdots \lambda_1 \in S$, then both $i_p \neq 0$ and $\lambda_p \neq 0$,

(2) if $i_1 \cdots i_p \lambda_p \cdots \lambda_1 \in S$ and $i_q \neq 0$ for some $q \in [1, p-1]$, then $i_1 \cdots i_q 1 \cdots 1 \in S$, and dually, if $i_1 \cdots i_p \lambda_p \cdots \lambda_1 \in S$ and $\lambda_q \neq 0$ for some $q \in [1, p-1]$, then $1 \cdots 1 \lambda_q \cdots \lambda_1 \in S$ (here, $i_1 \cdots i_q 1 \cdots 1$ and $1 \cdots 1 \lambda_q \cdots \lambda_1$ denote the elements from V_q), and

(3) either $i_p = 1$ for all $i_1 \cdots i_q \lambda_q \cdots \lambda_1 \in S$ with $q \geq p$ or $\lambda_p = 1$ for all $i_1 \cdots i_q \lambda_q \cdots \lambda_1 \in S$ with $q \geq p$.

Theorem 10.28. *Every semigroup from **P** is a projective in $\mathfrak{C}$.*

Proof. Let $S \in \mathbf{P}$. Let $f : S \to Q$ be a homomorphism of S into a finite semigroup Q and let $g : T \to Q$ be a continuous homomorphism of a compact Hausdorff right topological semigroup T onto Q. We adjoin the identities $\emptyset, 1_Q, 1_T$ to S, Q, T respectively and extend f, g in the obvious way. We shall inductively construct a homomorphism $h : S \to T$ such that $g \circ h = f$.

Let $l = \max\{p \in \mathbb{N} : S \cap V_p \neq \emptyset\}$. For each $p \in [1, l]$, let $S_p = S \cap V_p$ and define $e_p \in S_p$ by $e_p = 1 \cdots 1 1 \cdots 1$. Also put $S_0 = \{\emptyset\}$, $e_0 = \emptyset$, and $S_0^p = \bigcup_{q=0}^p S_q$. If $x = i_1 \cdots i_q 0 \cdots 0 i_p \lambda_p 0 \cdots 0 \lambda_r \cdots \lambda_1 \in S_p$ with $i_q, \lambda_r \neq 0$, we put $x^\star = i_1 \cdots i_q 1 \cdots 1 \in S_q$ and $x^{\star\star} = 1 \cdots 1 \lambda_r \cdots \lambda_1 \in S_r$ (if $q = 0$, $x^\star = \emptyset$, and if $r = 0$, $x^{\star\star} = \emptyset$).

Lemma 10.29.   (a) *For every $x \in S$, $x^* x = xx^{**} = x$.*

(b) *If $x \in S_p$ and $y \in S_q$, where $q < p$, then $(xy)^{**} = x^{**}y$ and $(yx)^* = yx^*$.*

(c) *If $x, y \in S_p$, then $x^{**}y^* = e_{p-1}$.*

Proof. (a) is obvious.

(b) Let $x = i_1 \cdots i_p \lambda_p 0 \cdots 0 \lambda_s \cdots \lambda_1$ and $y = j_1 \cdots j_q \rho_q \cdots \rho_1$, where $\lambda_s \neq 0$. Then $x^{**} = 1 \cdots 1 \lambda_s \cdots \lambda_1$ and $xy = i_1 \cdots i_p \lambda_p \cdots \lambda_{q+1} \rho_q \cdots \rho_1$. If $s = q$, then $x^{**}y = 1 \cdots 1 \rho_q \cdots \rho_1$ and $(xy)^{**} = 1 \cdots 1 \rho_q \cdots \rho_1$. If $s > q$, then $x^{**}y = 1 \cdots 1 \lambda_s \cdots \lambda_{q+1} \rho_q \cdots \rho_1$ and $(xy)^{**} = 1 \cdots 1 \lambda_s \cdots \lambda_{q+1} \rho_q \cdots \rho_1$. If $s < q$, then $j_{s+1} = \cdots = j_q = 1$, so $x^{**}y = 1 \cdots 1 \rho_q \cdots \rho_1$, and $(xy)^{**} = 1 \cdots 1 \rho_q \cdots \rho_1$.

The proof that $(yx)^* = yx^*$ is similar.

(c) Let $x = i_1 \cdots i_p \lambda_p 0 \cdots 0 \lambda_s \cdots \lambda_1$ and $y = j_1 \cdots j_t 0 \cdots 0 j_p \rho_p \cdots \rho_1$, where $\lambda_s, j_t \neq 0$. Then $x^{**} = 1 \cdots 1 \lambda_s \cdots \lambda_1$, $y^* = j_1 \cdots j_t 1 \cdots 1$, so

$$
x^{**}y^* = \begin{cases} 1 \cdots 1 1 \cdots 1 & \text{if } s = t \\ 1 \cdots 1 \lambda_s \cdots \lambda_{t+1} 1 \cdots 1 & \text{if } s > t \\ 1 \cdots 1 j_{s+1} \cdots j_t 1 \cdots 1 & \text{if } s < t. \end{cases}
$$

Now, if $s > t$, then $\lambda_s = \cdots = \lambda_{t+1} = 1$, and if $s < t$, then $j_{s+1} = \cdots = j_t = 1$. Furthermore, $s = p - 1$ or $t = p - 1$.   □

If $x = i_1 \cdots i_p \lambda_p \cdots \lambda_1 \in V_p$, we put $x' = i_1 \cdots i_p$ and $x'' = \lambda_p \cdots \lambda_1$, and for each $q = 1, \ldots, p$, put $x'_q = i_q$ and $x''_q = \lambda_q$. If $x \in S_p$, we put $R(x) = \{y \in S_p : y' = x'\}$ and $L(x) = \{y \in S_p : y'' = x''\}$. Note that these are respectively minimal right and minimal left ideals of S_p containing x.

Define h on S_0 by $h(\emptyset) = 1_T$. Suppose that h has been defined on S_0^{p-1}. We shall show that h can be extended to S_p.

Let $I_p = \{x'_p : x \in S_p\}$. For each $i \in I_p$, choose $z_i \in S_p$ such that $(z_i)'_p = i$ and $\min\{q \in [0, p-1] : (z_i)'_t = 0 \text{ for all } t \in [q+1, p-1]\}$ is as small as possible. Then choose a minimal right ideal $R_p(i)$ in $g^{-1}(f(S_p))$ with $g(R_p(i)) \subseteq f(R((z_i)))$. Note that for any $x \in S_p$ with $x'_p = i$, one has $x^* R((z_i)) \subseteq R(x)$, so $g(h(x^*)R_p(i)) \subseteq f(x^*)f(R((z_i))) \subseteq f(R(x))$. Consequently, for any $x \in S_p$, one has $g(h(x^*)R_p(x'_p)) \subseteq f(R(x))$. We define minimal left ideals $L_p(\lambda)$ in $g^{-1}(f(S_p))$ in the dual way. Now for every $x \in S_p$, we define $h(x)$ to be the idempotent of the group $h(x^*)R_p(x'_p)L_p(x''_p)h(x^{**})$. Since

$$
gh(x) \in g(h(x^*)R_p(x'_p))g(L_p(x''_p)h(x^{**})) \subseteq f(R(x))f(L(x)) = f(\{x\}),
$$

we have that $gh(x) = f(x)$.

Now we shall show that $h(x)h(y) = h(xy)$ for every $x \in S_p$ and $y \in S_0^p$. The proof that $h(y)h(x) = h(yx)$ is similar.

First let $y \in S_0^{p-1}$. We have that $h(x)h(y) \in h(x^\star)R_p(x_p')L_p(x_p'')h(x^{\star\star})h(y)$. But $x^\star = (xy)^\star$, $x_p' = (xy)_p'$, $x_p'' = (xy)_p''$, and $x^{\star\star}y = (xy)^{\star\star}$ by Lemma 10.29, so $h(x^{\star\star})h(y) = h(x^{\star\star}y) = h((xy)^{\star\star})$. It follows that $h(x)h(y)$ and $h(xy)$ belong to the same group in $g^{-1}(f(S_p))$. Therefore, it suffices to show that $h(x)h(y)$ is an idempotent. We show this by proving that $h(x)h(y)h(x) = h(x)$.

Write $h(x) = h(x^\star)zh(x^{\star\star})$ for some $z \in R_p(x_p')L_p(x_p'')$. Then

$$h(x)h(y)h(x) = h(x^\star)zh(x^{\star\star})h(y)h(x^\star)zh(x^{\star\star}) = h(x^\star)zh(x^{\star\star}yx^\star)zh(x^{\star\star}).$$

Since $x^{\star\star}yx^\star = (xy)^{\star\star}x^\star = e_{p-1} = x^{\star\star}x^\star$ by Lemma 10.29,

$$h(x)h(y)h(x) = h(x^\star)zh(x^{\star\star}x^\star)zh(x^{\star\star}) = h(x^\star)zh(x^{\star\star})h(x^\star)zh(x^{\star\star})$$
$$= h(x)h(x) = h(x).$$

Now let $y \in S_p$. Again, we have that $h(x)h(y) \in h(x^\star)R_p(x_p')L_p(y_p'')h(y^{\star\star})$ and also $x^\star = (xy)^\star$, $x_p' = (xy)_p'$, $y_p'' = (xy)_p''$, and $y^{\star\star} = (xy)^{\star\star}$. So $h(x)h(y)$ and $h(xy)$ belong to the same group. We again show that $h(x)h(y)$ is an idempotent by proving that $h(x)h(y)h(x) = h(x)$.

We know that either $z_p'' = 1$ for all $z \in S_p$ or $z_p' = 1$ for all $z \in S_p$. Suppose that the first possibility holds (considering the second is similar). Then $h(y) \in L_p(1)h(y^{\star\star})$ and $h(x) \in L_p(1)h(x^{\star\star})$. Consequently, $h(y)h(x^\star)$ and $h(x)h(y^\star)$ belong to the same minimal left ideal $L_p(1)h(e_{p-1})$ in $g^{-1}(f(S_p))$. We have seen that these elements are idempotents, so $h(x)h(y^\star)h(y)h(x^\star) = h(x)h(y^\star)$. Hence,

$$h(x)h(y)h(x) = h(x)h(y^\star)h(y)h(x^\star)h(x) = h(x)h(y^\star)h(x)$$
$$= h(x)h(x^{\star\star})h(y^\star)h(x) = h(x)h(e_{p-1})h(x).$$

This statement holds with y replaced by x, and so

$$h(x) = h(x)h(e_{p-1})h(x) = h(x)h(y)h(x). \qquad \square$$

Theorem 10.30. *Let S be a finite band. If S is an absolute coretract in $\mathfrak{F}$, then S is isomorphic to some semigroup from **P**.*

Proof. Let $k = |S|$ and $B = B(k, 1, 3)$. By Corollary 10.27, B is finite. We can define a surjective homomorphism $f : B \to S$. Then, since S is an absolute coretract in $\mathfrak{F}$, there exists a homomorphism $g : S \to B$ such that $f \circ g = \mathrm{id}_S$. Identifying S and $g(S)$, we may suppose that S is a subsemigroup of B and $f|_S = \mathrm{id}_S$. Let F be the free semigroup on a k-element alphabet A and let $h : F \to B$ be the canonical homomorphism. Note that $h(u) = h(v)$ if and only if v can be obtained from u by a succession of elementary operations in each of which a subword w of a word is replaced by w^3, or vice versa. By Lemma 10.24, $h(u)$ and $h(v)$ belong to the same completely simple component of B if and only if $\mathrm{ct}(u) = \mathrm{ct}(v)$. Recall that $\mathrm{ct}(v)$ denotes the set of letters from A appearing in v.

For any $w \in F$ and $C \subseteq A$, let $w|_C$ denote the word obtained from w by removing all letters from $A \setminus C$ and let $\alpha(w, C)$ and $\beta(w, C)$ denote the first and the last letters in $w|_C$, respectively. It is easy to see that if $h(u) = h(v)$, then $\alpha(u, C) = \alpha(v, C)$ and $\beta(u, C) = \beta(v, C)$.

For any $w \in F$, $C \subseteq A$ and $\rho \subseteq C^2$, let $\sigma(w, C, \rho)$ denote the number of pairs of neighboring letters in $w|_C$ belonging to ρ.

Lemma 10.31. *If $h(u) = h(v)$, then $\sigma(u, C, \rho) \equiv \sigma(v, C, \rho)$ (mod 2).*

Proof. It suffices to consider the case where $u = w_1 w w_2$, $v = w_1 w^3 w_2$. Put $\sigma(t) = \sigma(t, C, \rho)$. Then

$$\sigma(v) = \begin{cases} \sigma(u) + 2\sigma(w) + 2 & \text{if } w|_C \neq \emptyset \text{ and } (\beta(w, C), \alpha(w, C)) \in \rho \\ \sigma(u) + 2\sigma(w) & \text{otherwise.} \end{cases}$$

$\square$

Lemma 10.32. *S is a chain of its rectangular components.*

Proof. Assume the contrary. Then there exist $u, v \in h^{-1}(S)$ with $a \in \mathrm{ct}(u) \setminus \mathrm{ct}(v)$ and $b \in \mathrm{ct}(v) \setminus \mathrm{ct}(u)$. Put $\sigma(w) = \sigma(w, \{a, b\}, \{(a, b)\})$. Then $\sigma(uv) = 1$ and $\sigma(uvuv) = 2$, although $h(uvuv) = h(uv)$, a contradiction. $\square$

Let $S_1 > S_2 > \cdots > S_l$ be the rectangular components of S and for each $p \in [1, l]$, let

$$A_p = \{a \in A : fh(a) \in S_p\}.$$

Observe that for any $u \in h^{-1}(S)$, $h(u) \in S_p$ if and only if $p = \max\{q \leq l : \mathrm{ct}(u) \cap A_q \neq \emptyset\}$. Indeed, if $u = a_1 \cdots a_n$, then $h(u) = fh(u) = fh(a_1) \cdots fh(a_n)$.

Also if $1 \leq p \leq q \leq l$, let

$$A_p^q = \bigcup_{r=p}^{q} A_r \quad \text{and} \quad S_p^q = \bigcup_{r=p}^{q} S_r,$$

and let

$$M_p = \{\alpha(u, A_p^l) : u \in h^{-1}(S_p^l)\} \quad \text{and} \quad N_p = \{\beta(u, A_p^l) : u \in h^{-1}(S_p^l)\}.$$

Observe that $M_p \cap A_p \neq \emptyset$ and $N_p \cap A_p \neq \emptyset$.

Lemma 10.33. *For every $p \in [1, l]$, at least one of the sets M_p, N_p is a singleton.*

Proof. Choose $u \in h^{-1}(S_l)$. Let $a = \alpha(u, A_p^l)$ and $b = \beta(u, A_p^l)$. Put $\sigma(w) = \sigma(w, A_p^l, \{(b, a)\})$. Since $\sigma(uu) = 2\sigma(u) + 1 \equiv \sigma(u) \pmod 2$, $\sigma(u)$ is odd. Suppose that there exist $v_1, v_2 \in h^{-1}(S_p^l)$ with $\alpha(v_1, A_p^l) \neq a$ and $\beta(v_2, A_p^l) \neq b$. Put $v = v_1 v_2$. Since $\sigma(vv) = 2\sigma(v) \equiv \sigma(v) \pmod 2$, $\sigma(v)$ is even. Then $\sigma(uvu) = 2\sigma(u) + \sigma(v)$ is also even. On the other hand, in S, as in any chain of rectangular bands, the following statement holds: if $x, z \in S_q, y \in S_r$, and $r \leq q$, then $xyz = xz$. Therefore $h(uvu) = h(uu) = h(u)$, and so $\sigma(uvu) \equiv \sigma(u) \pmod 2$, a contradiction. $\square$

Lemma 10.34. *If $x \in S_p, y \in S_q, z \in S_r$, and $q \leq p, r$, then $xyz = xz$.*

Proof. Adjoin identities $\emptyset$, $1_B = 1_S$ to F, B, S and to extend h, f in the obvious way. Also put $S_0 = \{1_S\}$. Then the lemma is obviously true if $q = 0$. Fix $q > 0$ and assume that the lemma holds for all smaller values of q. Pick $u \in h^{-1}(x)$, $v \in h^{-1}(y)$ and $w \in h^{-1}(z)$. By Lemma 10.33, one of the sets M_q, N_q is a singleton. Suppose that $N_q = \{a\}$. Then we can write $u = u_1 a u_2$ and $v = v_1 a v_2$, where $\operatorname{ct}(u_2), \operatorname{ct}(v_2) \subseteq A_1^{q-1}$. Since $x = fh(u)$ and $y = fh(v)$, it follows from this that $x = x_1 s x_2$ and $y = y_1 s y_2$, where $s = fh(a) \in S_q, x_2 = fh(u_2), y_2 = fh(v_2) \in S_0^{q-1}$ and $y_1 = fh(v_1) \in S_0^q$. So $xyz = x_1 s x_2 y_1 s y_2 z$ and $xz = x_1 s x_2 z$. It is clear that $s x_2 y_1 s = s$. By our inductive assumption, $s y_2 z = sz$ and $s x_2 z = sz$. Hence $xyz = x_1 sz$ and $xz = x_1 sz$. The case $|M_q| = 1$ is similar. $\square$

Enumerate sets $M_p \cap A_p$ and $N_p \cap A_p$ without repetitions as $\{a_{pi} : 1 \leq i \leq m_p\}$ and $\{b_{p\lambda} : 1 \leq \lambda \leq n_p\}$ so that $a_{p1} = \alpha(u, A_p)$ and $b_{p1} = \beta(v, A_p)$ for some $u, v \in h^{-1}(S_p)$. Define functions φ_p and ψ_p on S_p^l as follows. Let $x \in S_p^l$. Pick $u \in h^{-1}(x)$ and put

$$\varphi_p(x) = \begin{cases} 0 & \text{if } \alpha(u, A_p^l) \notin A_p \\ i & \text{if } \alpha(u, A_p^l) = a_{pi} \end{cases} \quad \text{and} \quad \psi_p(x) = \begin{cases} 0 & \text{if } \beta(u, A_p^l) \notin A_p \\ \lambda & \text{if } \beta(u, A_p^l) = b_{p\lambda}. \end{cases}$$

We now define the mapping $\chi : S \to V$ by putting for every $x \in S_p$,

$$\chi(x) = \varphi_1(x)\varphi_2(x) \cdots \varphi_p(x)\psi_p(x)\psi_{p-1}(x) \cdots \psi_1(x).$$

It is clear that both $\varphi_p(x) \neq 0$ and $\psi_p(x) \neq 0$. By Lemma 10.33, either $\varphi_p(y) = 1$ for all $y \in S_p^l$ or $\psi_p(y) = 1$ for all $y \in S_p^l$.

Lemma 10.35. *χ is injective.*

Proof. Let $x \in S_p$ and pick $u \in h^{-1}(x)$. Let $p_1 < p_2 < \cdots < p_s = p$ be all $r \in [1, p]$ with $\varphi_r(x) \neq 0$, $q_1 < q_2 < \cdots < q_t = p$ all $r \in [1, p]$ with $\psi_r(x) \neq 0$, $\varphi_{p_j}(x) = i_j$ and $\psi_{q_k}(x) = \lambda_k$. Then

$$u = a_{p_1 i_1} u_1 a_{p_2 i_2} u_2 \cdots u_{s-1} a_{p_s i_s} w b_{q_t \lambda_t} v_t \cdots v_2 b_{q_2 \lambda_2} v_1 b_{q_1 \lambda_1},$$

where $ct(u_j) \subseteq A_1^{p_j}$ and $ct(v_k) \subseteq A_1^{q_k}$. But then, by Lemma 10.34,

$$x = fh(a_{p_1 i_1} a_{p_2 i_2} \cdots a_{p_s i_s} b_{q_t \lambda_t} \cdots b_{\lambda_2 q_2} b_{q_1 \lambda_1}),$$

and consequently, x is uniquely determined by $\chi(x)$. □

That χ is a homomorphism follows from the next lemma.

Lemma 10.36. *Let $x \in S_p$ and $y \in S_q$. Then*

(a) $\varphi_r(xy) = \varphi_r(x)$ *if* $r \le p$,

(b) $\varphi_r(xy) = \varphi_r(y)$ *if* $p < r \le q$,

(c) $\psi_r(xy) = \psi_r(y)$ *if* $r \le q$, *and*

(d) $\psi_r(xy) = \psi_r(x)$ *if* $q < r \le p$.

Proof. Let $u \in h^{-1}(x)$, $v \in h^{-1}(y)$, and $w = uv$. If $r \le p$, then $\alpha(w, A_r^l)$ occurs in u, because $ct(u) \cap A_p \ne \emptyset$, so $\varphi_r(xy) = \varphi_r(x)$. If $p < r \le q$, then $\alpha(w, A_r^l)$ occurs in v, because $ct(v) \cap A_r^l = \emptyset$, so $\varphi_r(xy) = \varphi_r(y)$. The check of (c) and (d) is similar. □

It remains to verify that the semigroup $\chi(S)$ satisfies condition (2) in the definition of the family **P**. Let $x \in S_p$ and let $\varphi_q(x) = a \ne 0$ for some $q \in [1, p]$ (the case $\psi(x) \ne 0$ is similar). Pick $u \in h^{-1}(x)$ and write it in the form $u = u_1 a u_2$, where $ct(u_1) \subseteq A_1^{q-1}$. Define $y \in S_q$ by $y = fh(u_1 a)$. Since $x = fh(u_1 a u_2)$, $yx = x$. By Lemma 10.36 (a), $\varphi_r(x) = \varphi_r(y)$ for each $r \le q$. By our choice of b_{r1}, there exists $v_r \in h^{-1}(S_r)$ such that $\beta(v_r, A_r) = b_{r1}$. Let $v = v_q v_{q-1} \cdots v_1$ and define $z \in S_q$ by $z = h(v)$. Then $yz \in S_q$ and by Lemma 10.36, $\chi(yz) = \varphi_1(x) \cdots \varphi_q(x) 1 \cdots 1$. □

From Theorem 10.28 and Theorem 10.30 we obtain the following result.

Theorem 10.37. *Let S be a finite band. Then the following statements are equivalent:*

(1) *S is isomorphic to some semigroup from* **P**,

(2) *S is an $\mathbb{H}$-projective,*

(3) *S is an absolute $\mathbb{H}$-coretract,*

(4) *S is a projective in $\mathfrak{C}$,*

(5) *S is an absolute coretract in $\mathfrak{C}$,*

(6) *S is a projective in $\mathfrak{F}$,*

(7) *S is an absolute coretract in $\mathfrak{F}$.*

Proof. We proceed by the circuits $(1) \Rightarrow (4) \Rightarrow (2) \Rightarrow (6) \Rightarrow (7) \Rightarrow (1)$ and $(1) \Rightarrow (4) \Rightarrow (5) \Rightarrow (3) \Rightarrow (7) \Rightarrow (1)$. The implications $(1) \Rightarrow (4)$ and $(7) \Rightarrow (1)$ are Theorem 10.28 and Theorem 10.28, $(2) \Rightarrow (6)$ and $(3) \Rightarrow (7)$ is Lemma 10.9, and the remaining implications are obvious. □

Using Theorem 10.37, we can summarize Theorem 10.10, Theorem 10.12 and Theorem 10.4 as follows:

Theorem 10.38. *The ultrafilter semigroup of any countable regular almost maximal left topological group is isomorphic to some semigroup from* **P**. *Assuming* $\mathfrak{p} = \mathfrak{c}$, *for every semigroup* $S \in$ **P**, *there is a group topology* $\mathcal{T}$ *on* $\bigoplus_\omega \mathbb{Z}_2$ *such that* $\mathrm{Ult}(\mathcal{T})$ *is isomorphic to* S.

From Theorem 10.38 and Proposition 7.7 we obtain the following.

Corollary 10.39. *Assuming* $\mathfrak{p} = \mathfrak{c}$, *for each combination of the properties of being extremally disconnected, irresolvable, and nodec, except for the combination* $(-,-,+)$, *there is a corresponding almost maximal topological group. There is no countable regular almost maximal left topological group corresponding to the combination* $(-,-,+)$.

Proof. A maximal topological group corresponds to the combination $(+,+,+)$. For the combinations $(-,+,+)$, $(+,-,+)$ and $(+,+,-)$, pick topological groups whose ultrafilter semigroups are the 2-element left zero semigroup, right zero semigroup and chain of idempotents, respectively. For the combinations $(+,-,-)$ and $(-,+,-)$, pick the semigroups $\{11, 1111, 1121\}$ and $\{11, 1111, 1211\}$ in **P**. These are 3-element semigroups consisting of 2 components with the second components being the 2-element right zero semigroup and left zero semigroup, respectively. For the combination $(-,-,-)$, pick the semigroup $\{11, 1111, 1110, 1211, 1210\}$ in **P**. This is a 5-element semigroup consisting of 2 components with the second component being the 2×2 rectangular band. Finally, every rectangular band in **P** is either a right zero semigroup or a left zero semigroup. Consequently, if a countable regular almost maximal left topological group is nodec, it is either extremally disconnected or irresolvable. □

As a consequence we also obtain from Theorem 10.37 the following result.

Theorem 10.40. *Let G be any infinite group, let $Q \in$ **P**, and let S be a subsemigroup of Q. Then there is in ZFC a Hausdorff left invariant topology $\mathcal{T}$ on G such that $\mathrm{Ult}(\mathcal{T}) \subseteq U(G)$ and $\mathrm{Ult}(\mathcal{T})$ is isomorphic to S.*

Proof. By Theorem 7.26, there is a zero-dimensional Hausdorff left invariant topology $\mathcal{T}_0$ on G such that $T = \mathrm{Ult}(\mathcal{T}_0)$ is topologically and algebraically isomorphic to

$\mathbb{H}_\kappa$ and $T \subseteq U(G)$. Pick a surjective continuous homomorphism $g : T \to Q$ (Theorem 7.24). By Theorem 10.37, Q is an absolute coretract in $\mathfrak{C}$. Consequently, there is an injective homomorphism $h : Q \to T$ (such that $g \circ h = \mathrm{id}_Q$). By Proposition 7.8, there is a left invariant topology $\mathcal{T}$ on G such that $\mathrm{Ult}(\mathcal{T}) = h(S)$. Since $\mathcal{T}_0 \subseteq \mathcal{T}$, $\mathcal{T}$ is Hausdorff.   □

It turns out that Theorem 10.37 remains to be true with 'finite band' replaced by 'finite semigroup'.

Let $\mathfrak{FR}$ denote the category of finite regular semigroups.

Theorem 10.41. *Let S be a finite semigroup. Then the following statements are equivalent:*

(1) *S is isomorphic to some semigroup from **P**,*

(2) *S is an $\mathbb{H}$-projective,*

(3) *S is an absolute $\mathbb{H}$-coretract,*

(4) *S is a projective in $\mathfrak{C}$,*

(5) *S is an absolute coretract in $\mathfrak{C}$,*

(6) *S is a projective in $\mathfrak{F}$,*

(7) *S is an absolute coretract in $\mathfrak{F}$,*

(8) *S is a projective in $\mathfrak{FR}$.*

Proof. See [95].   □

10.5   Topological Invariantness of Ult($\mathcal{T}$)

In this section we show that the ultrafilter semigroup of a countable regular almost maximal left topological group is its topological invariant. We start by pointing out a complete system of nonisomorphic representatives of **P**.

Let **M** denote the set of all matrices $M = (m_{p,q})_{p,q=0}^{l}$ without the main diagonal $(m_{p,p})_{p=0}^{l}$, where $l \in \mathbb{N}$ and $m_{p,q} \in \omega$, satisfying the following conditions for every $p \in [1, l]$:

(a) $m_{0,p} \leq m_{1,p} \leq \cdots \leq m_{p-1,p} \in \mathbb{N}$ and $m_{p,0} \leq m_{p,1} \leq \cdots \leq m_{p,p-1} \in \mathbb{N}$,

(b) either $m_{p-1,p} = 1$ and $m_{p-1,p+1} = \cdots = m_{p-1,l} = 0$ or $m_{p,p-1} = 1$ and $m_{p+1,p-1} = \cdots = m_{l,p-1} = 0$.

These are precisely matrices of the form

$$
\begin{pmatrix}
\blacksquare & 1 & 0 & \cdots & 0 & 0 & & & & & & & \\
+ & \blacksquare & 1 & \cdots & 0 & 0 & & & & & & & \\
* & + & \blacksquare & \cdots & 0 & 0 & & 0 & & & 0 & & 0 \\
\vdots & \vdots & \vdots & \ddots & \vdots & \vdots & & & & & & & \\
* & * & * & \cdots & \blacksquare & 1 & & & & & & & \\
* & * & * & \cdots & + & \blacksquare & + & * & \cdots & * & * & * & \\
* & * & * & \cdots & * & 1 & \blacksquare & + & \cdots & * & * & * & \\
& & & & & 0 & 1 & \blacksquare & \cdots & * & * & * & \\
& & 0 & & & \vdots & \vdots & \vdots & \ddots & \vdots & \vdots & \vdots & & 0 & & 0 \\
& & & & & 0 & 0 & 0 & \cdots & \blacksquare & + & * & \\
& & & & & 0 & 0 & 0 & \cdots & 1 & \blacksquare & 1 & 0 & \cdots & 0 & 0 \\
& & & & & & & & & & + & \blacksquare & 1 & \cdots & 0 & 0 \\
& & & & & & & & & & * & + & \blacksquare & \cdots & 0 & 0 \\
& & 0 & & & & 0 & & & \vdots & \vdots & \vdots & \vdots & \ddots & \vdots & \vdots & & 0 \\
& & & & & & & & & & * & * & * & \cdots & \blacksquare & 1 \\
& & & & & & & & & & * & * & * & \cdots & + & \blacksquare & + & * & \cdots \\
& & & & & & & & & & * & * & * & \cdots & * & 1 & \blacksquare & + & \cdots \\
& & & & & & & & & & & & & & 0 & 1 & \blacksquare & \cdots \\
& & 0 & & & & 0 & & & & 0 & & & & \vdots & \vdots & \vdots & \ddots
\end{pmatrix}
$$

and their transposes, where $+$ is a positive integer, $*$ is a nonnegative integer, and all rows and columns are nondecreasing up to the main diagonal.

Now, for every $M = (m_{p,q})_{p,q=0}^{l} \in \mathbf{M}$, let $V(M)$ denote the subsemigroup of V consisting of all words $i_1 i_2 \cdots i_p \lambda_p \lambda_{p-1} \cdots \lambda_1$, where $p \in [1,l]$, such that

(i) both $i_p \neq 0$ and $\lambda_p \neq 0$,

(ii) for every $q < r \leq p$, if $i_t = 0$ for all $t \in [q+1, r-1]$, then $i_r \leq m_{q,r}$, and dually, if $\lambda_t = 0$ for all $t \in [q+1, r-1]$, then $\lambda_r \leq m_{r,q}$.

It is obvious that for every $M \in \mathbf{M}$, $V(M) \in \mathbf{P}$.

Proposition 10.42. *For every $S \in \mathbf{P}$, there is a unique $M \in \mathbf{M}$ such that S is isomorphic to $V(M)$.*

Proof. Let $l = \max\{p \in \mathbb{N} : S \cap V_p \neq \emptyset\}$ and, for each $p \in [1,l]$, let $S_p = S \cap V_p$. For every $q < p \leq l$, let

$$I_{q,p} = \{i_p : i_1 \cdots i_q 0 \cdots 0 i_p \lambda_p \cdots \lambda_1 \in S_p\} \quad \text{and}$$
$$\Lambda_{q,p} = \{\lambda_p : i_1 \cdots i_p \lambda_p 0 \cdots 0 \lambda_q \cdots \lambda_1 \in S_p\}$$

and let $m_{q,p} = |I_{q,p}|$ and $m_{p,q} = |\Lambda_{p,q}|$. For every $p \in [1, l]$, choose bijections

$$f_p : I_{p-1,p} \to [1, m_{p-1,p}] \quad \text{and} \quad g_p : \Lambda_{p,p-1} \to [1, m_{p,p-1}]$$

such that

$$f_p(I_{q,p}) = [1, m_{q,p}] \quad \text{and} \quad g_p(\Lambda_{p,q}) = [1, m_{p,q}]$$

for each $q \le p - 1$. Also put $f_p(0) = 0$ and $g_p(0) = 0$. An easy check shows that $M = (m_{p,q})^l_{p,q=0} \in \mathbf{M}$ and

$$S \ni i_1 \cdots i_p \lambda_p \cdots \lambda_1 \mapsto f_1(i_1) \cdots f_p(i_p) g_p(\lambda_p) \cdots g_1(\lambda_1) \in V(M)$$

is an isomorphism.

Next, adjoin an identity $\emptyset$ to S and put $S_0 = \{\emptyset\}$. For every $p \in [0, l]$, let r_p denote the number of minimal right ideals of S_p, and for every different $p, q \in [0, l]$, let

$$S_{p,q} = \begin{cases} S_p S_q \setminus \bigcup_{k=p+1}^{q-1} S_k S_q & \text{if } p < q \\ S_p S_q \setminus \bigcup_{k=q+1}^{p-1} S_p S_k & \text{if } p > q. \end{cases}$$

Then the uniqueness of M follows from the next lemma. □

Lemma 10.43. *For every different $p, q \in [0, l]$, one has*

$$m_{p,q} = \frac{|S_{p,q}| \cdot r_q}{r_p \cdot |S_q|}.$$

Proof. To compute $|S_{p,q}|$, one may suppose that $S = V(M)$. Then

$$S_{p,q} = \begin{cases} \{i_1 \cdots i_p 0 \cdots 0 i_q \lambda_q \cdots \lambda_1 \in S_q : i_p \ne 0\} & \text{if } p < q \\ \{i_1 \cdots i_p \lambda_p 0 \cdots 0 \lambda_q \cdots \lambda_1 \in S_p : \lambda_q \ne 0\} & \text{if } p > q. \end{cases}$$

Since

$$|\{i_1 \cdots i_p : i_1 \cdots i_p \lambda_p \cdots \lambda_1 \in S_p\}| = r_p \quad \text{and}$$

$$|\{\lambda_q \cdots \lambda_1 : i_1 \cdots i_q \lambda_q \cdots \lambda_1 \in S_q\}| = \frac{|S_q|}{r_q},$$

it follows that

$$|S_{p,q}| = r_p \cdot m_{p,q} \cdot \frac{|S_q|}{r_q}. \qquad \square$$

Note that it also follows from Lemma 10.43 that the matrix M is uniquely determined by the numbers l and r_p and the sets S_p and $S_p S_q$, where $p, q \in [1, l]$ and $p \ne q$.

Theorem 10.44. *If countable regular almost maximal left topological groups are homeomorphic, then their ultrafilter semigroups are isomorphic.*

Proof. Let $(G, \mathcal{T})$ be a countable regular almost maximal left topological group and let $S = \mathrm{Ult}(\mathcal{T})$. By Theorem 10.38, S is isomorphic to some semigroup from $\mathbf{P}$. Let $S_1 > \cdots > S_l$ be the rectangular components of S. For every $p \in [1, l]$, let $\mathcal{T}_p$ be the left invariant topology on G with $\mathrm{Ult}(\mathcal{T}_p) = S_p$. For every different $p, q \in [1, l]$, let $\mathcal{T}_p \cdot \mathcal{T}_q$ be the left invariant topology on G with $\mathrm{Ult}(\mathcal{T}_p \cdot \mathcal{T}_q) = S_p \cdot S_q$. By Lemma 7.3, the number r_p of minimal right ideals of S_p is equal to the number of maximal open filters on $(G, \mathcal{T}_p)$ converging to the identity. Then by Lemma 10.43, in order to show that $\mathrm{Ult}(\mathcal{T})$ is a topological invariant of $(G, \mathcal{T})$, it suffices to show that topologies $\mathcal{T}_p$ and $\mathcal{T}_p \cdot \mathcal{T}_q$ are determined purely topologically.

For every $p = [1, l]$, let $\hat{\mathcal{T}}_p$ be the left invariant topology on G with $\mathrm{Ult}(\hat{\mathcal{T}}_p) = \bigcup_{k=1}^{p} S_k$. Then $\mathcal{T}_l = \mathcal{T}$ and by Proposition 7.7, for $p < l$, a nonprincipal ultrafilter $\mathcal{U}$ on G converges to a point $x \in G$ in $\hat{\mathcal{T}}_p$ if and only if $\mathcal{U}$ converges to x in $\hat{\mathcal{T}}_{p+1}$ and $\mathcal{U}$ is nowhere dense in $\hat{\mathcal{T}}_{p+1}$. Consequently, topologies $\hat{\mathcal{T}}_p$, $p = l, l-1, \ldots, 1$, are determined purely topologically. But then this holds for topologies $\mathcal{T}_p$ as well, since a nonprincipal ultrafilter $\mathcal{U}$ on G converges to a point $x \in G$ in $\mathcal{T}_p$ if and only if $\mathcal{U}$ converges to x in $\hat{\mathcal{T}}_p$ and $\mathcal{U}$ is dense in $\hat{\mathcal{T}}_p$.

Finally, a neighborhood base at a point $x \in G$ in the topology $\mathcal{T}_p \cdot \mathcal{T}_q$ consists of subsets of the form

$$\{x\} \cup \bigcup_{y \in U \setminus \{x\}} V_y \setminus \{y\}$$

where U is a neighborhood of x in $\mathcal{T}_p$ and V_y is a neighborhood of y in $\mathcal{T}_q$. Hence, topologies $\mathcal{T}_p \cdot \mathcal{T}_q$ are also determined purely topologically.   □

In Section 12.1 we will see that every countable homogeneous regular space admits a structure of a left topological group (Theorem 12.5).

Definition 10.45. For every countable homogeneous regular space X, pick a group operation $*$ on X with continuous left translations and let $\mathrm{Ult}(X)$ denote the ultrafilter semigroup of the left topological group $(X, *)$.

By Theorem 10.44, $\mathrm{Ult}(X)$ does not depend, up to isomorphism, on the choice of the operation $*$, so $\mathrm{Ult}(X)$ is a topological invariant of X.

References

Theorem 10.4 is a result from [99] and Corollary 10.6 from [84]. Theorem 10.10 was proved in [99], and Theorem 10.12 in [90]. Theorem 10.15 and Theorem 10.16 are from [87].

Theorem 10.19 is due to A. Clifford [9] and Corollary 10.20 to D. McLean [49]. Theorem 10.23 is a result of J. Green and D. Rees [31].

The definition of the family **P**, Theorem 10.28 and Theorem 10.30 are from [93]. Theorem 10.41 is a result from [95]. Its proof is based on Theorem 10.28, Theorem 10.30 and the fact that every projective in $\mathfrak{F}\mathfrak{R}$ is a band. The latter is a result of P. Trotter [74, 75] who also characterized projectives in $\mathfrak{F}\mathfrak{R}$. Theorem 10.41 tells us among other things that the semigroups from **P** are the same that those characterized by Trotter. Theorem 10.44 was proved in [99].

The exposition of this chapter is based on the treatment in [92].

Chapter 11
Almost Maximal Spaces

In this chapter we show that for every infinite group G and for every $n \in \mathbb{N}$, there is in ZFC a zero-dimensional Hausdorff left invariant topology $\mathcal{T}$ on G such that $\mathrm{Ult}(\mathcal{T})$ is a chain of n idempotents and $\mathrm{Ult}(\mathcal{T}) \subseteq U(G)$. As a consequence we obtain that for every infinite cardinal κ and for every $n \in \mathbb{N}$, there is a homogeneous zero-dimensional Hausdorff space of cardinality κ with exactly n nonprincipal ultrafilters converging to the same point, all of them being uniform. In particular, for every infinite cardinal κ, there is a homogeneous regular maximal space of dispersion character κ.

11.1 Right Maximal Idempotents in $\mathbb{H}_\kappa$

Recall that given an infinite cardinal κ, $\mathbb{H}_\kappa = \mathrm{Ult}(\mathcal{T}_0)$, where $\mathcal{T}_0$ denotes the group topology on $H = \bigoplus_\kappa \mathbb{Z}_2$ with a neighborhood base at 0 consisting of subgroups $H_\alpha = \{x \in H : x(\gamma) = 0 \text{ for each } \gamma < \alpha\}$, $\alpha < \kappa$. When working with $\mathbb{H}_\kappa$, the following two functions are also usful.

Definition 11.1. Define functions $\theta, \phi : H \setminus \{0\} \to \kappa$ by

$$\theta(x) = \min \mathrm{supp}(x) \quad \text{and} \quad \phi(x) = \max \mathrm{supp}(x)$$

and let $\overline{\theta}, \overline{\phi} : \beta H \setminus \{0\} \to \beta \kappa$ denote their continuous extensions.

The main properties of these functions are that for every $x \in \beta H \setminus \{0\}$ and $y \in \mathbb{H}_\kappa$,

$$\overline{\theta}(x + y) = \overline{\theta}(x) \quad \text{and} \quad \overline{\phi}(x + y) = \overline{\phi}(y).$$

In this section we show that for every $\kappa \geq \omega$, there is a right maximal idempotent $p \in \mathbb{H}_\kappa$ such that $C(p) = \{x \in \beta H \setminus \{0\} : x + p = p\}$ is a finite right zero semigroup, and if κ is not Ulam-measurable, every right maximal idempotent $p \in \mathbb{H}_\kappa$ enjoys this property.

Note that $\mathbb{H}_\kappa$ is left saturated in βH, so for every $p \in \mathbb{H}_\kappa$, one has $C(p) \subseteq \mathbb{H}_\kappa$.

The proof of the result about right maximal idempotents in $\mathbb{H}_\kappa$ involves right cancelable ultrafilters in $\mathbb{H}_\kappa$.

An element p of a semigroup S is called *right cancelable* if whenever $q, r \in S$ and $qp = rp$, one has $q = r$. Equivalently, p is right cancelable if the right translation by p is injective.

Theorem 11.2. *For every ultrafilter $p \in \mathbb{H}_\kappa$, the following statements are equivalent:*

(1) *p is right cancelable in βH,*

(2) *p is right cancelable in $\mathbb{H}_\kappa$,*

(3) *there is no idempotent $q \in H^*$ for which $p = q + p$,*

(4) *there is no $q \in H^*$ for which $p = q + p$.*

(5) *$H + p \subset \beta H$ is discrete,*

(6) *$H + p \subset \beta H$ is strongly discrete,*

(7) *p is strongly discrete.*

Proof. $(1) \Rightarrow (2)$ is obvious.

$(2) \Rightarrow (3)$ Assume on the contrary that there is an idempotent $q \in H^*$ for which $p = q + p$. Clearly $q \in \mathbb{H}_\kappa$. For every $\alpha < \kappa$, define $e_\alpha \in H$ by $\mathrm{supp}(e_\alpha) = \{\alpha\}$, and let $E_\alpha = \{e_\beta : \alpha \leq \beta < \kappa\}$. Pick any ultrafilter r on H extending the family of subsets E_α, where $\alpha < \kappa$. Then $r, r + q \in \mathbb{H}_\kappa$ and $r \neq r + q$. Indeed,

$$Y = \bigcup_{\alpha < \kappa} (e_\alpha + H_{\alpha+1} \setminus \{0\}) \in r + q$$

and $|\mathrm{supp}(y)| > 1$ for all $y \in Y$, but $|\mathrm{supp}(x)| = 1$ for all $x \in E_0$. On the other hand, it follows from $p = q + p$ that $r + p = r + q + p$ and, since p is right cancelable in $\mathbb{H}_\kappa$, we obtain that $r = r + q$, a contradiction.

$(3) \Rightarrow (4)$ Assume on the contrary that there is $q \in H^*$ for which $p = q + p$. Then $C(p) \neq \emptyset$. Since $C(p)$ is a closed subsemigroup of H^*, it has an idempotent, a contradiction.

$(4) \Rightarrow (5)$ Assume on the contrary that $H + p \subset \beta H$ is not discrete. Then there is $a \in H$ such that $a + p \in \mathrm{cl}((H \setminus \{a\}) + p)$. Since

$$\mathrm{cl}((H \setminus \{a\}) + p) = (\beta H \setminus \{a\}) + p,$$

we obtain that there is $r \in \beta H \setminus \{a\}$ such that $a + p = r + p$, so $-a + r + p = p$. Let $q = -a + r$. Then $q + p = p$, and since $r \neq a$, $q \in C(p) \subseteq \mathbb{H}_\kappa \subseteq H^*$, a contradiction.

$(5) \Rightarrow (6)$ Since $H + p \subset \beta H$ is discrete, for every $x \in H$, there is $B_x \in p$ such that $y + p \notin \overline{x + B_x}$ for all $y \in H \setminus \{x\}$, that is, $x + B_x \notin y + p$ for all $y \in H \setminus \{x\}$. For every $x \in H \setminus \{0\}$, let

$$F_x = \{0\} \cup \{y \in H \setminus \{0\} : \phi(y) < \phi(x) \quad \text{and} \quad \mathrm{supp}(y) \subset \mathrm{supp}(x)\}.$$

Put $A_0 = B_0$ and inductively for every $\alpha < \kappa$ and for every $x \in H$ with $\phi(x) = \alpha$, choose $A_x \in p$ such that

(i) $A_x \subseteq B_x \cap H_{\phi(x)+1}$, and

(ii) $(x + A_x) \cap (y + A_y) = \emptyset$ for all $y \in F_x$.

This can be done because F_x is finite and $y + A_y \subseteq y + B_y \not\subseteq x + p$ for all $y \in F_x$. We now claim that $(x + A_x) \cap (y + A_y) = \emptyset$ for all different $x, y \in H$.

Indeed, without loss of generality one may suppose that $x \neq 0$ and $\phi(y) \leq \phi(x)$ or $y = 0$. If $y \in F_x$, the statement holds by (ii). Otherwise $\text{supp}(y) \setminus \text{supp}(x) \neq \emptyset$ or $\phi(y) = \phi(x)$, in any case

$$(y + H_{\phi(y)+1}) \cap (x + H_{\phi(x)+1}) = \emptyset,$$

so the statement holds by (i).

For every $x \in H$, $\overline{x + A_x}$ is a neighborhood of $x + p \in \beta H$, and all these neighborhoods are pairwise disjoint. Hence $H + p \subset \beta H$ is strongly discrete.

(6) $\Rightarrow$ (7) Since $H + p \subset \beta H$ is strongly discrete, for every $x \in H$, there is $A_x \in p$ such that the subsets $\overline{x + A_x} \subset \beta H$, where $x \in H$, are pairwise disjoint. Then the subsets $x + A_x \subset H$, where $x \in H$, are pairwise disjoint. It follows that p is strongly discrete.

(7) $\Rightarrow$ (1) Since p is strongly discrete, for every $x \in H$, there is $A_x \in p$ such that the subsets $x + A_x$ are pairwise disjoint. Let $q, r \in \beta H$ and $q \neq r$. Choose disjoint $Q \in q$ and $R \in r$ and put

$$A = \bigcup_{x \in Q} x + A_x \quad \text{and} \quad B = \bigcup_{x \in R} x + A_x.$$

Then $A \in q + p$, $B \in r + p$ and $A \cap B = \emptyset$, so $q + p \neq r + p$. Hence p is right cancelable.   □

Recall that given a group G and $p \in \beta G$, $\mathcal{T}[p]$ is the largest left invariant topology on G in which p converges to 1, and C_p is the smallest closed subsemigroup of βG containing p.

Corollary 11.3. *Let p be a right cancelable ultrafilter in $\mathbb{H}_\kappa$. Then*

(1) *the topology $\mathcal{T}[p]$ is zero-dimensional, and*

(2) *there is a continuous homomorphism $\pi : \text{Ult}(\mathcal{T}[p]) \to \beta \mathbb{N}$ such that $\pi(p) = 1$ and $\pi(C_p) = \beta \mathbb{N}$.*

Proof. By Theorem 11.2, p is a strongly discrete ultrafilter on H. Then apply Theorem 4.18 and Theorem 7.29.   □

We now turn to the right maximal idempotents in $\mathbb{H}_\kappa$.

Proposition 11.4. *For every right maximal idempotent $p \in \mathbb{H}_\kappa$, $C(p)$ is a right zero semigroup.*

Proof. Let $C = C(p)$ and let $q \in C$. Suppose that q is not right cancelable in $\mathbb{H}_\kappa$. Then by Theorem 11.2, there is an idempotent $r \in H^*$ such that $r + q = q$. It follows that $r + q + p = q + p$, and so $r + p = p$. Thus, $p \leq_R r$, and since p is right maximal, $r \leq_R p$, that is, $p + r = r$. From this we obtain that

$$p + q = p + r + q = r + q = q$$

and

$$q + q = q + p + q = p + q = q,$$

so q is an idempotent. Hence, $p \leq_R q$ and $q \leq_R p$. It then follows that the elements of C which are not right cancelable in $\mathbb{H}_\kappa$ form a right zero semigroup.

Now we claim that no element of C is right cancelable in $\mathbb{H}_\kappa$. Indeed, assume on the contrary that some $q \in C$ is right cancelable in $\mathbb{H}_\kappa$. Then by Corollary 11.3, C_q admits a continuous homomorphism onto $\beta\mathbb{N}$. Taking any nontrivial finite left zero semigroup in $\beta\mathbb{N}$, we obtain, by the Lemma 10.2, that there is a nontrivial left zero semigroup in $C_q \subseteq C$, a contradiction.   □

Proposition 11.5. *Let C be a compact right zero semigroup in $\mathbb{H}_\kappa$ and let $\overline{\theta}(C) = \{u\}$. If u is countably incomplete, then C is finite.*

Note that for every right zero semigroup $C \subset \mathbb{H}_\kappa$, $\overline{\theta}(C)$ is a singleton. Indeed, if $x, y \in C$, then $y = x + y$, and so

$$\overline{\theta}(y) = \overline{\theta}(x + y) = \overline{\theta}(x).$$

Proof of Proposition 11.5. Assume on the contrary that C is infinite. Pick any countably infinite subset $X \subset C$ and pick $p \in (\text{cl } X) \setminus X$. Put $Y = (H \setminus \{0\}) + p$. Since cl $Y = (\beta H \setminus \{0\}) + p$ and $p = p + p$, $p \in$ cl Y. Consequently, $(\text{cl } X) \cap (\text{cl } Y) \neq \emptyset$. Also we have that for every $x \in X$, $x \notin$ cl Y. Indeed, otherwise $x = y + p$ for some $y \in \beta H$ and then

$$x + p = y + p + p = y + p = x.$$

But $x + p = p \neq x$, since $x \in X \subset C$, $p \in (\text{cl } X) \setminus X \subset C$ and C is a right zero semigroup.

Hence, in order to derive a contradiction, it suffices, by Corollary 2.24, to construct a partition $\{A_n : n < \omega\}$ of $H \setminus \{0\}$ such that $(\text{cl } X) \cap (\text{cl } Y_n) = \emptyset$ where $Y_n = A_n + p$.

Since u is countably incomplete, there is a partition $\{B_n : n < \omega\}$ of κ such that $B_n \notin u$ for all $n < \omega$, equivalently $u \notin \overline{B_n}$. Put $A_n = \theta^{-1}(B_n)$. Then for every $x \in$ cl X, $\overline{\theta}(x) = u$, and for every $y \in Y_n$, $\overline{\theta}(y) \in B_n$, so for every $y \in$ cl Y_n, $\overline{\theta}(y) \in \overline{B_n}$. Hence, $(\text{cl } X) \cap (\text{cl } Y_n) = \emptyset$.   □

Combining Proposition 11.4 and Proposition 11.5, we obtain the following result.

Theorem 11.6. *Let p be a right maximal idempotent in $\mathbb{H}_\kappa$. Then $C(p)$ is a compact right zero semigroup, and if $\overline{\theta}(p)$ is countably incomplete, $C(p)$ is finite.*

11.2   Projectivity of Ult($\mathcal{T}$)

Theorem 11.7. *Let $\mathcal{T}$ be a translation invariant topology on H such that $\mathcal{T}_0 \subseteq \mathcal{T}$ and let X be an open zero-dimensional neighborhood of 0 in $\mathcal{T}$. Then for every homomorphism $g : T \to Q$ of a semigroup T onto a semigroup Q and for every local homomorphism $f : X \to Q$, there is a local homomorphism $h : X \to T$ such that $f = g \circ h$.*

The proof of Theorem 11.7 is based on the following notion.

Definition 11.8. A *basis* in X is a subset $A \subseteq X \setminus \{0\}$ together with a partition $\{X(a) : a \in A\}$ of $X \setminus \{0\}$ such that for every $a \in A$, $X(a)$ is a clopen neighborhood of $a \in X \setminus \{0\}$ and $X(a) - a \subseteq X \cap H_{\phi(a)+1}$.

Lemma 11.9. *Whenever $\{U_x : x \in X \setminus \{0\}\}$ is a family of neighborhoods of $0 \in X$, there is a basis A in X such that for every $a \in A$, $X(a) - a \subseteq U_a$.*

Proof. Without loss of generality one may suppose that for every $x \in X \setminus \{0\}$, U_x is a clopen neighborhood of $0 \in X$ and $x + U_x \subseteq X \setminus \{0\}$. For every $x \in X \setminus \{0\}$, let

$$F_x = \{y \in X \setminus \{0\} : \phi(y) < \phi(x) \quad \text{and} \quad \text{supp}(y) \subset \text{supp}(x)\}.$$

Note that F_x is finite. For every $\alpha < \kappa$, let

$$X_\alpha = \{x \in X \setminus \{0\} : \phi(x) = \alpha\}.$$

Now put $A_{-1} = \emptyset$ and inductively, for every $\alpha < \kappa$, define a subset $A_\alpha \subseteq X_\alpha$ and for every $a \in A_\alpha$, a clopen neighborhood $X(a)$ of $a \in X \setminus \{0\}$ by

  (i) $A_\alpha = X_\alpha \setminus \bigcup_{b \in B_\alpha} X(b)$, where $B_\alpha = \bigcup_{\beta < \alpha} A_\beta$, and
  (ii) $X(a) = (a + U_a \cap H_{\phi(a)+1}) \setminus \bigcup_{b \in F_a} X(b)$.
We put $A = \bigcup_{\alpha < \kappa} A_\alpha$.

It follows from (i) that for every $a \in A_\alpha$, $a \notin \bigcup_{b \in B_\alpha} X(b)$. Then, since F_a is finite, we obtain from (ii) that $X(a)$ is indeed a clopen neighborhood of $a \in X \setminus \{0\}$. It is clear also that $X(a) - a \subseteq U_a \cap H_{\phi(a)+1}$ and that the subsets $X(a), a \in A$, cover $X \setminus \{0\}$. To see that they are disjoint, let $a \in A_\alpha$, $b \in A_\alpha \cup B_\alpha$ and $a \neq b$. If $b \in F_a$, then $X(a) \cap X(b) = \emptyset$ by (ii). Otherwise $\text{supp}(b) \setminus \text{supp}(a) \neq \emptyset$ or $b \in A_\alpha$, in any case $(a + H_{\phi(a)+1}) \cap (b + H_{\phi(b)+1}) = \emptyset$, so again $X(a) \cap X(b) = \emptyset$. □

Lemma 11.10. *Let A be a basis in X. Then*

  (1) *every $x \in X \setminus \{0\}$ can be uniquely written in the form $x = a_1 + \cdots + a_n$, where $a_1, \ldots, a_n \in A$ and $a_i + \cdots + a_n \in X(a_i)$ for each $i = 1, \ldots, n-1$, and*

  (2) *every mapping $h_0 : A \to S$ of A into a semigroup S extends to a local homomorphism $h : X \to S$ by $h(a_1 + \cdots + a_n) = h_0(a_1) \cdots h_0(a_n)$, where $a_1, \ldots, a_n \in A$ and $a_i + \cdots + a_n \in X(a_i)$ for each $i = 1, \ldots, n-1$.*

Proof. (1) Let $x \in X \setminus \{0\}$. Then $x \in X(a_1)$ for some $a_1 \in A$. If $x = a_1$, we are done. Otherwise $x = a_1 + x_1$, where $x_1 = x - a_1 \in X \setminus \{0\}$, and $\phi(a_1) < \theta(x_1)$.

Suppose that we have written x as

$$x = a_1 + \cdots + a_i + x_i$$

where $a_1, \ldots, a_i \in A$, $x_i \in X \setminus \{0\}$, and

(i) $\phi(a_j) < \theta(a_{j+1})$ for each $j = 1, \ldots, i - 1$ and $\phi(a_i) < \theta(x_i)$, and

(ii) $a_j + \cdots + a_i + x_i \in X(a_j)$ for each $j = 1, \ldots, i$.

Then $x_i \in X(a_{i+1})$ for some $a_{i+1} \in A$. Since $\theta(a_{i+1}) = \theta(x_i)$, $\phi(a_i) < \theta(a_{i+1})$. If $x_i = a_{i+1}$, we are done: $x = a_1 + \cdots + a_{i+1}$ and $a_j + \cdots + a_{i+1} \in X(a_j)$ for each $j = 1, \ldots, i$. Otherwise $x_i = a_{i+1} + x_{i+1}$ where $x_{i+1} = x_i - a_{i+1}$, then $x = a_1 + \cdots + a_{i+1} + x_{i+1}$ and (i) and (ii) are satisfied with i replaced by $i + 1$.

After $\leq |\mathrm{supp}(x)|$ steps we obtain the required decomposition.

Now suppose that x has two such decompositions, say

$$x = a_1 + \cdots + a_n = b_1 + \cdots + b_m.$$

We show that $n = m$ and $a_i = b_i$ for each $i = 1, \ldots, n$. We proceed by induction on $\min\{n, m\}$.

Let $\min\{n, m\} = 1$, say $n = 1$. We have that $a_1 \in X(a_1)$ and

$$a_1 = b_1 + \cdots + b_m \in X(b_1).$$

Since $\{X(a) : a \in A\}$ is disjoint, it follows that $a_1 = b_1$. But then also $m = 1$. Indeed, otherwise $b_2 + \cdots + b_m = 0$ which contradicts $b_2 + \cdots + b_m \in X(b_2)$.

Now let $\min\{n, m\} > 1$. Again we have that $a_1 + \cdots + a_n \in X(a_1)$ and

$$a_1 + \cdots + a_n = b_1 + \cdots + b_m \in X(b_1),$$

and so $a_1 = b_1$. But then $a_2 + \cdots + a_n = b_2 + \cdots + b_m$ and we can apply the inductive assumption.

(2) Let $x \in X \setminus \{0\}$. Write $x = a_1 + \cdots + a_n$, where $a_1, \ldots, a_n \in A$ and $a_i + \cdots + a_n \in X(a_i)$ for each $i = 1, \ldots, n$. Define the neighborhood U of $0 \in X$ by

$$U = \bigcap_{i=1}^{n} (X(a_i) - (a_i + \cdots + a_n)) \cap X.$$

Now let $y \in U \setminus \{0\}$. Then $a_i + \cdots + a_n + y \in X(a_i)$ for each $i = 1, \ldots, n$. Write $y = a_{n+1} + \cdots + a_{n+m}$, where $a_{n+1}, \ldots, a_{n+m} \in A$ and $a_{n+j} + \cdots + a_{n+m} \in X(a_{n+j})$ for each $j = 1, \ldots, m - 1$. We obtain that $a_i + \cdots + a_{n+m} \in X(a_i)$ for each $i = 1, \ldots, n + m - 1$ and

$$\begin{aligned} h(x + y) &= h(a_1 + \cdots + a_n + a_{n+1} \cdots + a_{n+m}) \\ &= h_0(a_1) \cdots h_0(a_n) h_0(a_{n+1}) \cdots h_0(a_{n+m}) \\ &= h(x) h(y). \end{aligned}$$

$\square$

Now we are in a position to prove Theorem 11.7.

Proof of Theorem 11.7. For every $x \in X \setminus \{0\}$, pick a neighborhood U_x of $0 \in X$ such that $f(xy) = f(x)f(y)$ for all $y \in U_x \setminus \{0\}$. By Lemma 11.9, there is a basis A in X such that for every $a \in A$, $X(a) - a \subseteq U_a$. By Lemma 11.10, every $x \in X \setminus \{0\}$ can be uniquely written as $x = a_1 + \cdots + a_n$ where $a_1, \ldots, a_n \in A$ and $a_i + \cdots + a_n \in X(a_i)$ for each $i = 1, \ldots, n-1$. Then $f(x) = f(a_1) \cdots f(a_n)$. Indeed, $a_1 + \cdots + a_n \in X(a_1)$ and $X(a_1) \subseteq a_1 + U_{a_1}$, consequently $a_2 + \cdots + a_n \in U_{a_1}$. It follows that $f(a_1 + a_2 + \cdots + a_n) = f(a_1)f(a_2 + \cdots + a_n)$ and then by induction $f(a_1 + \cdots + a_n) = f(a_1) \cdots f(a_n)$.

Now for every $a \in A$, pick $t_a \in T$ such that $g(t_a) = f(a)$. Define $h : X \to T$ by

$$h(a_1 + \cdots + a_n) = t_{a_1} \cdots t_{a_n}$$

where $a_1, \ldots, a_n \in A$ and $a_i + \cdots + a_n \in X(a_i)$ for each $i = 1, \ldots, n-1$. By Lemma 11.10, h is a local homomorphism, and since

$$g(t_{a_1} \cdots t_{a_n}) = g(t_{a_1}) \cdots g(t_{a_n}) = f(a_1) \cdots f(a_n),$$

it follows that $g \circ h = f$. $\square$

As a consequence we obtain from Theorem 11.7 the following result.

Corollary 11.11. *Let $\mathcal{T}$ be a locally zero-dimensional almost maximal translation invariant topology on H such that $\mathcal{T}_0 \subseteq \mathcal{T}$. Then* $\mathrm{Ult}(\mathcal{T})$ *is a projective in $\mathfrak{F}$.*

Proof. Let T and Q be finite semigroups, let $\alpha : \mathrm{Ult}(\mathcal{T}) \to Q$ be a homomorphism, and let $\beta : T \to Q$ be a surjective homomorphism. By Lemma 10.11, there are an open zero-dimensional neighborhood X of zero of $(H, \mathcal{T})$ and a local homomorphism $f : X \to Q$ such that $f^* = \alpha$, so α is proper. Now by Theorem 11.7, there is a local homomorphism $h : X \to T$ such that $f = \beta \circ h$. Define $\gamma : \mathrm{Ult}(\mathcal{T}) \to T$ by $\gamma = h^*$. Then $\alpha = \beta \circ \gamma$. $\square$

We will need also the following proposition.

Proposition 11.12. *Every projective S in $\mathfrak{F}$ is a chain of rectangular bands satisfying the following conditions:*

  (i) *whenever $x, y, z \in S$ and $y \mathcal{R} z$, $xy = xz$ implies $y = z$, and dually*

  (ii) *whenever $x, y, z \in S$ and $y \mathcal{L} z$, $yx = zx$ implies $y = z$.*

Proof. By Theorem 10.41, S is isomorphic to some semigroup from **P**. It is easy to see that every subsemigroup of V possesses these properties. $\square$

We conclude this section by noting that Theorem 11.7 can be used also to prove the following result.

Theorem 11.13. *For every infinite cardinal κ, the semigroup $\mathbb{H}_\kappa$ contains no nontrivial finite groups.*

Proof. Similar to the proof of Theorem 8.18 with Corollary 8.13 replaced by Theorem 11.7.   □

11.3   The Semigroup $C(p)$

Lemma 11.14. *Let $p \in \mathbb{H}_\kappa$ and let $C(p)$ be finite. Then for every $q, r \in \beta G$, the equality $q + p = r + p$ implies that $q \in r + C^1$ or $r \in q + C^1$, where $C^1 = C^1(p)$.*

Proof. Assume the contrary. Then, since C^1 is finite, there exist $A \in q$ and $B \in r$ such that

$$\overline{A} \cap (B + C^1) = \emptyset \quad \text{and} \quad \overline{B} \cap (A + C^1) = \emptyset.$$

By Theorem 7.18, there is a zero-dimensional translation invariant topology $\mathcal{T}$ on H with $\mathrm{Ult}(\mathcal{T}) = C(p)$. It follows that for every $x \in A \cup B$, there exists a clopen neighborhood U of $0 \in H$ in $\mathcal{T}$ such that

$$A \cap (x + U) = \emptyset \text{ if } x \in B, \quad \text{and} \quad B \cap (x + U) = \emptyset \text{ if } x \in A.$$

Enumerate $A \cup B$ as $\{x_\alpha : \alpha < \kappa\}$ so that the sequence $(\phi(x_\alpha))_{\alpha < \kappa}$ is nondecreasing. For each $\alpha < \kappa$, choose inductively a clopen neighborhood U_α of 0 in $\mathcal{T}$ so that the following conditions are satisfied:

(i) $U_\alpha \subseteq H_{\phi(x_\alpha)+1}$,

(ii) $A \cap (x_\alpha + U_\alpha) = \emptyset$ if $x_\alpha \in B$, and $B \cap (x_\alpha + U_\alpha) = \emptyset$ if $x_\alpha \in A$, and

(iii) $(x_\alpha + U_\alpha) \cap (x_\gamma + U_\gamma) = \emptyset$ for all $\gamma < \alpha$ such that $\mathrm{supp}(x_\gamma) \subseteq \mathrm{supp}(x_\alpha)$ and elements x_α, x_γ belong to different sets A, B.

To this end, fix $\alpha < \kappa$ and suppose that we have already chosen U_γ for all $\gamma < \alpha$ satisfying (i)–(iii). Without loss of generality one may suppose also that $x_\alpha \in A$. Let

$$F = \{\gamma < \alpha : \mathrm{supp}(x_\gamma) \subseteq \mathrm{supp}(x_\alpha) \text{ and } x_\gamma \in B\}.$$

It follows from (ii) that

$$x_\alpha \notin \bigcup_{\gamma \in F} (x_\gamma + U_\gamma).$$

Since F is finite and each U_γ is closed, there is a clopen neighborhood U_α of 0 such that

$$(x_\alpha + U_\alpha) \cap \left(\bigcup_{\gamma \in F} (x_\gamma + U_\gamma) \right) = \emptyset,$$

which means that (iii) is satisfied. Obviously, one can choose U_α to satisfy also (i) and (ii).

We now claim that $(x_\alpha + U_\alpha) \cap (x_\gamma + U_\gamma) = \emptyset$ whenever $\gamma < \alpha < \kappa$ and elements x_α, x_γ belong to different sets A, B.

Indeed, if $\text{supp}(x_\gamma) \subseteq \text{supp}(x_\alpha)$, then $(x_\alpha + U_\alpha) \cap (x_\gamma + U_\gamma) = \emptyset$ by (iii). If $\text{supp}(x_\gamma) \setminus \text{supp}(x_\alpha) \neq \emptyset$, then $(x_\alpha + H_{\phi(x_\alpha)+1}) \cap (x_\gamma + H_{\phi(x_\gamma)+1}) = \emptyset$, and consequently, $(x_\alpha + U_\alpha) \cap (x_\gamma + U_\gamma) = \emptyset$ by (i).

Thus, we have that

$$\Big(\bigcup_{x_\alpha \in A} (x_\alpha + U_\alpha) \Big) \cap \Big(\bigcup_{x_\gamma \in B} (x_\gamma + U_\gamma) \Big) = \emptyset,$$

so $q + p \neq r + p$, which is a contradiction. $\qquad\square$

Theorem 11.15. *Let $p \in \mathbb{H}_\kappa$ and let $C(p)$ be finite. Then*

(1) *$C(p)$ is a projective in $\mathfrak{F}$, and*

(2) *$C(p)$ is a chain of right zero semigroups.*

Proof. (1) By Theorem 7.18, there is a zero-dimensional translation invariant topology $\mathcal{T}$ on H such that $\text{Ult}(\mathcal{T}) = C(p)$. Then apply Corollary 11.11.

(2) Let $C = C(p)$. By (1) and Proposition 11.12, C is a chain of rectangular bands. We have to show that for every $x, y \in C$, $x \mathcal{L} y$ implies $x = y$. Let $K = K(C)$. Pick any $z \in K$. Then $x + z, y + z \in K$ and $(x + z)\mathcal{L}(y + z)$. We have also that $x + z + p = y + z + p$. It follows from this and Lemma 11.14 that either $x + z \in y + z + C^1$ or $y + z \in x + z + C^1$, where $C^1 = C^1(p)$. Both $y + z + C^1$ and $x + z + C^1$ are $\mathcal{R}$-classes of K. Therefore in any case, $(x + z)\mathcal{R}(y + z)$. Since also $(x+z)\mathcal{L}(y+z)$, we obtain that $x+z = y+z$ and then, by Proposition 11.12 (ii), $x = y$. $\qquad\square$

In the rest of this section, we show that for every $n \in \mathbb{N}$, there is an idempotent $p \in \mathbb{H}_\kappa$ such that $C(p)$ is a chain of n finite right zero semigroups.

We first prove several auxiliary statements.

Lemma 11.16. *Let $p \in \mathbb{H}_\kappa$ and let $C(p)$ be finite. Then for every $q \in H^*$,*

$$|\{x \in \beta H : x + p = q\}| \leq |C^1(p)|.$$

Proof. Let $X = \{x \in \beta H : x + p = q\}$ and let $C^1 = C^1(p)$. Choose $y \in X$ with maximally possible $|y + C^1|$. For every $z \in C^1$, one has $y + z + p = y + p = q$, so $y + C^1 \subseteq X$. We claim that $X = y + C^1$.

To see this, let $x \in X$. We have that $x + p = y + p$. Then by Lemma 11.14, either $x \in y + C^1$ or $y \in x + C^1$. The first possibility is what we wish to show. The second implies that $y + C^1 \subseteq x + C^1$. Since $|y + C^1|$ is maximally possible, we obtain that $y + C^1 = x + C^1$, so again $x \in y + C^1$.

It follows from $X = y + C^1$ that $|X| \leq |C^1|$. $\qquad\square$

Lemma 11.17. *Let $p, q \in \mathbb{H}_\kappa$ and let $C(p), C(q)$ be finite. Then*

$$|C^1(p+q)| \le |C^1(p)| \cdot |C^1(q)|.$$

Proof. We have that

$$C^1(p+q) = \{x \in \beta H : x + p + q = p + q\}$$
$$= \{x \in \beta H : x + p \in \{y \in \beta H : y + q = p + q\}\}.$$

Let $Y = \{y \in \beta H : y + q = p + q\}$ and for each $y \in Y$, let $X(y) = \{x \in \beta H : x + p = y\}$. Then

$$C^1(p+q) = \bigcup_{y \in Y} X(y)$$

and, by Lemma 11.16, $|Y| \le |C^1(q)|$ and $|X(y)| \le C^1(p)$. $\square$

Lemma 11.18. *Let G be a group, let $p \in G^*$, and let $Q = \mathrm{Ult}(\mathcal{T}[p])$. Then $Q = (Qp) \cup \{p\}$.*

Proof. Clearly $(Qp) \cup \{p\} \subseteq Q$. We have to show that for every $q \in G^* \setminus ((Qp) \cup \{p\})$, one has $q \notin Q$. Pick $A \in q$ such that $1 \notin A$ and $\overline{A} \cap ((Qp) \cup \{p\}) = \emptyset$. It suffices to construct a neighborhood U of 1 in $\mathcal{T}[p]$ such that $U \cap A = \emptyset$. To this end, we use Theorem 4.8.

Since $\overline{A} \cap ((Qp) \cup \{p\}) = \emptyset$, there is an open neighborhood V of 1 in $\mathcal{T}[p]$ such that $A \cap (\overline{V}p) = \emptyset$. For every $x \in V$, pick $M(x) \in p$ such that $xM(x) \subseteq V$ and $(xM(x)) \cap A = \emptyset$. Put $U = [M]_1$. Then $U \subseteq \{1\} \cup \bigcup_{x \in V} (xM(x))$. It follows that $U \cap A = \emptyset$. $\square$

Lemma 11.19. *For every $p \in \mathbb{H}_\kappa$ and $q \in \mathrm{Ult}(\mathcal{T}[p])$, one has $\overline{\theta}(q) = \overline{\theta}(p)$.*

Proof. Let $A \in p$. Choose $M : H \to p$ such that $M(0) \subseteq A$ and for every $x \in H \setminus \{0\}$ and $y \in M(x)$, $\phi(x) < \theta(y)$. Then whenever $0 \ne z \in [M]_0$, $\theta(z) \in \theta(A)$. $\square$

Proposition 11.20. *For every idempotent $p \in \mathbb{H}_\kappa$ with finite $C(p)$, there is a right maximal idempotent $q \in \mathbb{H}_\kappa$ with finite $C(q)$ such that for each $x \in C(q)$, $x <_L p$.*

Proof. Choose a right cancelable ultrafilter $r \in \mathbb{H}_\kappa$ such that $\overline{\phi}(r) \notin \overline{\phi}(C(p))$ and $\overline{\theta}(r)$ is countably incomplete.

To this end, pick a partition $\{A_n : n < \omega\}$ of κ such that $|A_n| = \kappa$ for all $n < \omega$ and $\overline{\phi}(C(p)) \subseteq A_0$. For every $\alpha < \kappa$, let e_α denote the element of H with $\mathrm{supp}(e_\alpha) = \{\alpha\}$. Then any ultrafilter r on G extending the family of subsets

$$E_{\alpha,n} = \{e_\gamma : \gamma \ge \alpha \text{ and } \gamma \in A_m \text{ for some } m \ge n\},$$

where $\alpha < \kappa$ and $n < \omega$, is as required.

We claim that $r + p$ is right cancelable and $p \notin (\beta H) + r + p$.

To see that $r + p$ is right cancelable, suppose that $u + r + p = v + r + p$ for some $u, v \in \beta H$. Then by Lemma 11.14, either $u + r \in v + r + C^1$ or $v + r \in u + r + C^1$, where $C^1 = C^1(p)$. But

$$\overline{\phi}(u + r) = \overline{\phi}(v + r) = \overline{\phi}(r)$$

and for every $x \in C(p)$,

$$\overline{\phi}(u + r + x) = \overline{\phi}(v + r + x) = \overline{\phi}(x) \in \overline{\phi}(C(p)).$$

It follows that $u + r = v + r$ and, since r is right cancelable, $u = v$.

To see that $p \notin (\beta H) + r + p$, assume on the contrary that $p = u + r + p$ for some $u \in \beta H$. Then $u + r \in C(p)$, which is a contradiction, since $\overline{\phi}(u + r) = \overline{\phi}(r)$ and $\overline{\phi}(r) \notin \overline{\phi}(C(p))$.

Now let $Q = \mathrm{Ult}(\mathcal{T}[r + p])$. Pick any right maximal idempotent $q \in Q$. By Corollary 11.3 and Lemma 7.12, Q is left saturated. Consequently, $C(q) \subset Q$ and q is right maximal in H^*. By Lemma 11.19, $\overline{\theta}(q) = \overline{\theta}(r + p) = \overline{\theta}(r)$, so $\overline{\theta}(q)$ is countably incomplete. Then by Theorem 11.6, $C(q)$ is finite. By Lemma 11.18, $Q \subseteq (\beta H) + r + p$. For every $x \in (\beta H) + r + p$, one has $x + p = x$. Indeed, $x = u + r + p$ for some $u \in \beta H$ and then $x + p = u + r + p + p = u + r + p = x$. It follows that for every $x \in C(q)$, $x + p = x$, so $x \leq_L p$. But $p + x \neq p$, since $p + x \in (\beta H) + r + p$ and $p \notin (\beta H) + r + p$. Hence $x <_L p$.   □

Lemma 11.21. *Let $p \in \mathbb{H}_\kappa$ be an idempotent with finite $C(p)$, and let $q \in \mathbb{H}_\kappa$ be a right maximal idempotent such that for each $x \in C(q)$, $x <_L p$. Then $p + q$ is an idempotent and $p + q < p$. Furthermore, if $C(q)$ is finite, then $C(p + q) \setminus C(p)$ is a finite right zero semigroup.*

Proof. It follows from $q \leq_L p$ that $p + q + p + q = p + q + q = p + q$, so $p + q$ is an idempotent. Also $p + q + p = p + q$ and $p + p + q = p + q$, so $p + q \leq p$. Since $q <_L p$, $p + q \neq p$, and consequently, $p + q < p$.

Now let $C(q)$ be finite. Then, by Lemma 11.17, $C(pq)$ is finite as well. By Theorem 11.15, $C(p + q)$ is a chain of right zero semigroups. Let $r \in C(p + q) \setminus C(p)$ and let $K = K(C(p + q))$. Clearly $p + q \in K$. It suffices to show that $r \in K$.

We have that $r + p + q = p + q$. Then by Lemma 11.14, either $r + p \in p + C^1(q)$ or $p \in r + p + C^1(q)$. The second possibility cannot hold, since it implies that $r + p = p$ and then $r \in C(p)$, which is a contradiction. Consequently, the first possibility holds. Since $r + p = p$ gives a contradiction, it follows that $r + p \in p + C(q)$. For every $x \in C(q)$, we have that $p + x + p + q = p + x + q = p + q$ and $p + q + p + x = p + q + x = p + x$, so $p + C(q) \subseteq K$. Hence $r + p \in K$ and then $r \in K$. The latter follows from the fact that $C(p + q)$ is a chain and so $C(p + q) \setminus K$ is a subsemigroup.   □

We now come to the main result of this section.

Theorem 11.22. *For every right maximal idempotent $q \in \mathbb{H}_\kappa$ with finite $C(q)$ and for every $n \in \mathbb{N}$, there is an idempotent $p \in \mathbb{H}_\kappa$ such that $q \in C(p)$ and $C(p)$ is a chain of n finite right zero semigroups.*

Proof. If $n = 1$, put $p = q$. By Theorem 11.6, $C(p)$ is a finite right zero semigroup.

Now let $n > 1$ and suppose that we have found an idempotent $p' \in \mathbb{H}_\kappa$ such that $q \in C(p')$ and $C(p')$ is a chain of $n - 1$ finite right zero semigroups, say $C_1 > \cdots > C_{n-1}$. By Proposition 11.20, there is a right maximal idempotent $q' \in \mathbb{H}_\kappa$ with finite $C(q')$ such that for each $x \in C(q')$, $x <_L p'$. Put $p = p' + q'$. By Lemma 11.21, p is an idempotent, $p < p'$ and $C_n = C(p) \setminus C(p')$ is a finite right zero semigroup. It follows that $C(p)$ is the chain $C_1 > \cdots > C_n$. $\square$

Now, having proved Theorem 11.22, we can show that

Theorem 11.23. *For every $n \in \mathbb{N}$, there is a locally zero-dimensional translation invariant topology $\mathcal{T}$ on H such that $\mathcal{T}_0 \subseteq \mathcal{T}$ and $\mathrm{Ult}(\mathcal{T})$ is a chain of n idempotents.*

Proof. By Theorem 11.22, there is an idempotent $p \in \mathbb{H}_\kappa$ such that $C(p)$ is a chain of n finite right zero semigroups, say $C_1 > \cdots > C_n$. Inductively for each $i = 1, \ldots, n$, pick $q_i \in C_i$ and define $p_i \in C_i$ by $p_1 = q_1$ and, for $i > 1$,

$$p_i = p_{i-1} + q_i + p_{i-1}.$$

Then $p_1 > \cdots > p_n$. Let $C = C(p)$ and $S = \{p_1, \ldots, p_n\}$. We claim that the subsemigroup $S \subseteq C$ possesses the following property:

For every $q \in C \setminus S$, $(S + q) \cap S = \emptyset$.

Indeed, let $q \in C_i$. Then $q \mathcal{R} p_i$ and $q \neq p_i$. It follows that for every $r \in C$, one has $(r + q)\mathcal{R}(r + p_i)$ and, by Theorem 11.15 (1) and Proposition 11.12 (i), $r + q \neq r + p_i$. If $r \in S$, then $r + p_i \in S$, so $r + q \notin S$, since no different elements of S are $\mathcal{R}$-related. Hence $(S + q) \cap S = \emptyset$.

Now let $\mathcal{T}$ be the translation invariant topology on H such that $\mathrm{Ult}(\mathcal{T}) = S$. It follows from the property above and Proposition 7.21 that $\mathcal{T}$ is locally regular. Being a chain of idempotents, S has only one minimal right ideal. Hence by Proposition 7.7, $\mathcal{T}$ is extremally disconnected. Let X be an open regular neighborhood of $0 \in H$ in $\mathcal{T}$. Since extremal disconnectedness is preserved by open subsets and a regular extremally disconnected space is zero-dimensional, we obtain that X is zero-dimensional. $\square$

11.4   Local Monomorphisms

Given a local left group X and a semigroup S with identity, a *local monomorphism* is an injective local homomorphism $f : X \to S$ with $f(1_X) = 1_S$.

Lemma 11.24. *Let X be a local left group, let S be a left cancellative semigroup with identity, and let $f : X \to S$ be a local monomorphism. Then there is a left invariant T_1-topology $\mathcal{T}^f$ on S with a neighborhood base at $s \in S$ consisting of subsets $sf(U)$, where U runs over neighborhoods of 1_X. Furthermore, let $Y = f(X) \subseteq (S, \mathcal{T}^f)$ and let $\overline{f} : \beta X_d \to \beta S$ be the continuous extension of f. Then f homeomorphically maps X onto Y and $\overline{f}$ isomorphically maps $\mathrm{Ult}(X)$ onto $\mathrm{Ult}(\mathcal{T}^f)$.*

Proof. Let $\mathcal{B}$ be an open neighborhood base at 1_X. For every $U \in \mathcal{B}$ and $x \in U$, there is $V \in \mathcal{B}$ such that $xV \subseteq U$ and $f(xy) = f(x)f(y)$ for all $y \in V$. Then $f(x)f(V) = f(xV) \subseteq f(U)$. Since f is injective, we have also that $\bigcap f(\mathcal{B}) = \{1\}$. Consequently by Corollary 4.4, there is a left invariant T_1-topology $\mathcal{T}^f$ on S in which for each $s \in S$, $sf(\mathcal{B})$ is an open neighborhood base at s.

To see that f homeomorphically maps X onto Y, let $x \in X$. Choose a neighborhood U of 1_X such that $xU \subseteq X$ and $f(xy) = f(x)f(y)$ for all $y \in U$. Then whenever V is a neighborhood of 1_X and $V \subseteq U$, one has $f(xV) = f(x)f(V)$.

Finally, by Lemma 8.4, $\overline{f}$ isomorphically maps $\mathrm{Ult}(X)$ onto $\mathrm{Ult}(\mathcal{T}^f)$. $\qquad\square$

Definition 11.25. Let $\mathcal{T}$ be a translation invariant topology on H such that $\mathcal{T}_0 \subseteq \mathcal{T}$ and let X be an open neighborhood of 0 in $\mathcal{T}$. Denote by $P(X)$ the set of all $x \in X \setminus \{0\}$ which cannot be decomposed into a sum $x = y + z$ where $y, z \in X \setminus \{0\}$ and $\phi(y) < \theta(z)$. Note that $|P(X)| = \kappa$. We say that X satisfies the *P-condition* if there is a neighborhood W of $0 \in X$ such that $|P(X) \setminus W| = \kappa$.

It follows from the next lemma that $P(X)$ is a strongly discrete subset of X with at most one limit point 0.

Lemma 11.26. *Let $x \in P(X)$ and $y \in X \setminus \{0\}$. If $x \in y + H_{\phi(y)+1} \cap X$, then $x = y$.*

Proof. Otherwise $x = y + z$ for some $z \in H_{\phi(y)+1} \cap (X \setminus \{0\})$ and then $\phi(y) < \theta(z)$, which contradicts $x \in P(X)$. $\qquad\square$

Suppose that X has the property that, whenever D is a strongly discrete subset of X with exactly one limit point 0, there is $A \subset D$ such that $|A| = \kappa$ and 0 is not a limit point of A. Then, obviously, X satisfies the P-condition. In particular, X satisfies the P-condition if $\mathcal{T}$ is almost maximal.

In fact, the P-condition is always satisfied in the following sense.

Lemma 11.27. *Let $x \in X \setminus \{0\}$, let $F_x = \{y \in X \setminus \{0\} : \mathrm{supp}(y) \subseteq \mathrm{supp}(x)\}$, and let $Y = X \setminus F_x$. Then Y satisfies the P-condition.*

Proof. Choose a subset $A \subset Y$ with $|A| = \kappa$ such that whenever $y, z \in A$ and $y \neq z$, one has $\mathrm{supp}(y) \cap \mathrm{supp}(z) = \mathrm{supp}(x)$. For every $y \in A$, there is $z_y \in P(Y)$ such that $\mathrm{supp}(z_y) \subseteq \mathrm{supp}(z)$ and $\mathrm{supp}(z_y) \cap \mathrm{supp}(x) \neq \emptyset$. Since $F_x \cap Y = \emptyset$, we have that $\mathrm{supp}(z_y) \setminus \mathrm{supp}(x) \neq \emptyset$ for all $y \in A$. Put $B = \{z_y : y \in A\}$. Then $B \subseteq P(Y)$, $B \cap H_{\phi(x)+1} = \emptyset$ and $|B| = \kappa$. $\square$

We now come to the main result about local monomorphisms.

Theorem 11.28. *Let $\mathcal{T}$ be a translation invariant topology on H such that $\mathcal{T}_0 \subseteq \mathcal{T}$, let X be an open neighborhood of 0 in $\mathcal{T}$, and let G be a group of cardinality κ. Suppose that X is zero-dimensional and satisfies the P-condition. Then there is a local monomorphism $f : X \to G$ such that the topology $\mathcal{T}^f$ is zero-dimensional. If $G = H$, then f can be chosen to be continuous with respect to $\mathcal{T}_0$.*

Proof. Using the P-condition, choose a clopen neighborhood W of $0 \in X$ such that $|P(X) \setminus W| = \kappa$. By Lemma 11.9, there is a basis A in X such that

(1) for each $a \in A$, $X(a) - a \subseteq W$, and

(2) for each $a \in A \setminus W$, $X(a) \cap W = \emptyset$.

Let F be the free semigroup on the alphabet A including the empty word $\emptyset$. Define $h : X \to F$ by putting $h(0) = \emptyset$ and

$$h(a_1 + \cdots + a_n) = a_1 \cdots a_n$$

where $a_1, \ldots, a_n \in A$ and $a_i + \cdots + a_n \in X(a_i)$ for each $i = 1, \ldots, n-1$. By Lemma 11.10, h is a local monomorphism. By Lemma 11.24, h induces a left invariant T_1-topology $\mathcal{T}^h$ on F. We have that $Y = h(X)$ is an open neighborhood of the identity of $(F, \mathcal{T}^h)$ and h homeomorphically maps X onto Y, so Y is zero-dimensional.

Lemma 11.29. *$\mathcal{T}^h$ is zero-dimensional.*

Proof. It suffices to show that Y is closed in $\mathcal{T}^h$. Let $a_1 \cdots a_n \in F \setminus Y$. Then $a_i + \cdots + a_n \notin X(a_i)$ for some $i = 1, \ldots, n-1$, so $a_{i+1} + \cdots + a_n \notin X(a_i) - a_i$. Taking the biggest such i we obtain that $a_{i+1} + \cdots + a_n \in X(a_{i+1}) \subseteq X$. It follows that there is a neighborhood U of $0 \in X$ such that $(a_{i+1} + \cdots + a_n + U) \cap (X(a_i) - a_i) = \emptyset$, so

$$(a_i + \cdots + a_n + U) \cap X(a_i) = \emptyset.$$

We claim that $(a_1 \cdots a_n h(U)) \cap Y = \emptyset$.

Indeed, assume on the contrary that $a_1 \cdots a_n h(y) \in Y$ for some $y \in U \setminus \{0\}$. Let $h(y) = a_{n+1} \cdots a_{n+m} \in Y$. Since $a_1 \cdots a_n h(y) = a_1 \cdots a_n a_{n+1} \cdots a_{n+m} \in Y$, we obtain that

$$a_i + \cdots + a_n + y = a_i + \cdots + a_n + a_{n+1} + \cdots + a_{n+m} \in X(a_i),$$

which is a contradiction. $\square$

Denote $I = A \setminus W$. Then $|I| = \kappa$ and for every $a_1 \cdots a_n \in Y$,

$$a \in \{a_1, \ldots, a_n\} \cap I \quad \text{implies } a = a_1.$$

Indeed, by the construction of A and Lemma 11.26, $P(X) \subseteq A$, and by the choice of W, $|P(X) \setminus W| = \kappa$, so the first statement holds. And since

$$a_i + \cdots + a_n \in X(a_{i-1}) - a_{i-1} \subseteq W$$

for each $i = 2, \ldots, n$, the second one holds as well.

Now let Z denote the subset of F consisting of all words $a_1 \cdots a_n$ such that $\phi(a_i) < \theta(a_{i+1})$ for each $i = 1, \ldots, n-1$ and

$$a \in \{a_1, \ldots, a_n\} \cap I \quad \text{implies } a = a_1.$$

Clearly Z is a neighborhood of the identity of $(F, \mathcal{T}^h)$ containing Y. Furthermore, for every $\alpha < \kappa$,

$$Z_\alpha = \{b_1 \cdots b_m \in h(W) : \theta(b_1) \geq \alpha\} \cup \{\emptyset\}$$

is a neighborhood of the identity, and for every $a_1 \cdots a_n \in Z$,

$$a_1 \cdots a_n Z_{\phi(a_n)+1} \subseteq Z,$$

so Z is open. In addition, and as distinguished from Y, Z has the property that, whenever $a_1 \cdots a_n \in Z$ and $i = 1, \ldots, n-1$, one has $a_1 \cdots a_i \in Z$.

Lemma 11.30. *There is a bijective local monomorphism $g : Z \to G$. If $G = H$, then g can be chosen to be continuous with respect to $\mathcal{T}_0$.*

Proof. We shall construct a bijection $g : Z \to G$ such that $g(\emptyset) = 1$ and

$$g(a_1 \cdots a_n) = g(a_1) \cdots g(a_n)$$

for every $a_1 \cdots a_n \in Z$. That such g is a local homomorphism follows from the last but one sentence preceding the lemma.

It suffices to define g on A so that

(i) whenever $a_1 \cdots a_n$ and $b_1 \cdots b_m$ are different elements of Z, $g(a_1) \cdots g(a_n)$ and $g(b_1) \cdots g(b_m)$ are different elements of G, and

(ii) for each $s \in G \setminus \{1\}$, there is $a_1 \cdots a_n \in Z$ such that $g(a_1) \cdots g(a_n) = s$.

To this end, enumerate A without repetitions as $\{c_\alpha : \alpha < \kappa\}$ so that if $a, b \in A$, $\phi(a) < \phi(b)$, $a = c_\alpha$ and $b = c_\beta$, then $\alpha < \beta$. This defines $\nu : A \to \kappa$ by $\nu(c_\alpha) = \alpha$. Note that whenever $a_1 \cdots a_n \in Z$, one has $\nu(a_1) < \cdots < \nu(a_n)$. Also enumerate $G \setminus \{1\}$ as $\{s_\alpha : \alpha < \kappa\}$.

Fix $\alpha < \kappa$ and suppose that values $g(c_\beta)$ have already been defined for all $\beta < \alpha$ so that, whenever $a_1 \cdots a_n$ and $b_1 \cdots b_m$ are different elements of Z with $\nu(a_n)$, $\nu(b_m) < \alpha$, $g(a_1) \cdots g(a_n)$ and $g(b_1) \cdots g(b_m)$ are different elements of G. Let

$$G_\alpha = \{g(a_1) \cdots g(a_n) \in G : a_1 \cdots a_n \in Z \text{ and } \nu(a_n) < \alpha\} \cup \{1\}.$$

Consider two cases.

Case 1: $c_\alpha \notin I$. Pick as $g(c_\alpha)$ any element of $G \setminus (G_\alpha^{-1} G_\alpha)$. This can be done because $|G_\alpha^{-1} G_\alpha| \leq |G_\alpha|^2 < \kappa$. Then whenever $a_1 \cdots a_n \in Z$ and $a_n = c_\alpha$, one has $g(a_1) \cdots g(a_n) \notin G_\alpha$. Indeed, otherwise

$$g(c_\alpha) = g(a_n) \in (g(a_1) \cdots g(a_{n-1}))^{-1} G_\alpha \subseteq G_\alpha^{-1} G_\alpha.$$

Also if $a_1 \cdots a_n$ and $b_1 \cdots b_m$ are different element of Z with $a_n = b_m = c_\alpha$, then $g(a_1) \cdots g(a_n) \neq g(b_1) \cdots g(b_m)$, by the inductive hypothesis.

Case 2: $c_\alpha \in I$. Then whenever $a_1 \cdots a_n \in Z$ and $a_n = c_\alpha$, one has $n = 1$. Put $g(c_\alpha)$ to be the first element in the sequence $\{s_\beta : \beta < \kappa\} \setminus G_\alpha$.

It is clear that the mapping $g : A \to G$ so constructed satisfies (i), and since $|I| = \kappa$, (ii) is satisfied as well.

If $G = H$, the construction remains the same with the only correction in Case 1: we pick $g(c_\alpha)$ so that $\phi(g(c_\beta)) < \theta(g(c_\alpha))$ for all $\beta < \alpha$. To see that such g is continuous, let $\alpha < \kappa$ be given. Choose $\beta < \kappa$ such that $\phi(g(c_\beta)) \geq \alpha$ and put

$$U_\alpha = \{a_1 \cdots a_n \in h(W) : \nu(a_1) > \beta\} \cup \{\varnothing\}.$$

Then U_α is a neighborhood of identity of Z and $g(U_\alpha) \subseteq H_{\alpha+1}$. $\square$

The bijective local monomorphism $g : Z \to G$ induces a left invariant topology $\mathcal{T}^g$ on G. Since g homeomorphically maps Z onto $(G, \mathcal{T}^g)$, $\mathcal{T}^g$ is zero-dimensional and Hausdorff. Let $f = g \circ h$. Then $f : X \to G$ is a local monomorphism and $\mathcal{T}^f = \mathcal{T}^g$. If $G = H$, choose g to be continuous with respect to $\mathcal{T}_0$, then so is f. $\square$

We will use Theorem 11.28 in a very special situation when the topology $\mathcal{T}$ is almost maximal. But it is also interesting in the general case.

Corollary 11.31. *Let $\mathcal{T}_1$ be a locally zero-dimensional translation invariant topology on H such that $\mathcal{T}_0 \subseteq \mathcal{T}_1$ and let G be a group of cardinality κ. Then there is a zero-dimensional left invariant topology $\mathcal{T}$ on G such that*

(1) *$(G, \mathcal{T})$ and $(H, \mathcal{T}_1)$ are locally homeomorphic,*

(2) *$\mathrm{Ult}(\mathcal{T})$ is topologically and algebraically isomorphic to $\mathrm{Ult}(\mathcal{T}_1)$, and*

(3) *$\mathrm{Ult}(\mathcal{T})$ is left saturated in βG.*

If $G = H$, then $\mathcal{T}$ can be chosen to be stronger then $\mathcal{T}_0$.

Proof. Let X be an open zero-dimensional neighborhood of zero of $(H, \mathcal{T}_1)$. By Lemma 11.27, one may suppose that X satisfies the P-condition. Then by Theorem 11.28, there is a local monomorphism $f : X \to G$ such that the topology $\mathcal{T}^f$ on G is zero-dimensional. Put $\mathcal{T} = \mathcal{T}^f$. By Lemma 11.24, conditions (1) and (2) are satisfied, and by Lemma 7.12, (3) is satisfied as well. If $G = H$, choose f to be continuous with respect to $\mathcal{T}_0$.   □

Now, using Theorem 11.23 and Theorem 11.28, we prove the main result of this chapter.

Theorem 11.32. *For every infinite group G and for every $n \in \mathbb{N}$, there is a zero-dimensional Hausdorff left invariant topology $\mathcal{T}$ on G such that $\mathrm{Ult}(\mathcal{T})$ is a chain of n idempotents and $\mathrm{Ult}(\mathcal{T}) \subseteq U(G)$.*

Proof. Let G be a group of cardinality $\kappa = |H|$ and let $n \in \mathbb{N}$. By Theorem 11.23, there is a locally zero-dimensional translation invariant topology $\mathcal{T}_1$ on H such that $\mathcal{T}_0 \subseteq \mathcal{T}_1$ and $\mathrm{Ult}(\mathcal{T}_1)$ is a chain of n idempotents. Pick an open zero-dimensional neighborhood X of $0 \in H$. By Theorem 11.28, there is a local monomorphism $f : X \to G$ such that the topology $\mathcal{T}^f$ on G is zero-dimensional. Put $\mathcal{T} = \mathcal{T}^f$.   □

The *dispersion character* of a space is the minimum cardinality of a nonempty open set.

Corollary 11.33. *For every infinite cardinal κ and for every $n \in \mathbb{N}$, there is a homogeneous zero-dimensional Hausdorff space of cardinality κ with exactly n nonprincipal ultrafilters converging to the same point, all of them being uniform. In particular, for every infinite cardinal κ, there is a homogeneous regular maximal space of dispersion character κ.*

Remark 11.34. If $G = H$, the topology $\mathcal{T}$ in Theorem 11.32 can be chosen to be stronger than $\mathcal{T}_0$, and if $G = \mathbb{R}$, stronger than the natural topology of the real line.

To see the second, apply Theorem 11.32 to the circle group $\mathbb{T}$. This gives us a left saturated chain of n uniform idempotents in $\beta\mathbb{T}_d$. Since $\mathbb{T}$ is a compact group, every idempotent converges to $1 \in \mathbb{T}$ (Lemma 7.10). Then identifying $\mathbb{T}$ with the subset $[-\frac{1}{2}, \frac{1}{2}) \subset \mathbb{R}$, we obtain a left saturated chain of n uniform idempotents in $\beta\mathbb{R}_d$ converging to $0 \in \mathbb{R}$.

One can show also that if $G = \mathbb{R}$, the topology $\mathcal{T}$ in Theorem 11.32 can be chosen to be stronger than the Sorgenfrey topology.

References

The question of whether there exists a regular maximal space was raised by M. Katětov [42]. A countable example of such a space was constructed by E. van Douwen

[78, 80] and that of arbitrary dispersion character by A. El'kin [19]. The first consistent example of a homogeneous regular maximal space was produced by V. Malykhin [46].

Right cancelable and right maximal idempotent ultrafilters on a countable group G have been studied by N. Hindman and D. Strauss [37, Sections 8.2, 8.5, and 9.1]. In particular, they showed that for every right cancelable ultrafilter p on G, the semigroup C_p admits a continuous homomorphism onto $\beta\mathbb{N}$ [37, Theorem 8.51], and for every right maximal idempotent p in G^*, $C(p)$ is a finite right zero semigroup [37, Theorem 9.4]. That every countably infinite group admits in ZFC a regular maximal left invariant topology was proved by I. Protasov [60].

Theorem 11.2 and Theorem 11.6 are from [107]. Theorem 11.7 and Theorem 11.13 were proved in [101]. The results of Section 11.3 are from [108] and Theorem 11.28 is from [107]. Theorem 11.32 was proved in [107] for $n = 1$ and in [108] for any n.

Chapter 12

Resolvability

In this chapter we prove a structure theorem for a broad class of homeomorphisms of finite order on countable regular spaces. Using this, we show that every countable nondiscrete topological group not containing an open Boolean subgroup is ω-resolvable. We also show that every infinite Abelian group not containing an infinite Boolean subgroup is absolutely ω-resolvable, and consequently, can be partitioned into ω subsets such that every coset modulo infinite subgroup meets each subset of the partition.

12.1 Regular Homeomorphisms of Finite Order

Definition 12.1. Let X be a space with a distinguished point $1 \in X$ and let $f : X \to X$ be a homeomorphism with $f(1) = 1$. We say that f is *regular* if for every $x \in X \setminus \{1\}$, there is a homeomorphism g_x of a neighborhood of 1 onto a neighborhood of x such that $fg_x|_U = g_{f(x)}f|_U$ for some neighborhood U of 1.

Note that if a space X admits a regular homeomorphism, then for any two points $x, y \in X$, there is a homeomorphism g of a neighborhood of x onto a neighborhood of y with $g(x) = y$, and if in addition X is zero-dimensional and Hausdorff, then g can be chosen to be a homeomorphism of X onto itself. Hence, a zero-dimensional Hausdorff space admitting a regular homeomorphism is homogeneous.

The notion of a regular homeomorphism generalizes that of a local automorphism on a local left group. To see this, let X be a local left group and let $f : X \to X$ be a local automorphism. For every $x \in X \setminus \{1\}$, choose a neighborhood U_x of 1 such that xy is defined for all $y \in U_x$, xU_x is a neighborhood of x and $\lambda_x : U_x \ni y \mapsto xy \in xU_x$ is a homeomorphism, and put $g_x = \lambda_x$. Clearly $g_x(1) = x$. Choose a neighborhood V_x of 1 such that $V_x \subseteq U_x$, $f(V_x) \subseteq U_{f(x)}$ and $f(xy) = f(x)f(y)$ for all $y \in V_x$. Then for every $y \in V_x$, $fg_x(y) = f(xy) = f(x)f(y) = g_{f(x)}f(y)$.

We now show that, likewise in the case of a local automorphism, the spectrum of a spectrally irreducible regular homeomorphism of finite order is a finite subset of $\mathbb{N}$ closed under taking the least common multiple.

Lemma 12.2. *Let X be a Hausdorff space with a distinguished point 1 and let $f : X \to X$ be a spectrally irreducible regular homeomorphism of finite order. Let $x_0 \in X \setminus \{1\}$ with $|O(x_0)| = s$ and let U be a spectrally minimal neighbourhood of x_0.*

Then

$$\text{spec}(f, U) = \{\text{lcm}(s, t) : t \in \{1\} \cup \text{spec}(f)\}.$$

Proof. The proof is similar to that of Lemma 8.27.

For each $x \in O(x_0)$, let g_x be a homeomorphism of a neighborhood U_x of 1 onto a neighborhood of x such that $fg_x|V_x = g_{f(x)}f|V_x$ for some neighborhood $V_x \subseteq U_x$ of 1. Choose a neighborhood V of 1 such that

$$V \subseteq \bigcap_{x \in O(x_0)} V_x,$$

$g_{x_0}(V) \subseteq U$, and the subsets $g_x(V)$, where $x \in O(x_0)$, are pairwise disjoint. Let n be the order of f. Choose a neighborhood W of 1 such that $f^i(W) \subseteq V$ for all $i < n$. Then clearly this inclusion holds for all $i < \omega$, and furthermore, for every $y \in W$, $f^i g_{x_0}(y) = g_{f^i(x_0)}f^i(y)$. Indeed, it is trivial for $i = 0$, and further, by induction, we obtain that

$$f^i g_{x_0}(y) = ff^{i-1}g_{x_0}(y) = fg_{f^{i-1}(x_0)}f^{i-1}(y) = g_{f^i(x_0)}f^i(y).$$

Now let $y \in W$, $|O(y)| = t$ and $k = \text{lcm}(s, t)$. We claim that $|O(g_{x_0}(y))| = k$. Indeed,

$$f^k(g_{x_0}(y)) = g_{f^k(x_0)}(f^k(y)) = g_{x_0}(y).$$

On the other hand, suppose that $f^i(g_{x_0}(y)) = g_{x_0}(y)$ for some i. Then

$$g_{f^i(x_0)}(f^i(y)) = g_{x_0}(y).$$

Since the subsets $g_x(V)$, $x \in O(x_0)$, are pairwise disjoint, it follows from this that $f^i(x_0) = x_0$, so $s|i$. But then also $f^i(y) = y$, as g_{x_0} is injective, and so $t|i$. Hence $k|i$. $\square$

The next theorem is the main result of this section.

Theorem 12.3. *Let X be a countable nondiscrete regular space with a distinguished point $1 \in X$, let $f : X \to X$ be a spectrally irreducible regular homeomorphism of finite order, let $S = \text{spec}(f)$, and let $m = 1 + \sum_{s \in S} s$. Let π be the standard permutation on $\bigoplus_\omega \mathbb{Z}_m$ of spectrum S, and for every $a \in \bigoplus_\omega \mathbb{Z}_m$, define $\sigma_a : \bigoplus_\omega \mathbb{Z}_m \to \bigoplus_\omega \mathbb{Z}_m$ by $\sigma_a(x) = a + x$. Then there is a continuous bijection $h : X \to \bigoplus_\omega \mathbb{Z}_m$ with $h(1) = 0$ such that*

(1) $f = h^{-1}\pi h$, *and*

(2) *for every $x \in X$, $\lambda_x = h^{-1}\sigma_{h(x)}h$ is a homeomorphism of X onto itself.*

Furthermore, if X is a local left group and f is a local automorphism, then h can be chosen so that

(3) $\lambda_x(y) = xy$, *whenever $\max \text{supp}(h(x)) + 1 < \min \text{supp}(h(y))$.*

Recall that the topology of $\bigoplus_\omega \mathbb{Z}_m$ is generated by taking as a neighborhood base at 0 the subgroups

$$H_n = \left\{ x \in \bigoplus_\omega \mathbb{Z}_m : x(i) = 0 \text{ for all } i < n \right\}$$

where $n < \omega$.

The conclusion of Theorem 12.3 can be rephrased as follows:

One can define the operation of the group $\bigoplus_\omega \mathbb{Z}_m$ on X in such a way that $0 = 1$, the topology of $\bigoplus_\omega \mathbb{Z}_m$ is weaker than that of X and

(1) $\pi = f$, and

(2) for every $x \in X$, $\sigma_x : X \ni y \mapsto x + y \in X$ is a homeomorphism.

Furthermore, if X is a local left group and f is a local automorphism, then the operation can be defined so that

(3) $x + y = xy$, whenever $\max \operatorname{supp}(x) + 1 < \min \operatorname{supp}(y)$.

Actually, Theorem 12.3 characterizes spectrally irreducible regular homeomorphisms of finite order on countable regular spaces. If $f : X \to X$ is a spectrally irreducible homeomorphism and for some m there is a continuous bijection $h : X \to \bigoplus_\omega \mathbb{Z}_m$ with $h(1) = 0$ such that

(1) hfh^{-1} is a coordinatewise permutation on $\bigoplus_\omega \mathbb{Z}_m$, and

(2) for every $x \in X$, $h^{-1}\sigma_{h(x)}h$ is a homeomorphism of X onto itself,

then f is regular.

To see this, let $\pi = hfh^{-1}$. For every $x \in X \setminus \{1\}$, let $n(x) = \max \operatorname{supp}(h(x)) + 1$, $U_x = h^{-1}(H_{n(x)})$ and $g_x = h^{-1}\sigma_{h(x)}h|_{U_x}$. Then for every $y \in U_x$, $f(y) \in U_x = U_{f(x)}$ and

$$
\begin{aligned}
fg_x(y) &= h^{-1}\pi h h^{-1}\sigma_{h(x)}h(y) \\
&= h^{-1}\pi \sigma_{h(x)}h(y) \\
&= h^{-1}\pi (h(x) + h(y)) \\
&= h^{-1}(\pi(h(x)) + \pi(h(y))) \\
&= h^{-1}(h(f(x)) + h(f(y))) \\
&= h^{-1}\sigma_{h(f(x))}h(f(y)) \\
&= g_{f(x)}f(y).
\end{aligned}
$$

Proof of Theorem 12.3. Let $W = W(\mathbb{Z}_m)$. The permutation π_0 on $\mathbb{Z}_m$, which induces the standard permutation π on $\bigoplus_\omega \mathbb{Z}_m$, also induces the permutation π_1 on W. If $w = \xi_0 \cdots \xi_n$, then $\pi_1(w) = \pi_0(\xi_0) \cdots \pi_0(\xi_n)$. We will write π instead of π_0 and π_1. For each $x \in X \setminus \{1\}$, choose a homeomorphism g_x of a neighborhood of 1 onto a neighborhood of x with $g_x(1) = x$ such that $fg_x = g_{f(x)}f|_U$ for some neighborhood U of 1. Also put $g_1 = \mathrm{id}_X$. If X is a local left group and f is a local automorphism, choose g_x so that $g_x(y) = xy$. Enumerate X as $\{x_n : n < \omega\}$ with $x_0 = 1$.

We shall assign to each $w \in W$ a point $x(w) \in X$ and a clopen spectrally minimal neighborhood $X(w)$ of $x(w)$ such that

  (i) $x(0^n) = 1$ and $X(\emptyset) = X$,

  (ii) $\{X(w^\frown\xi) : \xi \in \mathbb{Z}(m)\}$ is a partition of $X(w)$,

  (iii) $x(w) = g_{x(w_0)} \cdots g_{x(w_{k-1})}(x(w_k))$ and $X(w) = g_{x(w_0)} \cdots g_{x(w_{k-1})}(X(w_k))$, where $w = w_0 + \cdots + w_k$ is the canonical decomposition,

  (iv) $f(x(w)) = x(\pi(w))$ and $f(X(w)) = X(\pi(w))$,

  (v) $x_n \in \{x(v) : v \in W \text{ and } |v| = n\}$.

Enumerate S as $s_1 < \cdots < s_t$ and for each $i = 1, \ldots, t$, pick a representative ζ_i of the orbit in $\mathbb{Z}_m \setminus \{0\}$ of length s_i. Choose a clopen invariant neighborhood U_1 of 1 such that $x_1 \notin U_1$ and $\mathrm{spec}(f, X \setminus U_1) = \mathrm{spec}(f)$. Put $x(0) = 1$ and $X(0) = U_1$. Then choose points $a_i \in X \setminus U_1$, $i = 1, \ldots, t$, with pairwise disjoint orbits of lengths s_i such that $x_1 \in \bigcup_{i=1}^t O(a_i)$. For each $i = 1, \ldots, t$ and $j < s_i$, put

$$x(\pi^j(\zeta_i)) = f^j(a_i).$$

By Lemma 8.30, there is an invariant partition $\{X(\xi) : \xi \in \mathbb{Z}_m \setminus \{0\}\}$ of $X \setminus U_1$ such that $X(\xi)$ is a clopen spectrally minimal neighborhood of $x(\xi)$.

Fix $n > 1$ and suppose that $X(w)$ and $x(w)$ have been constructed for all $w \in W$ with $|w| < n$ so that conditions (i)–(v) are satisfied.

Note that the subsets $X(w)$, $|w| = n-1$, form a partition of X. So one of them, say $X(u)$, contains x_n. Let $u = u_0 + \cdots + u_q$ be the canonical decomposition. Then $X(u) = g_{x(u_0)} \cdots g_{x(u_{q-1})}(X(u_q))$ and $x_n = g_{x(u_0)} \cdots g_{x(u_{q-1})}(y_n)$ for some $y_n \in X(u_q)$. Choose a clopen invariant neighborhood U_n of 1 such that for all basic w with $|w| = n-1$,

  (a) $g_{x(w)}(U_n) \subset X(w)$,

  (b) $fg_{x(w)}|_{U_n} = g_{f(x(w))}f|_{U_n}$, and

  (c) $\mathrm{spec}(f, X(w) \setminus g_{x(w)}(U_n)) = \mathrm{spec}(X(w))$.

If $y_n \neq x(u_q)$, choose U_n in addition so that

  (d) $y_n \notin g_{x(u_q)}(U_n)$.

Put $x(0^n) = 1$ and $X(0^n) = U_n$.

Let $w \in W$ be an arbitrary nonzero basic word with $|w| = n-1$ and let $O(w) = \{w_j : j < s\}$, where $w_{j+1} = \pi(w_j)$ for $j < s-1$ and $\pi(w_{s-1}) = w_0$. Put $Y_j = X(w_j) \setminus g_{x(w_j)}(U_n)$. Using Lemma 12.2, choose points $b_i \in Y_0$, $i = 1, \ldots, t$, with pairwise disjoint orbits of lengths $\mathrm{lcm}(s_i, s)$. If $u_q \in O(w)$, choose b_i in addition so that

$$y_n \in \bigcup_{i=1}^{t} O(b_i).$$

For each $i = 1, \ldots, t$ and $j < s_i$, put

$$x(\pi^j(w^\frown \xi_i)) = f^j(b_i).$$

Then, using Lemma 8.30, inscribe an invariant partition

$$\{X(v^\frown \xi) : v \in O(w), \xi \in \mathbb{Z}_m \setminus \{0\}\}$$

into the partition $\{Y_j : j < s\}$ such that $X(v^\frown \xi)$ is a clopen spectrally minimal neighborhood of $x(v^\frown \xi)$.

For nonbasic $w \in W$ with $|w| = n$, define $x(w)$ and $X(w)$ by condition (iii).

To check (ii) and (iv), let $|w| = n - 1$ and let $w = w_0 + \cdots + w_k$ be the canonical decomposition. Then

$$X(w^\frown 0) = g_{x(w_0)} \cdots g_{x(w_k)}(X(0^n)) = g_{x(w_0)} \cdots g_{x(w_{k-1})}(g_{x(w_k)}(X(0^n)))$$

and

$$X(w^\frown \xi) = g_{x(w_0)} \cdots g_{x(w_{k-1})}(X(w_k^\frown \xi)),$$

so (ii) is satisfied. Next,

$$
\begin{aligned}
f(x(w)) &= f g_{x(w_0)} \cdots g_{x(w_{k-1})}(x(w_k)) \\
&= g_{f(x(w_0))} f g_{x(w_1)} \cdots g_{x(w_{k-1})}(x(w_k)) \\
&\vdots \\
&= g_{f(x(w_0))} \cdots g_{f(x(w_{k-1}))} f(x(w_k)) \\
&= g_{x(\pi(w_0))} \cdots g_{x(\pi(w_{k-1}))}(x(\pi(w_k))) \\
&= x(\pi(w_0) \cdots \pi(w_{k-1})\pi(w_k)) \\
&= x(\pi(w)),
\end{aligned}
$$

so (iv) is satisfied as well.

To check (v), suppose that $x_n \notin \{x(w) : |w| = n - 1\}$. Then

$$
\begin{aligned}
x_n &= g_{x(u_0)} \cdots g_{x(u_{q-1})}(y_n) \\
&= g_{x(u_0)} \cdots g_{x(u_{q-1})}(u_q^\frown \xi) \\
&= x(u^\frown \xi).
\end{aligned}
$$

Now, for every $x \in X$, there is $w \in W$ with nonzero last letter such that $x = x(w)$, so $\{v \in W : x = x(v)\} = \{w^\frown 0^n : n < \omega\}$. Hence, we can define $h : X \to \bigoplus_\omega \mathbb{Z}_m$ by putting for every $w = \xi_0 \cdots \xi_n \in W$,

$$h(x(w)) = \overline{w} = (\xi_0, \ldots, \xi_n, 0, 0, \ldots).$$

It is clear that h is bijective and $h(1) = 0$. Since for every $z = (\xi_i)_{i<\omega} \in \bigoplus_\omega \mathbb{Z}_m$,

$$h^{-1}(z + H_n) = X(\xi_0 \cdots \xi_n),$$

h is continuous.

To see (1), let $x = x(w)$. Then

$$
\begin{aligned}
h(f(x(w))) &= h(x(\pi(w))) \\
&= \overline{\pi(w)} \\
&= \pi(\overline{w}) \\
&= \pi(h(x(w))).
\end{aligned}
$$

To see (2), let $x = x(w)$, $w = w_0 + \cdots + w_k$ and $n = \max \text{supp}(h(x)) + 1$. We first show that

$$\lambda_x|_{h^{-1}(H_n)} = g_{x(w_0)} \cdots g_{x(w_k)}|_{h^{-1}(H_n)}.$$

Let $y \in h^{-1}(H_n)$, $y = x(v)$ and $v = v_0 + \cdots + v_l$. Then

$$
\begin{aligned}
h g_{x(w_0)} \cdots g_{x(w_k)}(y) &= h g_{x(w_0)} \cdots g_{x(w_k)} g_{x(v_0)} \cdots g_{x(v_{l-1})}(x(v_l)) \\
&= h(x(w + v)) \\
&= \overline{w + v} \\
&= \overline{w} + \overline{v} \\
&= h(x(w)) + h(x(v)) \\
&= \sigma_{h(x)} h(y).
\end{aligned}
$$

It follows from (iii) that $g_{x(w_0)} \cdots g_{x(w_k)}$ homeomorphically maps $X(0^n)$, a neighborhood of 1, onto $X(w^\frown 0)$, a neighborhood of x, and so λ_x does.

Now, to see that λ_x homeomorphically maps a neighborhood of an arbitrary point $y \in X$ onto a neighborhood of $z = \lambda_x(y)$, it suffices to check that $\lambda_x = \lambda_z(\lambda_y)^{-1}$. Indeed, $z = h^{-1}\sigma_{h(x)}h(y) = h^{-1}(h(x) + h(y))$ and then

$$
\begin{aligned}
\lambda_z(\lambda_y)^{-1} &= h^{-1}\sigma_{h(x)+h(y)}h(h^{-1}\sigma_{h(y)}h)^{-1} \\
&= h^{-1}\sigma_{h(x)+h(y)}hh^{-1}(\sigma_{h(y)})^{-1}h \\
&= h^{-1}\sigma_{h(x)+h(y)}\sigma_{-h(y)}h \\
&= h^{-1}\sigma_{h(x)}h \\
&= \lambda_x.
\end{aligned}
$$

To see (3), let $x = x(w)$ and $w = w_0 + \cdots + w_k$. If $k = 0$, then $\lambda_x(y) = g_x(y) = xy$. Continuing, by induction on k, we obtain that

$$\lambda_x(y) = g_{x(w_0)} \cdots g_{x(w_k)}(y)$$
$$= g_{x(w_0)} \cdots g_{x(w_{k-1})}(x(w_k) \cdot y)$$
$$= x(w_0 + \cdots + w_{k-1}) \cdot (x(w_k) \cdot y)$$
$$= (x(w_0 + \cdots + w_{k-1}) \cdot x(w_k)) \cdot y$$
$$= (g_{x(w_0)} \cdots g_{x(w_{k-1})}(x(w_k))) \cdot y$$
$$= x(w) \cdot y. \qquad \qquad \square$$

The second part of Theorem 12.3, the case where X is a local left group and f is a local automorphism, is Theorem 8.29. The first part with $f = \mathrm{id}_X$ tells us that

Corollary 12.4. *Every countably infinite homogeneous regular space admits a Boolean group operation with continuous translations.*

We conclude this section by deducing from Corollary 12.4 and Corollary 8.12 the following result.

Theorem 12.5. *Let X be a countably infinite homogeneous regular space and let G be a countably infinite group. Then there is a group operation $*$ on X such that $(X, *)$ is a left topological group algebraically isomorphic to G.*

Proof. One may suppose that X is nondiscrete. By Corollary 12.4, there is a Boolean group operation $+$ on X with continuous translations. Endowing G with any nondiscrete regular left invariant topology and applying Corollary 8.12, we obtain that there is a bijective local homomorphism $f : (X, +) \to G$. For any $x, y \in X$, define $x * y = f^{-1}(f(x)f(y))$. Obviously, $(X, *)$ is a group isomorphic to G. Now given any $x \in X$, we can choose a neighborhood U of 0 such that $f(x + z) = f(x)f(z)$ for all $z \in U$, and then $x * z = f^{-1}(f(x)f(z)) = f^{-1}(f(x + z)) = x + z$. It follows from this that the left translations of $(X, *)$ are continuous and open at the identity. Consequently, the left translations of $(X, *)$ are continuous. $\qquad \square$

12.2 Resolvability of Topological Groups

Theorem 12.6. *If a countable regular space admits a nontrivial regular homeomorphism of finite order, then it is ω-resolvable.*

Proof. Let X be a countable regular space with a distinguished point $1 \in X$ and let $f : X \to X$ be a nontrivial regular homeomorphism of finite order. By Corollary 8.25 and Corollary 3.27, one may suppose that f is spectrally irreducible. Let $h : X \to$

$\bigoplus_\omega \mathbb{Z}_m$ be a bijection guaranteed by Theorem 12.3. Pick an orbit C in $\mathbb{Z}_m$ (with respect to π_0) of a smallest possible length $s > 1$. For every $x \in X$, consider the sequence of coordinates of $h(x)$ belonging to C and define $v(x)$ to be the number of pairs of distinct neighbouring elements in this sequence. Denote also by $\alpha(x)$ and $\gamma(x)$ the first and the last elements in the sequence (if nonempty). Then whenever $x, y \in X$ and $\max \operatorname{supp}(h(x)) + 1 < \min \operatorname{supp}(h(y))$,

$$v(\lambda_x(y)) = \begin{cases} v(x) + v(y) & \text{if } \gamma(x) = \alpha(y) \\ v(x) + v(y) + 1 & \text{otherwise.} \end{cases}$$

We define a disjoint family $\{X_n : n < \omega\}$ of subsets of X by

$$X_n = \{x \in X : v(x) \equiv 2^n \ (\operatorname{mod} 2^{n+1})\}.$$

To see that every X_n is dense in X, let $x \in X$ and let U be an open neighbourhood of 1. We have to show that $\lambda_x(U) \cap X_n \neq \emptyset$. Put $k = 2^{n+1}$ and choose inductively $x_1, \ldots, x_k \in U$ such that

(i) $|O(x_j)| = s$,

(ii) $\qquad\qquad\qquad \max \operatorname{supp}(h(x_j)) + 1 < \min \operatorname{supp}(h(x_{j+1}))$,

and if $x \neq 0$, then

$$\max \operatorname{supp}(h(x)) + 1 < \min \operatorname{supp}(h(x_1)),$$

(iii) $\lambda_{y_1} \cdots \lambda_{y_k}(1) \in U$ whenever $y_j \in O(x_j)$.

Without loss of generality one may suppose that $\gamma(x_j) = \alpha(x_{j+1})$, and that if $x \in X$, then $\gamma(x) = \alpha(x_1)$. For every $l = 0, 1, \ldots, k - 1$, define $z_l \in U$ by

$$z_l = \lambda_{x_1} \lambda_{f(x_2)} \cdots \lambda_{f^l(x_{l+1})} \lambda_{f^l(x_{l+2})} \cdots \lambda_{f^l(x_k)}(1)$$

(in particular, $z_0 = \lambda_{x_1} \lambda_{x_2} \cdots \lambda_{x_k}(1)$). Then

$$h(\lambda_x(z_l)) = h(x) + h(x_1) + \pi h(x_2) + \cdots + \pi^l h(x_{l+1}) + \pi^l h(x_{l+2}) + \cdots + \pi^l h(x_k).$$

It follows that $v(\lambda_x(z_0)) = v(x) + v(x_1) + \cdots + v(x_k)$ and $v(\lambda_x(z_l)) = v(z_0) + l$. Hence, for some l, $v(\lambda_x(z_l)) \equiv 2^n \ (\operatorname{mod} 2^{n+1})$, so $\lambda_x(z_l) \in X_n$. $\qquad\square$

The next proposition says that every nondiscrete topological group not containing an open Boolean subgroup admits a nontrivial regular homeomorphism of order 2.

Proposition 12.7. *Let G be a nondiscrete topological group not containing an open Boolean subgroup. Suppose that for every element $x \in G$ of order 2, the conjugation $G \ni y \mapsto xyx^{-1} \in G$ is a trivial local automorphism. Then the inversion $G \ni y \mapsto y^{-1} \in G$ is a nontrivial regular homeomorphism.*

In order to prove Proposition 12.7, we need the following lemma.

Lemma 12.8. *Let X be a homogeneous space with a distinguished point $1 \in X$ and let $f : X \to X$ be a homeomorphism of finite order n with $f(1) = 1$. Suppose that for every $x \in X \setminus \{1\}$ with $|O(x)| = s < n$, there is a homeomorphism g_x of a neighborhood U of 1 onto a neighborhood of x with $g_x(1) = x$ such that $f^s g_x(y) = g_x f^s(y)$ for all $y \in U$. Then f is regular. In particular, if for every $x \in X \setminus \{1\}$, $|O(x)| = n$, then f is regular.*

Proof. Consider an arbitrary orbit in X distinct from $\{1\}$ and enumerate it as $\{x_i : i < s\}$, where $x_{i+1} = f(x_i)$ for $i = 0, \ldots, s-2$ and $f(x_{s-1}) = x_0$. If $s = n$, choose as g_{x_0} any homeomorphism of a neighborhood U of 1 onto a neighborhood of x_0 with $g_{x_0}(e) = x_0$. If $s < n$, choose g_{x_0} in addition such that $f^s g_{x_0}(y) = g_{x_0} f^s(y)$ for all $y \in U$. For every $i = 1, \ldots, s-1$, put $g_{x_i} = f^i g_{x_0} f^{-i}|_U$. Then for every $i = 0, \ldots, s-1$ and $y \in U$,

$$f g_{x_i}(y) = f f^i g_{x_0} f^{-i}(y) = f^{i+1} g_{x_0} f^{-(i+1)} f(y).$$

If $i < s - 1$, then $f^{i+1} g_{x_0} f^{-(i+1)} f(y) = g_{x_{i+1}} f(y)$, so $f g_{x_i}(y) = g_{x_{i+1}} f(y)$. Hence, it remains only to check that $f g_{x_{s-1}}(y) = g_{x_0} f(y)$. If $s = n$, then

$$f g_{x_{s-1}}(y) = f^s g_{x_0} f^{-s} f(y) = \mathrm{id}_X g_{x_0} \mathrm{id}_X f(y) = g_{x_0} f(y).$$

If $s < n$, then

$$f g_{x_{s-1}}(y) = f^s g_{x_0} f^{-s} f(y) = g_{x_0} f^s f^{-s} f(y) = g_{x_0} f(y). \qquad \square$$

Proof of Proposition 12.7. Let f denote the inversion and let $B = B(G)$. We have that f is a homeomorphism of order ≤ 2 and B is the set of fixed points of f, in particular, $f(1) = 1$. By Lemma 5.3, B is not a neighborhood of 1, so f is nontrivial.

To see that f is regular, let $x \in G \setminus \{1\}$ and $|O(x)| < 2$. Then $x \in B$. But then there is a neighborhood U of 1 such that $x y x^{-1} = y$ for all $y \in U$, that is, $xy = yx$. Define $g_x : U \to xU$ by $g_x(y) = xy$. We have that

$$f g_x(y) = (xy)^{-1} = (yx)^{-1} = x^{-1} y^{-1} = xy^{-1} = g_x f(y).$$

Hence, by Lemma 12.8, f is regular. $\qquad \square$

Combining Theorem 12.6 and Proposition 12.7, we obtain that

Theorem 12.9. *Every countable nondiscrete topological group not containing an open Boolean subgroup is ω-resolvable.*

Note that in the Abelian case Theorem 12.9 can be proved easier. If a topological group is Abelian, then the inversion is a local automorphism. Therefore, it suffices to use Theorem 8.29 instead of Theorem 12.3. Also in the Abelian case the restriction 'countable' is redundant.

Theorem 12.10. *Every nondiscrete Abelian topological group not containing a count-able open Boolean subgroup is ω-resolvable.*

Since every Abelian group can be isomorphically embedded into a direct sum of countable groups, Theorem 12.10 is immediate from the Abelian case of Theorem 12.9 and the following result.

Theorem 12.11. *Let $\kappa > \omega$. For every $\alpha < \kappa$, let G_α be a countable group and let G be an uncountable subgroup of $\bigoplus_{\alpha < \kappa} G_\alpha$. Then G can be partitioned into ω subsets dense in every group topology of uncountable dispersion character.*

The proof of Theorem 12.11 involves the functions $\theta_2, \phi_2 : \mathbb{N} \to \omega$ (see Definition 6.35).

Lemma 12.12. *Let $m \in \mathbb{N}$ and $n < \omega$. Let I denote the integer interval*

$$[\phi_2(m) + 1, \phi_2(m) + 2^{n+1}].$$

Then for every $k \in \mathbb{N}$, there is $i \in I$ such that $\theta_2(\phi_2(m + k \cdot 2^i)) = n$.

Proof. For every $i \geq \phi_2(m) + 1$, one has $\theta_2(k \cdot 2^i) \geq \phi_2(m) + 1$. It follows that

$$\phi_2(m + k \cdot 2^i) = \phi_2(k \cdot 2^i) = \phi_2(k) + i.$$

Consequently, $J = \{\phi_2(m + k \cdot 2^i) : i \in I\}$ is an integer interval of length 2^{n+1}, and so there is $j \in J$ such that $j \equiv 2^n \pmod{2^{n+1}}$. □

Proof of Theorem 12.11. Define a disjoint family $\{Y_n : n < \omega\}$ of subsets of G by

$$Y_n = \{x \in G : \theta_2(\phi_2(|\operatorname{supp}(x)|)) = n\}.$$

Let G be endowed with any group topology of uncountable dispersion character, let U be a neighborhood of $1 \in G$, and let $x \in G$. We show that $(xU) \cap Y_n \neq \emptyset$.

Without loss of generality one may suppose that $x \neq 1$. Let $m = |\operatorname{supp}(x)|$. Put $s = 2^{\phi_2(m)+2^{n+1}}$ and pick a neighborhood V of 1 such that $V^s \subseteq U$. Since G has uncountable dispersion character, it follows that for every countable $A \subseteq \kappa$, there is $y \in V \setminus \{1\}$ such that $\operatorname{supp}(y) \cap A = \emptyset$.

To see this, pick a neighborhood W of 1 such that $WW^{-1} \subseteq V$. Since $|W| > \omega$ and $|\bigoplus_{\alpha \in A} G_\alpha| \leq \omega$, there exist distinct $v, w \in W$ such that $v(\alpha) = w(\alpha)$ for each $\alpha \in A$. Put $y = vw^{-1}$. Then $1 \neq y \in WW^{-1} \subseteq V$ and $\operatorname{supp}(y) \cap A = \emptyset$.

Now construct an ω_1-sequence $(y_\alpha)_{\alpha < \omega_1}$ in $V \setminus \{1\}$ such that

(i) $\operatorname{supp}(y_\alpha) \cap \operatorname{supp}(x) = \emptyset$ for all $\alpha < \omega_1$, and

(ii) $\operatorname{supp}(y_\alpha) \cap \operatorname{supp}(y_\beta) = \emptyset$ for all distinct $\alpha, \beta < \omega_1$.

Passing to a cofinal subsequence, one may suppose that there is $k \in \mathbb{N}$ such that $|\mathrm{supp}(y_\alpha)| = k$ for all $\alpha < \kappa$. Let i be a natural number guaranteed by Lemma 12.12 and let $y = y_1 \cdots y_{2^i}$. Then $y \in V^{2^i} \subseteq U$, $|\mathrm{supp}(xy)| = m + k \cdot 2^i$, and $\theta_2(\phi_2(|\mathrm{supp}(xy)|)) = n$. $\qquad\square$

We conclude this section by showing that the existence of a countable nondiscrete ω-irresolvable topological group cannot be established in ZFC.

Theorem 12.13. *The existence of a countable nondiscrete ω-irresolvable topological group implies the existence of a P-point in ω^*.*

Proof. Let $(G, \mathcal{T})$ be a countable ω-irresolvable topological group. By Theorem 12.9, one may suppose that $G = \bigoplus_\omega \mathbb{Z}_2$. Let $\mathcal{T}_0$ denote the topology on G induced by the product topology on $\prod_\omega \mathbb{Z}_2$. Then $\mathcal{T} \vee \mathcal{T}_0$ is a nondiscrete ω-irresolvable group topology. Therefore, one may suppose that $\mathcal{T}_0 \subseteq \mathcal{T}$. By Theorem 3.33, every discrete subset of $(G, \mathcal{T})$ is closed. Let $\mathcal{F}$ be the neighborhood filter of 0 in $\mathcal{T}$. It is easy to see that $\{\overline{\phi}^{-1}(u) : u \in \overline{\phi(\mathcal{F})}\}$ is a partition of $\overline{\mathrm{Ult}(\mathcal{T})}$ into right ideals. It then follows from Theorem 3.35 and Proposition 7.7 that $\overline{\phi(\mathcal{F})}$ is finite. Hence by Theorem 5.19, each point of $\overline{\phi(\mathcal{F})}$ is a P-point. $\qquad\square$

Combining Theorem 12.13 and Theorem 2.38 gives us that

Corollary 12.14. *It is consistent with ZFC that there is no countable nondiscrete ω-irresolvable topological group.*

12.3   Absolute Resolvability

Definition 12.15. Let G be a group. A subset $A \subseteq G$ is *absolutely dense* if A is dense in every nondiscrete group topology on G. Given a cardinal $\kappa \geq 2$, G is *absolutely κ-resolvable* (*absolutely resolvable* if $\kappa = 2$) if G can be partitioned into κ absolutely dense subsets.

Lemma 12.16. *Let G be an Abelian group and let $A \subseteq G$. If A is absolutely dense, then $G \setminus A$ contains no coset modulo infinite subgroup.*

Proof. Suppose that there exist an infinite subgroup H of G and $g \in G$ such that $g + H \subseteq G \setminus A$. Pick any nondiscrete group topology $\mathcal{T}_H$ on H and extend it to the group topology $\mathcal{T}$ on G by declaring H to be an open subgroup. Then $g + H$ is an open subset of $(G, \mathcal{T})$ disjoint from A, so A is not dense. Hence, A is not absolutely dense. $\qquad\square$

The next proposition is a consequence of Hindman's Theorem.

Proposition 12.17. *Whenever an infinite Boolean group is partitioned into finitely many subsets, there is a coset modulo infinite subgroup contained in one subset of the partition.*

Proof. Let B be an infinite Boolean group and let $\{A_i : i < r\}$ be a finite partition of B. By Hindman's Theorem, there exist $i < r$ and a one-to-one sequence $(x_n)_{n<\omega}$ in B such that $\mathrm{FS}((x_n)_{n<\omega}) \subseteq A_i$. Let $H = \mathrm{FS}((x_n)_{1\leq n<\omega}) \cup \{0\}$. Then H is a subgroup of B, since B is Boolean, and $x_0 + H \subseteq \mathrm{FS}((x_n)_{n<\omega}) \subseteq A_i$.   □

From Proposition 12.17 and Lemma 12.16 we obtain that

Corollary 12.18. *If an Abelian group contains an infinite Boolean subgroup, then it is not absolutely resolvable.*

Definition 12.19. Let G be a group written additively. Given a sequence $(x_n)_{n<\omega}$ in G, the set of *finite sums with inverses* of the sequence is defined by

$$\mathrm{FSI}((x_n)_{n<\omega}) = \left\{ \sum_{n \in F} \varepsilon_n^F x_n : F \in \mathcal{P}_f(\omega) \text{ and } \varepsilon_n^F \in \{1, -1\} \text{ for all } n \in F \right\}.$$

In this section we prove the following result.

Theorem 12.20. *Let G be an Abelian group and let $A = G \setminus B(G)$ be infinite. Then there is a disjoint family $\{A_m : m < \omega\}$ of subsets of A such that whenever $(x_n)_{n<\omega}$ is a one-to-one sequence in A, $g \in G$ and $m < \omega$, one has*

$$(g + \mathrm{FSI}((x_n)_{n<\omega})) \cap A_m \neq \emptyset.$$

As a consequence we obtain from Theorem 12.20 that

Corollary 12.21. *Every infinite Abelian group not containing an infinite Boolean subgroup is absolutely ω-resolvable.*

Proof. Let $\{A_m : m < \omega\}$ be a disjoint family of subsets of $A = G \setminus B(G)$ guaranteed by Theorem 12.20 and let $\mathcal{T}$ be a nondiscrete group topology on G. We have to show that each A_m is dense in $\mathcal{T}$. Let $g \in G$ and let U be an open neighborhood of zero of $(G, \mathcal{T})$. By the assumption, $B(G)$ is finite, so pick $x_0 \in U \setminus B(G)$ with $-x_0 \in U$. Fix $k < \omega$ and suppose that we have constructed a one-to-one sequence $(x_n)_{n \leq k}$ in $U \setminus B(G)$ with $\mathrm{FSI}((x_n)_{n\leq k}) \subseteq U$. Let $H_k = \mathrm{FSI}((x_n)_{n\leq k}) \cup \{0\}$. Pick

$$x_{k+1} \in U \setminus (B(G) \cup \{x_n : n \leq k\})$$

such that $H_k + \varepsilon x_{k+1} \subseteq U$ for each $\varepsilon \in \{1, -1\}$. As a result, we obtain a one-to-one sequence $(x_n)_{n<\omega}$ in $U \setminus B(G)$ with $\mathrm{FSI}((x_n)_{n<\omega}) \subseteq U$. Since $(g+\mathrm{FSI}((x_n)_{n<\omega}))\cap A_m \neq \emptyset$, $(g + U) \cap A_m \neq \emptyset$ as well.   □

Corollary 12.21 and Lemma 12.16 give us in turn that

Corollary 12.22. *Every infinite Abelian group not containing an infinite Boolean subgroup can be partitioned into ω subsets such that every coset modulo infinite subgroup meets each subset of the partition.*

Note that the cardinal number ω in Theorem 12.20, Corollary 12.21 and Corollary 12.22 is maximally possible.

We first prove Theorem 12.20 in the case where G is a direct sum of finite groups, not necessarily Abelian.

Theorem 12.23. *Let $\kappa \geq \omega$. For each $\alpha < \kappa$, let G_α be a finite group written additively, let $G = \bigoplus_{\alpha<\kappa} G_\alpha$, and let $A = G \setminus B(G)$ be infinite. Then there is a disjoint family $\{A_m : m < \omega\}$ of subsets of A such that whenever $(x_n)_{n<\omega}$ is a one-to-one sequence in A, $g \in G$ and $m < \omega$, one has*

$$(g + \mathrm{FSI}((x_n)_{n<\omega})) \cap A_m \neq \emptyset.$$

Proof. For every $x \in G$, let

$$\mathrm{supp}(x) = \{\alpha < \kappa : x(\alpha) \neq 0_\alpha\} \text{ and } \mathrm{supp}_0(x) = \{\alpha < \kappa : x(\alpha) \notin B(G_\alpha)\}.$$

Since $B(G) = \bigoplus_{\alpha<\kappa} B(G_\alpha)$, we have that $x \in A$ if and only if $\mathrm{supp}_0(x) \neq \emptyset$. For every $\alpha < \kappa$, $G_\alpha \setminus B(G_\alpha)$ is a disjoint union of 2-element subsets of the form $\{y, -y\}$ and we put $\mathrm{sgn}(y) = 1$ and $\mathrm{sgn}(-y) = -1$. Thus, there is assigned to every $x \in A$ a nonempty finite sequence of $\mathrm{sgn}(x(\alpha))$, where $\alpha \in \mathrm{supp}_0(x)$.

The crucial idea of the construction is contained in the following lemma.

Lemma 12.24. *Let $(c_n)_{n<\omega}$ be an increasing sequence in $\mathbb{N}$ such that $c_0 = 1$ and $c_{n+1} - c_n \to \infty$. Then there is a function $\varphi : \mathbb{N} \to [\mathbb{N}]^{<\omega}$ with the following properties:*

(1) *if $a \in [c_n, c_{n+1})$, then $\varphi(a) = \{a_0, \ldots, a_{n-1}\}$, where $a_k \in [c_k, c_{k+1})$ for all $k \leq n - 1$ (in particular, if $a \in [c_0, c_1)$, then $\varphi(a) = \emptyset$), and*

(2) *for every $d \in \mathbb{N}$, there is $c \in \mathbb{N}$ such that whenever $a, b \in \mathbb{N}$, $0 < |a - b| \leq d$, $u \in \varphi(a)$, $v \in \varphi(b)$ and $u, v \geq c$, one has $|u - v| > d$.*

Proof. Without loss of generality one may suppose that $c_{n+1} - c_n \geq 5$ for all $n < \omega$. For each $n < \omega$, pick an integer $d_n \geq 2$ such that $d_n^2 + d_n - 1 \leq c_{n+1} - c_n$ and $d_n \to \infty$. To define φ, let $k < \omega$ and let $a \geq c_{k+1}$. Choose the largest integer $l \geq 0$ for which $c_{k+1} + l d_k^2 \leq a$, then choose the largest integer $i \geq 0$ for which $c_{k+1} + l d_k^2 + i d_k \leq a$, and then put $j = a - c_{k+1} - l d_k^2 - i d_k$. Thereby we write a in the form

$$a = c_{k+1} + l d_k^2 + i d_k + j,$$

where l is a nonnegative integer and $i, j \in \{0, \ldots, d_k - 1\}$. Put

$$a_k = c_k + jd_k + i.$$

Since $a_k \in [c_k, c_k + d_k^2)$, the function φ so defined satisfies the condition (1). To check (2), let $d \in \mathbb{N}$ be given. Choose $n_0 < \omega$ such that $d_n \geq d + 2$ for all $n \geq n_0$ and put $c = c_{n_0}$. Now let $a, b \in \mathbb{N}$, $0 < |a - b| \leq d$, $u \in \varphi(a)$, $v \in \varphi(b)$ and $u, v \geq c$. Since

$$
\begin{aligned}
c_{k+1} - a_k &= c_{k+1} - c_k - jd_k - i \\
&\geq d_k^2 + d_k - 1 - jd_k - i \\
&\geq d_k,
\end{aligned}
$$

one may suppose that $u = a_k$ and $v = b_k$ for some $k \geq n_0$. Let

$$
\begin{aligned}
a &= c_{k+1} + ld_k^2 + id_k + j, \\
b &= c_{k+1} + l'd_k^2 + i'd_k + j'.
\end{aligned}
$$

Then

$$
\begin{aligned}
a_k &= jd_k + i, \\
b_k &= j'd_k + i'.
\end{aligned}
$$

Thus, we have that

$$
\begin{aligned}
a - b &= [(l - l')d_k + (i - i')]d_k + (j - j'), \\
a_k - b_k &= (j - j')d_k + (i - i').
\end{aligned}
$$

Notice that $|i - i'| < d_k$ and $|j - j'| < d_k$. Then it follows from the second equality that if $|j - j'| > 1$, $|a_k - b_k| > d_k$. Therefore one may suppose that $|j - j'| \leq 1$. But then it follows from the first equality that $a - b$ differs from a multiple of d_k by $0, 1$ or -1. Since $|a - b| \leq d \leq d_k - 2$, this multiple of d_k must be zero, so

$$(l - l')d_k + (i - i') = 0.$$

This gives us $l = l'$ and $i = i'$. Consequently, $|j - j'| = 1$, and we obtain that $|a_k - b_k| = d_k$.   □

Let $\varphi : \mathbb{N} \to [\mathbb{N}]^{<\omega}$ be a function guaranteed by Lemma 12.24. For every $x \in A$, define the subset $\text{supp}_\varphi(x) \subseteq \text{supp}_0(x)$ and the nonnegative integer $\nu(x)$ as follows. Let $|\text{supp}_0(x)| = l$ and let $l \in [c_n, c_{n+1})$. Then $\text{supp}_0(x) = \{\alpha_0, \ldots, \alpha_{l-1}\}$ for some $\alpha_0 < \cdots < \alpha_{l-1} < \kappa$ and $\varphi(l) = \{l_0, \ldots, l_{n-1}\}$ for some $l_k \in [c_k, c_{k+1})$, $k \leq n - 1$. Put

$$\text{supp}_\varphi(x) = \{\alpha_{l_0}, \ldots, \alpha_{l_{n-1}}\}$$

and define $\nu(x)$ to be the number of pairs of distinct neighbouring elements in the sequence

$$\mathrm{sgn}(x(\alpha_{l_0})), \ldots, \mathrm{sgn}(x(\alpha_{l_{n-1}})).$$

(If $l \in [c_0, c_1)$, then $\varphi(l) = \emptyset$, and then $\mathrm{supp}_\varphi(x) = \emptyset$ and $\nu(x) = 0$.)

We define the disjoint family $\{A_m : m < \omega\}$ of subsets of A by

$$A_m = \{x \in A : \nu(x) \equiv 2^m \ (\mathrm{mod}\ 2^{m+1})\}.$$

Now let $(x_n)_{n<\omega}$ be a one-to-one sequence in A, $g \in G$ and $m < \omega$. We shall show that

$$(g + \mathrm{FSI}((x_n)_{n<\omega})) \cap A_m \neq \emptyset.$$

Note that if $(y_n)_{n<\omega}$ is a sum subsystem of $(x_n)_{n<\omega}$, then

$$\mathrm{FSI}((y_n)_{n<\omega}) \subseteq \mathrm{FSI}((x_n)_{n<\omega}).$$

Lemma 12.25. *Let G be a totally bounded topological group and let $(x_n)_{n<\omega}$ be a sequence in G. Then whenever U and U_x, $x \in U$, are neighborhoods of $0 \in G$, there is a sum subsystem $(y_n)_{n<\omega}$ of $(x_n)_{n<\omega}$ such that $y_0 \in U$ and for every $n < \omega$, $y_{n+1} \in \bigcap_{i \leq n} U_{y_i}$.*

Proof. We first show that for every $m < \omega$ and for every neighborhood V of $0 \in G$,

$$\mathrm{FS}((x_n)_{m\leq n<\omega}) \cap V \neq \emptyset.$$

To this end, define a closed subsemigroup $T \subseteq \beta G_d$ by

$$T = \bigcap_{m<\omega} \overline{\mathrm{FS}((x_n)_{m\leq n<\omega})}$$

(Proposition 6.39) and pick an idempotent $p \in T$ (Theorem 6.12). We then have that $\mathrm{FS}((x_n)_{m\leq n<\omega}) \in p$ for every $m < \omega$. But since G is totally bounded, we also have by Lemma 7.10, that $V \in p$ for every neighborhood V of $0 \in G$. It follows that $\mathrm{FS}((x_n)_{m\leq n<\omega}) \cap V \neq \emptyset$ for every $m < \omega$ and for every neighborhood V of $0 \in G$.

Now pick $y_0 \in \mathrm{FS}((x_n)_{n<\omega}) \cap U$. Fix $k < \omega$ and suppose that we have constructed a sum subsystem $(y_n)_{n\leq k}$ of $(x_n)_{n<\omega}$ such that $y_0 \in U$ and for every $n < k$, $y_{n+1} \in \bigcap_{i \leq n} U_{y_i}$. Let $y_k = \sum_{n \in H_k} x_n$ and let $n_{k+1} = \max H_k + 1$. Pick

$$y_{k+1} \in \mathrm{FS}((x_n)_{n_{k+1}\leq n<\omega}) \cap \bigcap_{i \leq k} U_{y_i}.$$

The sequence $(y_n)_{n<\omega}$ so constructed is as required. $\square$

By Lemma 12.25, one may suppose that

$$\text{supp}(g) \cap \text{supp}(x_i) = \emptyset \quad \text{and} \quad \text{supp}(x_i) \cap \text{supp}(x_j) = \emptyset$$

for all $i < j < \omega$. Also one may suppose that the sequence $(\min \text{supp}_0(x_n))_{n<\omega}$ is increasing. Let

$$\gamma = \sup\{\min \text{supp}_0(x_n) : n < \omega\}.$$

Every $x \in G$ can be uniquely written in the form $x = x' + x''$, where $x', x'' \in G$,

$$\text{supp}(x') = \text{supp}(x) \cap \gamma \quad \text{and} \quad \text{supp}(x'') = \text{supp}(x) \setminus \text{supp}(x').$$

Put $k = 2^{m+1} - 1$. Choose inductively a sum subsystem $(y_n)_{n<k}$ of $(x_n)_{n<\omega}$ such that

(a) $\max \text{supp}_0(g') < \min \text{supp}_0(y'_0)$ and $\max \text{supp}_0(y'_i) < \min \text{supp}_0(y'_{i+1})$ for all $i < k - 1$, and

(b) each of the intervals $(|\text{supp}_0(g')|, |\text{supp}_0(g' + y'_0)|]$ and

$$(|\text{supp}_0(g' + y'_0 + \cdots + y'_i)|, |\text{supp}_0(g' + y'_0 + \cdots + y'_i + y'_{i+1})|],$$

where $i < k - 1$, contains some interval $[c_n, c_{n+1})$.

Let $h = g + y_0 + \cdots + y_{k-1}$, $y_{k-1} = \sum_{n \in H} x_n$ and $n_0 = \max H + 1$. We claim that there is $y_k \in \text{FS}((x_n)_{n_0 \le n < \omega})$ such that

(i) $\max \text{supp}_0(h') < \min \text{supp}_0(y_k)$,

(ii) the interval $(|\text{supp}_0(h')|, |\text{supp}_0(h + y_k)|]$ contains some $[c_n, c_{n+1}]$, and

(iii) $\text{supp}_\varphi(h + y_k) \cap \text{supp}_0(h'') = \emptyset$.

Assume the contrary. Then by Corollary 6.38, there exist $\delta \in \text{supp}_0(h'')$ and a sum subsystem $(z_n)_{n<\omega}$ of $(x_n)_{n_0 \le n < \omega}$ such that

$$\max \text{supp}_0(h') < \min \text{supp}_0(z_0), \quad \text{and}$$
$$\delta \in \text{supp}_\varphi(h + z) \quad \text{for all } z \in \text{FS}((z_n)_{n<\omega}).$$

Put $d = |\text{supp}_0(h + z_0)|$ and let c be a constant guaranteed by Lemma 12.16. Pick $z \in \text{FS}((z_n)_{1 \le n < \omega})$ such that

$$\max \text{supp}_0(h' + z'_0) < \min \text{supp}_0(z), \quad \text{and}$$
$$|\text{supp}_0(h' + z')| \ge c.$$

Let $a = |\text{supp}_0(h + z_0 + z)|$ and $b = |\text{supp}_0(h + z)|$. Then

$$0 < a - b = |\text{supp}_0(z_0)| \leq |\text{supp}_0(h + z_0)| = d.$$

Let $u = |\text{supp}_0(h + z_0 + z) \cap (\delta + 1)|$, $v = |\text{supp}_0(h + z) \cap (\delta + 1)|$, $w = |\text{supp}_0(h + z_0) \cap (\delta + 1)|$ and $t = |\text{supp}_0(h) \cap (\delta + 1)|$. Then $u = v + w - t$, and so

$$u - v = w - t,$$

which is a contradiction, since $u \in \varphi(a)$, $v \in \varphi(b)$, $u, v \geq c$ and $w - t \leq w \leq d$.

Now consider the element

$$g_0 = g + y_0 + y_1 + \cdots + y_k.$$

By the construction,

$$\text{supp}_\varphi(g_0) \cap \text{supp}_0(y_k) \neq \emptyset,$$

and for each $i \leq k - 1$,

$$\emptyset \neq \text{supp}_\varphi(g_0) \cap \text{supp}_0(y_i) \subseteq \text{supp}_0(y_i').$$

Therefore we have that

$$\beta_0 < \alpha_1 \leq \beta_1 < \alpha_2 \leq \beta_2 < \cdots < \alpha_{k-1} \leq \beta_{k-1} < \alpha_k,$$

where

$$\alpha_i = \min(\text{supp}_\varphi(g_0) \cap \text{supp}_0(y_i)),$$
$$\beta_i = \max(\text{supp}_\varphi(g_0) \cap \text{supp}_0(y_i)).$$

Put $\varepsilon_0 = 1$ and, by induction on $i = 1, \ldots, k$, choose $\varepsilon_i \in \{1, -1\}$ so that

$$\varepsilon_{i-1} y_{i-1}(\beta_{i-1}) = \varepsilon_i y_i(\alpha_i).$$

Without loss of generality one may suppose that all $\varepsilon_i = 1$. For each $i \leq k$, define $g_i \in g + \text{FSI}((y_n)_{n \leq k})$ by

$$g_i = g + y_0 - y_1 + \cdots + (-1)^i y_i + (-1)^i y_{i+1} + \cdots + (-1)^i y_k.$$

Then

$$v(g_i) = v(g_0) + i.$$

Hence, there is $j \leq k = 2^{m+1} - 1$ such that

$$v(g_j) \equiv 2^m \pmod{2^{m+1}},$$

and so $g_j \in A_m$. $\square$

In order to to prove Theorem 12.20 in the general case, we need in addition the following result.

Theorem 12.26. *Let G be an infinite Abelian group endowed with the largest totally bounded group topology and let $|G| = \kappa$. Then there is a continuous injection f : $G \to \bigoplus_\kappa \mathbb{Z}_4$ such that*

(1) $f(-x) = -f(x)$ *for all $x \in G$,*

(2) $f(x+y) = f(x) + f(y)$ *for all $x, y \in G$ with $S(f(x)) \cap S(f(y)) = \emptyset$, where for each $a \in \bigoplus_\kappa \mathbb{Z}_4$, $S(a) = \operatorname{supp}(a) \cup (\operatorname{supp}(a) + 1)$.*

Here, $\bigoplus_\kappa \mathbb{Z}_4$ is endowed with the topology induced by the product topology on $\prod_\kappa \mathbb{Z}_4$.

Theorem 12.26 allows us to identify an arbitrary Abelian group G of cardinality κ with a subset of $\bigoplus_\kappa \mathbb{Z}_4$ so that

(a) for every $x \in G$, $-x \in \bigoplus_\kappa \mathbb{Z}_4$ is the inverse of x in G,

(b) for every $x, y \in G$ with $S(x) \cap S(y) = \emptyset$, $x + y \in \bigoplus_\kappa \mathbb{Z}_4$ is the sum of x and y in G, and

(c) the largest totally bounded group topology on G is stronger than that induced from $\bigoplus_\kappa \mathbb{Z}_4$.

And that is all what we need to extend the proof of Theorem 12.23 to Theorem 12.20.

We deduce Theorem 12.26 from the following fact.

Proposition 12.27. *Let G be a countable Abelian topological group. Suppose that $B(G)$ is neither discrete nor open. Then there is a continuous bijection $f : G \to \bigoplus_\omega \mathbb{Z}_4$ such that*

(i) $f(-x) = -f(x)$ *for all $x \in G$,*

(ii) $f(x + y) = f(x) + f(y)$ *for all $x, y \in G \setminus \{0\}$ with* $\max \operatorname{supp}(f(x)) + 2 \leq \min \operatorname{supp}(f(y))$.

Proposition 12.27 is an immediate consequence of Theorem 8.29.

Before proving Theorem 12.26 we show that condition (ii) in Proposition 12.27 can be replaced by (2) from Theorem 12.26.

Indeed, every nonzero element $a \in \bigoplus_\omega \mathbb{Z}_4$ has a unique canonical decomposition

$$a = a_1 + \cdots + a_n,$$

where for each $i = 1, \ldots, n$, $\operatorname{supp}(a_i)$ is a nonempty interval in ω, and for each $i = 1, \ldots, n - 1$, $\max \operatorname{supp}(a_i) + 2 \leq \min \operatorname{supp}(a_{i+1})$. And it follows from (ii) that if $x_i = f^{-1}(a_i)$ for each $i = 1, \ldots, n$, then

$$f(x_1 + \cdots + x_n) = a_1 + \cdots + a_n.$$

Let $a = f(x)$, $b = f(y)$ and $S(a) \cap S(b) = \emptyset$. Let $a = a_1 + \cdots + a_n$ and $b = b_1 + \cdots + b_m$ be the canonical decompositions. Since $S(a) \cap S(b) = \emptyset$, there is a permutation $c_1, \ldots, c_{n+m}$ of $a_1, \ldots, a_n, b_1, \ldots, b_m$ such that

$$a + b = c_1 + \cdots + c_{n+m}$$

is the canonical decomposition. Let $x_i = f^{-1}(a_i)$, $y_j = f^{-1}(b_j)$ and $z_k = f^{-1}(c_k)$. Then $x = x_1 + \cdots + x_n$, $y = y_1 + \cdots + y_m$ and, since G is Abelian, $x + y = z_1 + \cdots + z_{n+m}$. Finally, we obtain that

$$f(x + y) = f(z_1 + \cdots + z_{n+m})$$

$$= c_1 + \cdots + c_{n+m}$$

$$= (a_1 + \cdots + a_n) + (b_1 + \cdots + b_m)$$

$$= f(x) + f(y).$$

Proof of Theorem 12.26.   Without loss of generality one may suppose that

$$G = \bigoplus_{\alpha < \kappa} G_\alpha,$$

where for each $\alpha < \kappa$,

$$|G_\alpha| = |B(G_\alpha)| = |G_\alpha : B(G_\alpha)| = \omega.$$

For each $\alpha < \kappa$, endow G_α with a totally bounded group topology and let

$$f_\alpha : G_\alpha \to \bigoplus_\omega \mathbb{Z}_4$$

be a continuous bijection guaranteed by Proposition 12.27. We define

$$f : G \to \bigoplus_{\alpha < \kappa} \bigoplus_{[\alpha, \alpha + \omega)} \mathbb{Z}_4$$

by

$$f = \bigoplus_{\alpha < \kappa} f_\alpha. \qquad \qquad \square$$

References

The study of resolvability for topological groups was initiated by W. Comfort and J. van Mill [13]. They showed that if G is an infinite Abelian group not containing an infinite Boolean subgroup, then every nondiscrete group topology on G is resolvable. Theorem 12.3 and Theorem 12.6 are results from [105]. Theorem 12.5 was proved in [96]. Theorem 12.9 is a result from [94] and Theorem 12.10 from [89]. Theorem 12.11 is due to I. Protasov [57]. He also proved Theorem 12.13 in the Abelian case [59].

The question of characterizing absolutely resolvable groups was raised in [13]. Corollary 12.18 is due to I. Protasov [57]. Theorem 12.20 was proved in [102]. That every infinite Abelian group not containing an infinite Boolean subgroup is absolutely resolvable was proved in [91]. Corollary 12.22 and Proposition 12.17 complement the Graham-Rothschild Theorem [29] (see also [30, Section 2.4]) which can be stated as follows. If an infinite Abelian group of finite exponent is partitioned into finitely many subsets, then there are arbitrarily large finite cosets contained in one subset of the partition.

Chapter 13
Open Problems

In this chapter we list several open questions related to the topic of this book.

(1) (Old question) Is there any element of finite order in $\beta\mathbb{N}$ other than idempotents?

(2) Let $\kappa > \omega$ and let $\mathcal{T}$ denote the topology on $\bigoplus_\kappa \mathbb{Z}_2$ induced by the product topology on $\prod_\kappa \mathbb{Z}_2$. Are finite groups in $\mathrm{Ult}(\mathcal{T})$ trivial?

(3) Is the structure group of $K(\mathbb{H})$ a free group?

(4) (Old question) Does every point of $\mathbb{Z}^*$ lie in a maximal proper principal left ideal of $\beta\mathbb{Z}$?

(5) (Old question) Is there an infinite increasing sequence of principal left ideals of $\beta\mathbb{Z}$?

(6) (N. Hindman and D. Strauss) Is there an infinite increasing ($\leq_L$-increasing) sequence of idempotents in $\beta\mathbb{Z}$?

(7) (N. Hindman and D. Strauss) Is there in ZFC a left maximal idempotent in $\mathbb{Z}^*$?

(8) Can every compact Hausdorff right topological semigroup be topologically and algebraically embedded into a compact Hausdorff right topological semigroup with a dense topological center?

(9) Let $b\mathbb{Z}$ denote the Bohr compactification of the discrete group $\mathbb{Z}$ and let $\mathcal{T}$ be the topology on $\mathbb{N}$ induced by $b\mathbb{Z}$. Is the largest semigroup compactification of $(\mathbb{N}, \mathcal{T})$ different from $b\mathbb{Z}$?

(10) (A. Arhangel'skiĭ, 1967) Is there in ZFC a (countable) nondiscrete extremally disconnected topological group?

(11) (W. Comfort and J. van Mill) Is there in ZFC a nondiscrete irresolvable (ω-irresolvable, almost maximal) topological group?

(12) (V. Malykhin) Is there an irresolvable (ω-irresolvable, almost maximal) topological group of uncountable dispersion character?

(13) Is there a (countable) nondiscrete ω-irresolvable topological group different from almost maximal?

(14) (I. Protasov and Y. Zelenyuk) Does every maximal (almost maximal) topological group have a neighborhood base at zero consisting of subgroups?

(15) (A. Arhangel'skiĭ and P. Collins) Is there in ZFC a countable nondiscrete nodec topological group?

(16) (I. Protasov) Is there in ZFC a countable nondiscrete topological group in which every discrete subset is closed?

(17) Is it true that for every projective S in $\mathfrak{F}$, there exists in ZFC a countable homogeneous regular almost maximal space X with $\mathrm{Ult}(X)$ isomorphic to S?

(18) Let $(G_1, \mathcal{T}_1)$ and $(G_2, \mathcal{T}_2)$ be countable regular left topological groups and suppose that they are homeomorphic. Are $\mathrm{Ult}(\mathcal{T}_1)$ and $\mathrm{Ult}(\mathcal{T}_2)$ topologically and algebraically isomorphic?

Bibliography

[1] S. Adian, *Classification of periodic words and their application in group theory*, Burnside groups (Berlin) (J. Mennicke, ed.), Springer-Verlag, 1980, pp. 1–40.

[2] R. Arens, *The adjoint of a bilinear operation*, Proc. Amer. Math. Soc. 2 (1951), pp. 839–848.

[3] A. Arhangel'skiĭ, *Groupes topologiques extremalement discontinus*, C.R. Acad. Sci. Paris 265 (1967), pp. 822–825.

[4] A. Arhangel'skiĭ and M. Tkachenko, *Topological groups and related structures*, World Scientific, Amsterdam-Paris, 2008.

[5] V. Arnautov, *Nondiscrete topologizability of countable rings*, Dokl. AN SSSR 191 (1970), pp. 747–750.

[6] D. Booth, *Ultrafilters on a countable set*, Ann. Math. Logic 2 (1970), pp. 1–24.

[7] N. Bourbaki, *Elements of Mathematics: General Topology, Chapters 1-4*, Springer-Verlag, Berlin, 1989.

[8] P. Civin and B. Yood, *The second conjugate space of a Banach algebra as an algebra*, Pacific J. Math. 11 (1961), pp. 847–870.

[9] A. Clifford, *Semigroups admitting relative inverses*, Ann. Math. 42 (1941), pp. 1037–1049.

[10] W. Comfort, *Ultrafilters: some old and some new results*, Bull. Amer. Math. Soc. 83 (1977), pp. 417–455.

[11] W. Comfort, *Topological groups*, Handbook of Set-theoretic Topology (Amsterdam) (K. Kunen and J. Vaughan, eds.), Elsevier, 1984, pp. 1143–1263.

[12] W. Comfort and S. Negrepontis, *The Theory of Ultrafilters*, Springer-Verlag, Berlin, 1974.

[13] W. Comfort and J. van Mill, *Groups with only resolvable group topologies*, Proc. Amer. Math. Soc. 120 (1994), pp. 687–696.

[14] D. Davenport and N. Hindman, *A proof of van Douwen's right ideal theorem*, Proc. Amer. Math. Soc. 113 (1991), pp. 573–580.

[15] M. Day, *Amenable semigroups*, Illinois J. Math. 1 (1957), pp. 509–544.

[16] N. de Bruijn and P. Erdős, *A colour problem for infinite graphs and a problem in the theory of relations*, Indag. Math. 13 (1951), pp. 369–373.

[17] D. Dikranjan, I. Prodanov, and L. Stoyanov, *Topological groups: characters, dualities and minimal group topologies*, Marcel Dekker Inc., New York-Basel, 1982.

[18] A. El'kin, *Ultrafilters and undecomposable spaces*, Moscow Univ. Math Bull. 24 (1969), pp. 37–40.

[19] A. El'kin, *Regular maximal spaces*, Mat. Zametki 27 (1980), pp. 301–305, 320.

[20] R. Ellis, *Distal transformation groups*, Pacific J. Math. 8 (1958), pp. 401–405.

[21] R. Ellis, *Lectures on Topological Dynamics*, Benjamin, New York, 1969.

[22] R. Engelking, *General Topology*, Heldermann Verlag, Berlin, 1989.

[23] S. Ferri, N. Hindman, and D. Strauss, *Digital representation of of semigroups and groups*, Semigroup Forum 77 (2008), pp. 36–63.

[24] M. Filali and P. Salmi, *Slowly oscillating functions in semigroup compactifications and convolution algebras*, J. Functional Analysis 250 (2007), pp. 144–166.

[25] D. Fremlin, *Consequences of Martin's axiom*, Cambridge Univ. Press, Cambridge, 1984.

[26] Z. Frolík, *Homogeneity problems for extremally disconnected spaces*, Comment. Math. Univ. Carolinae 8 (1967), pp. 757–763.

[27] Z. Frolík, *Fixed points of maps of* $\beta\mathbb{N}$, Bull. Amer. Math. Soc. 74 (1968), pp. 187–191.

[28] L. Fuchs, *Infinite Abelian Groups I*, Academic Press, New York and London, 1970.

[29] R. Graham and B. Rothschild, *Ramsey's theorem for n-parameter sets*, Trans. Amer. Math. Soc. 159 (1971), pp. 257–292.

[30] R. Graham, B. Rothschild, and J. Spencer, *Ramsey Theory*, Wiley, New York, 1990.

[31] J. Green and D. Rees, *On semigroups in which* $x^r = x$, Proc. Camb. Philos. Soc. 48 (1952), pp. 35–40.

[32] G. Hesse, *Zur Topologisierbarkeit von Gruppen*, Dissertation, Hannover, 1979.

[33] E. Hewitt, *A problem of set-theoretic topology*, Duke Math. J. 10 (1943), pp. 309–333.

[34] E. Hewitt and K. Ross, *Abstract Harmonic Analysis I*, Springer-Verlag, Berlin, 1963.

[35] N. Hindman, *Finite sums from sequences within cells of a partition of* $\mathbb{N}$, J. Combin. Theory, Ser. A 17 (1974), pp. 1–11.

[36] N. Hindman and J. Pym, *Free groups and semigroups in* $\beta\mathbb{N}$, Semigroup Forum 30 (1984), pp. 177–193.

[37] N. Hindman and D. Strauss, *Algebra in the Stone–Čech Compactification*, de Gruyter, Berlin, 1998.

[38] N. Hindman, D. Strauss, and Y. Zelenyuk, *Large rectangular semigroups in Stone–Čech Compactifications*, Trans. Amer. Math. Soc. 355 (2003), pp. 2795–2812.

[39] K. Hofmann, J. Lawson, and J. Pym (eds.), *The analytical and topological theory of semigroups*, de Gruyter, Berlin, 1990.

[40] A. Illanes, *Finite and* ω-*resolvability*, Proc. Amer. Math. Soc. 124 (1996), pp. 1243–1246.

[41] T. Jech, *The Axiom of Choice*, North Holland, Amsterdam, 1973.

[42] M. Katětov, *On nearly discrete spaces*, Časopis Pěst. Mat. Fys. 75 (1950), pp. 69–78.

[43] K. Kunen, *Set theory: an introduction to independence proofs*, North Holland, Amsterdam, 1980.

[44] A. Lau and J. Pym, *The topological centre of a compactification of a locally compact group*, Math. Z. 219 (1995), pp. 567–579.

[45] A. Louveau, *Sur un article de S. Sirota*, Bull. Sci. Math. (France) 96 (1972), pp. 3–7.

[46] V. Malykhin, *Extremally disconnected and similar groups*, Soviet Math. Dokl. 16 (1975), pp. 21–25.

[47] V. Malykhin, *On extremally disconnected topological groups*, Uspekhi Math. Nauk 34 (1979), pp. 59–66.

[48] A. Markov, *On unconditionally closed sets*, Mat. Sbornik 18 (1946), pp. 3–28.

[49] D. McLean, *Idempotent semigroups*, Amer. Math. Monthly 61 (1954), pp. 110–113.

[50] S. Morris, *Pontryagin duality and the structure of locally compact Abelian groups*, Cambridge Univ. Press, Cambridge, 1977.

[51] P. Novikov and S. Adian, *Infinite periodic groups I, II, III*, Izvestia Acad. Nauk SSSR Ser. Mat. 32 (1968), pp. 212–244, 351–524, 709–731.

[52] K. Numakura, *On bicompact semigroups*, Math. J. Okayama Univ. 1 (1952), pp. 99–108.

[53] A. Ol'šanskiĭ, *A note on a countable nontopologizable group*, Vesnyk Moscow Univ. (mat., meh.) (1980), p. 103.

[54] T. Papazyan, *Extremal topologies on a semigroup*, Topology and its Applications 39 (1991), pp. 229–243.

[55] L. Pontryagin, *Topological Groups*, Princeton Univ. Press, Princeton, NJ, 1939.

[56] I. Protasov, *Filters and topologies on semigroups*, Matematychni Studii 3 (1994), pp. 15–28.

[57] I. Protasov, *Partitions of direct products of groups*, Ukr. Math. J. 49 (1997), pp. 1385–1395.

[58] I. Protasov, *Finite groups in βG*, Matematychni Studii 10 (1998), pp. 17–22.

[59] I. Protasov, *Irresolvable topologies on groups*, Ukr. Math. J. 50 (1998), pp. 1646–1655.

[60] I. Protasov, *Maximal topologies on groups*, Sibirsk. Math. J. 39 (1998), pp. 1368–1381.

[61] I. Protasov, *Coronas of balleans*, Topology Appl. 149 (2005), pp. 149–160.

[62] I. Protasov and Y. Zelenyuk, *Topologies on groups determined by sequences*, Lviv: VNTL Publishers, 1998.

[63] D. Rees, *On semi-groups*, Math. Proc. Cambridge Philos. Soc. 36 (1940), pp. 387–400.

[64] F. Riesz, *Stetigkeitsbegriff und abstrakte Mengenlehre*, Atti del Congr. Intern. Mat., Roma, vol. 2, R. Accademia dei Lincei, 1908, pp. 18–24.

[65] D. Ross and Y. Zelenyuk, *Topological groups and $P_{\kappa^+}^{-}$-points*, Topology Appl. 157 (2010), pp. 2467–2475.

[66] W. Rudin, *Homogeneity problems in the theory of Čech compactifications*, Duke Math. J. 23 (1956), pp. 409–419.

[67] W. Ruppert, *Compact semitopological semigroups: an intrinsic theory*, Lecture Notes in Math. (Berlin), vol. 1079, Springer-Verlag, 1984.

[68] S. Shelah, *On a problem of Kurosh, Jonsson groups, and applications*, Word problems II (Amsterdam) (W. Boone S. Adian and eds. G. Higman, eds.), North-Holland, 1980, pp. 373–394.

[69] S. Shelah, *Proper forcing*, Springer-Verlag, Berlin, 1982.

[70] O. Shuungula, Y. Zelenyuk, and Yu. Zelenyuk, *The closure of the smallest ideal of an ultrafilter semigroup*, Semigroup Forum 79 (2009), pp. 531–539.

[71] S. Sirota, *The product of topological groups and extremal disconnectedness*, Math. USSR Sbornik 8 (1969), pp. 169–180.

[72] M. Stone, *Applications of the theory of Boolean rings to general topology*, Trans. Amer. Math. Soc. 41 (1937), pp. 375–481.

[73] A. Suschkewitsch, *Über die endlichen Gruppen ohne das Gesetz der eindeutigen Umkehrbarkeit*, Math. Ann. 99 (1928), pp. 30–50.

[74] P. Trotter, *Projectives in some categories of semigroups*, Semigroups (New York/London) (J. Preston eds. T. Hall, P. Jones, ed.), Academic Press, 1980, pp. 133–144.

[75] P. Trotter, *Projectives of finite regular semigroups*, J. Algebra 77 (1982), pp. 49–61.

[76] S. Ulam, *Concerning functions of sets*, Fund. Math. 14 (1929), pp. 231–233.

[77] S. Ulam, *Zur Masstheorie in der allgemeinen Mengenlehre*, Fund. Math. 16 (1930), pp. 140–150.

[78] E. van Douwen, *Simultaneous extensions of continuous functions*, Dissertation, Vrije Universiteit, 1975.

[79] E. van Douwen, *The integers and topology*, Handbook of Set-theoretic Topology (Amsterdam) (K. Kunen and J. Vaughan, eds.), Elsevier, 1984, pp. 111–167.

[80] E. van Douwen, *Applications of maximal topologies*, Topology Appl. 51 (1993), pp. 125–139.

[81] J. van Mill, *An introduction to $\beta\omega$*, Handbook of Set-theoretic Topology (Amsterdam) (K. Kunen and J. Vaughan, eds.), Elsevier, 1984, pp. 503–567.

[82] E. Čech, *On bicompact spaces*, Ann. Math. (2) 38 (1937), pp. 823–844.

[83] Y. Zelenyuk, *Topologies on Abelian groups*, C. Sc. Dissertation, Kyiv University, 1990, 80 pp.

[84] Y. Zelenyuk, *Topological groups with finite semigroups of ultrafilters*, Matematychni Studii 6 (1996), pp. 41–52.

[85] Y. Zelenyuk, *Finite groups in $\beta\mathbb{N}$ are trivial*, Ukranian Nat. Acad. Sci. Inst. Mat., Preprint 96.3, 1996, 12 pp.

[86] Y. Zelenyuk, *Finite groups in* $\beta\mathbb{N}$ *are trivial*, Semigroup Forum 55 (1997), pp. 131–132.

[87] Y. Zelenyuk, *On topological groups with finite semigroup of ultrafilters*, Matematychni Studii 7 (1997), pp. 139–144.

[88] Y. Zelenyuk, *Topological groups in which each nowhere dense subset is closed*, Math. Zametki 64 (1998), pp. 207–211.

[89] Y. Zelenyuk, *Resolvability of topological groups*, Ukr. Math. J. 51 (1999), pp. 41–47.

[90] Y. Zelenyuk, *Extremal ultrafilters and topologies on groups*, Matematychni Studii 14 (2000), pp. 121–140.

[91] Y. Zelenyuk, *Partitions of groups into absolutely dense subsets*, Math. Zametki 67 (2000), pp. 706–711.

[92] Y. Zelenyuk, *Ultrafilters and topologies on groups*, D. Sc. Dissertation, Kyiv University, 2000, 260 pp.

[93] Y. Zelenyuk, *On subsemigroups of* $\beta\mathbb{N}$ *and absolute coretracts*, Semigroup Forum 63 (2001), pp. 457–465.

[94] Y. Zelenyuk, *On partitions of groups into dense subsets*, Topology Appl. 126 (2002), pp. 327–339.

[95] Y. Zelenyuk, *Weak projectives of finite semigroups*, J. Algebra 266 (2003), pp. 77–86.

[96] Y. Zelenyuk, *On group operations on homogeneous spaces*, Proc. Amer. Math. Soc. 132 (2004), pp. 1219–1222.

[97] Y. Zelenyuk, *On extremally disconnected topological groups*, Topology Appl. 153 (2006), pp. 2382–2385.

[98] Y. Zelenyuk, *On the ultrafilter semigroup of a topological group*, Semigroup Forum 73 (2006), pp. 301–307.

[99] Y. Zelenyuk, *Almost maximal spaces*, Topology Appl. 154 (2007), pp. 339–357.

[100] Y. Zelenyuk, *On topologizing groups*, J. Group Theory 10 (2007), pp. 235–244.

[101] Y. Zelenyuk, *Finite groups in Stone–Čech compactifications*, Bull. London Math. Soc. 40 (2008), pp. 337–346.

[102] Y. Zelenyuk, *Partitions and sums with inverses in Abelian groups*, J. Combin. Theory, Ser. A 115 (2008), pp. 331–339.

[103] Y. Zelenyuk, *Topologies on groups determined by discrete subsets*, Topology Appl. 155 (2008), pp. 1332–1339.

[104] Y. Zelenyuk, *The number of minimal right ideals of* βG, Proc. Amer. Math. Soc. 137 (2009), pp. 2483–2488.

[105] Y. Zelenyuk, *Regular homeomorphisms of finite order on countable spaces*, Canad. J. Math. 61 (2009), pp. 708–720.

[106] Y. Zelenyuk, *The smallest ideal of* βS *is not closed*, Topology Appl. 156 (2009), pp. 2670–2673.

[107] Y. Zelenyuk, *Regular idempotents in* βS, Trans. Amer. Math. Soc. 362 (2010), pp. 3183–3201.

[108] Y. Zelenyuk, *Almost maximal topologies on semigroups*, Proc. Amer. Math. Soc., to appear (2011).

[109] Y. Zelenyuk, *Closed left ideal decompositions of* $U(G)$, Canad. Math. Bull., to appear (2011).

[110] Y. Zelenyuk, *Metrizable refinements of group topologies and the pseudo-intersection number*, to appear (2011).

[111] Y. Zelenyuk and Yu. Zelenyuk, *Free groups in the smallest ideal of* βG, Semigroup Forum 78 (2009), pp. 360–367.

Index